Hybrid Imaging and Visualization

Joseph Awange • Béla Paláncz
Lajos Völgyesi

Hybrid Imaging and Visualization

Employing Machine Learning with *Mathematica - Python*

Springer

Joseph Awange
Spatial Sciences
Curtin University
Perth, WA, Australia

Béla Paláncz
Department of Photogrammetry & Geoinformatics
Budapest University of Technology & Economics
Budapest, Hungary

Lajos Völgyesi
Department of Geodesy & Surveying
Budapest University of Technology & Economics
Budapest, Hungary

ISBN 978-3-030-26155-9 ISBN 978-3-030-26153-5 (eBook)
https://doi.org/10.1007/978-3-030-26153-5

This Springer imprint is published by the registered company Springer Nature Switzerland AG
The registered company address is: Gewerbestrasse 11, 6330 Cham, Switzerland

Preface

Computer vision is a subfield of artificial intelligence that enables the understanding of the content of digital images, such as photographs. Currently, machine learning is making impressive inroads in tackling challenges posed by computer vision related tasks, promising further impressive advances.

Speaking of computer vision, two modes of books frequently appear (i) reference-based textbooks written by experts, who often are academics, targeting students and practitioners, and (ii), programming oriented books (i.e., play books) written by experts, who often are developers and engineers, and designed to be used as references by practitioners. Whereas the former mainly focus on general methods and theory (Maths) and not on the practical aspects of the problems and the applications of methods (code), the latter focuses mainly on the techniques and practical concerns of the problem solving, where the focus is placed on examples of codes and standard libraries.

Although programme-based books briefly describe techniques with relevant theory (Maths), they probably do not predispose themselves for use as primary reference. In this regard, Dr. Jason Brownlee, a machine learning specialist who teaches developers how to obtain results from modern machine learning methods via hands-on tutorials recommends the work of Richard Szeliski (2010; Computer Vision: Algorithms and Applications) since it provides a short, focused, and readable introduction to computer vision complete with relevant theory, without getting too bogged down. For programmers, he suggests Jan Erik Solem's (2012; Programming Computer Vision with *Python*) since it focuses on real computer vision techniques with standard (or close enough) *Python* libraries. It is an excellent starting point for those who want to get their hands dirty with computer vision.

Our contribution, therefore, intends to be a go between these two types of books. On the one hand, it is like a programmer's book presenting many different techniques illustrated by a large number of examples, accompanied by detailed discussions on the Mathematics behind the different methods. The codes of the algorithms are given in *Python* as well as in *Mathematica* form. In the book the version of *Mathematica* 11 integrates *Python*, most of the codes are blended as hybrid codes, however in the electronic supplement the recent version of Mathematica 12 is employed and uploaded to Researchgate. *Mathematica* is an incredibly powerful platform with a fun and intellectually pleasing language, but is expensive and closed source. *Python* is a convenient, powerful language with a lot of support from the developer community. For as long as the two have existed people have been trying to tie them together, so that one can utilize the integrated advantages of both languages.

This book is divided into five chapters. The first one deals with dimension reduction techniques of visual objects where besides the standard methods; it includes Independent Component Analysis, AutoEncoding and Fractal Compression. The second chapter discusses classification methods that include Support Vector Classification. In the third chapter, different clustering techniques are demonstrated, like Hierarchical Clustering, Density-Based Spatial Clustering of Applications with Noise and Spectral Clustering. The fourth chapter presents different regression techniques, where different robust regression models such as Expectation Maximization, RANSAC and Symbolic Regression are also discussed. The last chapter provides a deep insight into applications of neural networks in computer vision. Besides the standard network types, Deep Learning and Convolutional Networks are also discussed. At the end of every chapter, the considered methods are compared and qualified from different practical points of views.

The authors:

Joseph L. Awange
Curtin University
Australia

Béla Paláncz
Budapest University of Technology and Economics
Hungary

Lajos Völgyesi
Budapest University of Technology and Economics
Hungary

Budapest – Perth – Karlsruhe
May 2019

Acknowledgements

This work was partially funded by National Research, Development and Innovation Office – NKFIH No. 124286. The first author also wishes to express his sincere thanks to *Prof. B. Heck* and *Prof. Hansjörg Kutterer* (Karlsruhe Institute of Technology (KIT), Germany) for hosting him during the period of his Alexander von Humboldt Fellowship (July–September 2019). Parts of this book were written during this period. He is also grateful to *Prof. I. Fitzsimons* (Head of School, Earth and Planetary Sciences, Curtin University, Australia) for the support and motivation that enabled the preparation of this edition. He also wishes to acknowledge the support of Alexander von Humboldt that facilitated his stay at KIT. To all, he says, "ahsante sana" (Swahili for thank you very much). Special thanks go to *Dr Cecilia Xia* for her inputs during the preparation of this book, especially on the *Python* code.

The authors highly appreaciate are greatly thankful for the time and efforts taken by *Dr. Daniel Lichtblau*, Wolfram Research in reading the draft version of the book and providing valuable comments and advice.

Contents

Introduction

1 Computer Vision and Machine Learning

Computer vision (also known as machine vision; Jain et al., 1995), a multidisciplinary field that is broadly a subfield of artificial intelligence and machine learning has as one of its goals the extraction of useful information from images. A basic problem in computer vision, therefore, is to try to understand, i.e., "*see*" the structure of the real world from a given set of images through use of specialized methods and general learning algorithms (e.g., Hartley and Zisserman 2003; see Fig. 1). Its applications are well documented in Jähne and Haußecker (2000), where it finds use e.g., in human motion capture (Moeslund and Granum 2001). With the plethora of unmanned aircraft vehicles (UAVs) or drones (see Awange 2018; Awange and Kiema 2019), computer vision is stamping its authority in the UAV field owing to its intelligent capability (Al-Kaff et al., 2018). Several publications abound on computer vision, e.g., on algorithms for image processing (e.g., Parker 2011, Al-Kaff et al., 2018), pattern recognition/languages in computer vision (e.g., Chen 2015), feature extraction (Nixon and Aguado 2012) and among others.

On its part, *Machine Learning* (ML) is the employment of statistical techniques by computers to learn specific and complex tasks from given data that are discriminated into learnt and defined classes (Anantrasirichai et al., 2018, 2019). They have widely been used, e.g., for landslides studies (Yilmaz, 2010), vegetations (Brown et al., 2008), earthquakes (Adeli and Panakkat, 2009), land surface classificatios (Li et al., 2014) and for classification of volcanic deformation (Anantrasirichai et al., 2018, 2019). Lary et al., (2016) provides a good exposition of its application.

Traditionally, computer vision's contributions are largely grouped into two categories; *textbook-based* that focus on methods and theory rather than on the practicality, and *programming-based* that focus on the techniques and the practicality of solving the problems. There is hardly any book that tries to bring the two together; i.e., methods/theory on the one hand, and techniques/practicality (i.e., codes) on the other hand. This present book attempts to fill this missing gap by treating computer vision as a machine-learning problem (Fig. 1), and disregarding everything we know about the creation of an image. For example, it does not exploit our understanding of perspective projection.

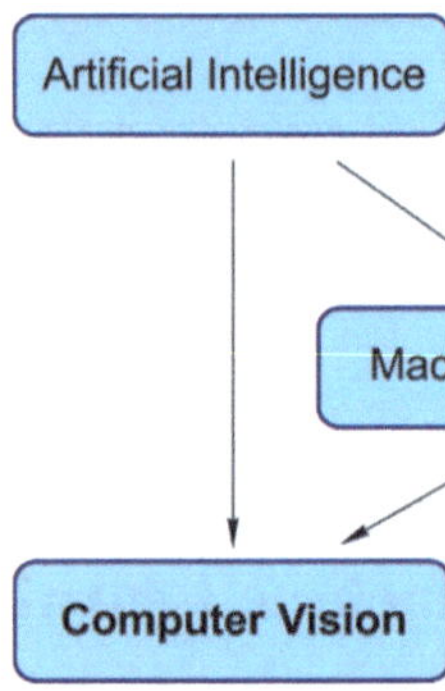

Fig. 1 Relationship between Computer Vision, Artificial Intelligent and Machine Learning.

In general the image processing chain contains five different tasks: reprocessing, data reduction, segmentation, object recognition and image understanding. Optimisation techniques are used as a set of auxiliary tools that are available in all steps of the image processing chain, see Fig. 2.

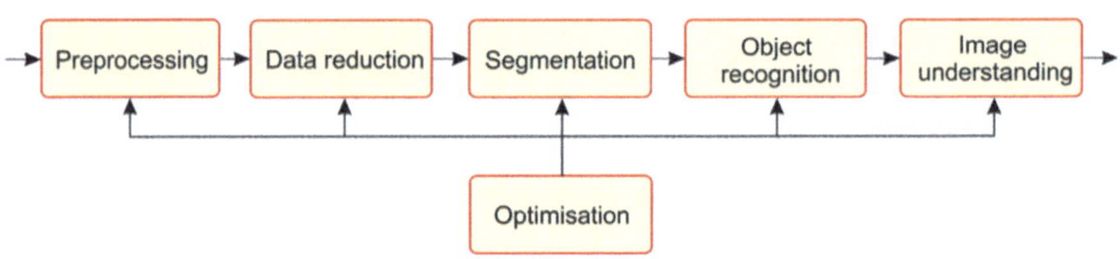

Fig. 2 The image processing chain containing the five different tasks.

Many popular computer vision applications involve trying to recognize things in photographs; for example:

1. Object Classification: What broad category of objects are in this photograph?
2. Object Identification: Which type of a given object is in this photograph?
3. Object Verification: Is the object in the photograph?
4. Object Detection: Where are the objects in the photograph?
5. Object Landmark Detection: What are the key points for the object in the photograph?
6. Object Segmentation: What pixels belong to the object in the image? and
7. Object Recognition: What objects are in this photograph and where are they?

Let us consider some examples, where Machine Learning techniques are applied to solving these computer vision problems.

Example 1 (*Segmentation as Clustering*)

Segmentation is any operation that partitions an image into regions that are coherent with respect to some criterion. One example is the segregation of different textures. The following is an image that highlights bacteria. Now, we would like to remove the background in order to get clear information about the size and form of the bacteria.

⇒ `img=` 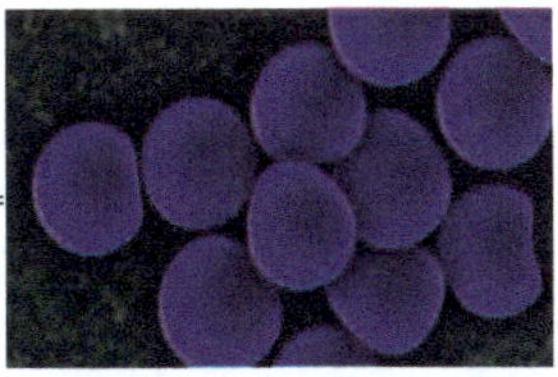 `;`

Fig. 3. Original image.

```
⇒ RemoveBackground[img, {"Foreground", "Uniform"}]
```

⇐ 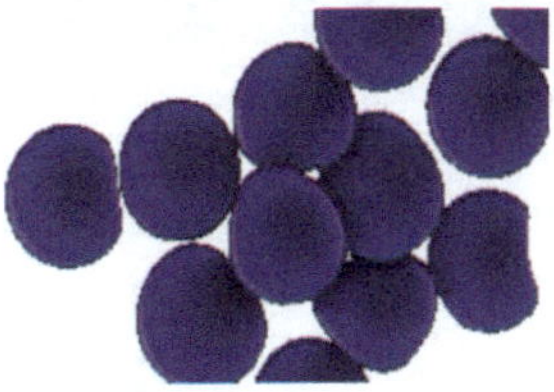

Fig. 4. Image of bacterias after background removal.

Example 2 (*Object Recognition as Classification*)

Object detection and recognition determine the position and, possibly, also the orientation and scale of specific objects in an image, and classify them. In the image below, we can get information on the type of the image object (car) and the possible subclasses probability.

⇒ `img=`  `;`

Fig. 5. Original image.

```
⇒ data=ImageIdentifyn[img, car(WORD), 10, "Probability"]
⇐ <|convertible → 0.725076,saloon → 0.150854,coupe → 0.0760313,
   station wagon → 0.0414005,hatchback → 0.00342106,
   limousine → 0.00153371,automobile → 1.|>
```

Example 3 (*Image Understanding as Landmark Detection*)

Key points of an image can characterize the main feature locations of an image. This information can be employed for further image processing operations like image transform, classification and clustering. Consider the image in Fig. 6 below.

```
⇒ img=
```

```
;
```

Fig. 6. Original image of a cathedral.

Let us find the first thirty most important keypoints of the image.

```
⇒ HighlightImage[img, ImageKeypoints[img, "MaxFeatures" → 30]]
⇐
```

Fig. 7. The first thirty most important keypoints.

2 *Python* and *Mathematica*

In this book, we employ *Python* (see e.g., Lutz 2001; Oliphant 2007) and *Mathematica* (e.g., Maeder 1991) as well as their blending, since *Python* code can be run from *Mathematica* directly. It is therefore appropriate to provide a brief discussion on them. This section is thus dedicated to their exposition.

Python is now undoubtedly the most popular language for data science projects, while the *Wolfram Language* is rather a niche language in this concern. Consequently, *Python* is probably well-known to the reader compared to *Mathematica.* Given that *Wolfram Language,* widely used in academia (especially in physics, mathematics and financial analytics) has been around for over 30 years, it is actually older than both *R* and *Python.*

The general principle of the *Wolfram Language* is that each function is very high level and automated as much as possible. For example the Classify[ ] function chooses the method automatically for the user. However, the user can also set it manually to something like Method → "RandomForest". The neural network function employed in *Mathematica* uses MxNet as a backend and is similar in its use to *Keras* in *Python*, although nicer to view. In general, the *Machine Learning* (ML) functions in *Wolfram* have a black box feeling to them, although there are lower level functions as well. One should therefore not blindly trust that the automatic solutions provided by the Predict and Classify functions are the only optimal solutions. They are often far from that and at best give baseline solutions on which to rely upon. One can then always use lower level functions to build one's own customized ML solution with *Wolfram* or *Python.* However this ability of *Mathematica* has been improved considerably in the last version released in 2019.

Mathematica has a very good system for documentation with all built-in functions. Also, the documentation itself is in notebooks so that one can quickly try something directly inside the documentation. The documentation in *Mathematica* is really good, but *Python* has a much bigger community with a widened network of support such that it is very likely that one finds an answer to a given problem. Also, one can learn a lot through sites like *Kaggle*. The *Wolfram Mathematica* community in comparison is small and therefore it is harder to find relevant information, although the *Mathematica* community (https://mathematica.stackexchange.com) on stack exchange is really helpful.

So let us talk about the elephant in the room: the price. *Mathematica* is not free, it is actually quite expensive. Since *Mathematica* comes with all functions from the start, there is no need to buy additional "Toolboxes" like in *Matlab.* Now, we present some bullet points for both languages in no particular order.

Wolfram Mathematica

- natural language interpretation
- pattern matching is powerful and prominent, for example in function declaration
- interactive and very good documentation
- consistent

- symbolic, one can pass everything into a function (has a lot of advantages but also makes it harder to debug)
- more advanced notebooks
- no virtual environments and dependencies
- works the same in every OS
- most of the time there is only one obvious way to do things, for example plotting
- Dynamic and manipulates functions for more interactivity
- built-in knowledge
- indices start at 1
- instant Application Programming Interface (API) (although only in the *Wolfram Cloud* or one's own *Wolfram Enterprise Cloud*)
- hard to find a job / hard to recruit people who know *Wolfram*

Python

- "There is a package for that"
- closer to state of the art
- codes are easier to read and to maintain
- debug messages are usually more helpful
- free
- learn from *Kaggle*
- lots of possibilities to deploy a trained model
- a lot of online courses, podcasts and other resources
- use of google-colab or *Kaggle* for learning ML without a local GPU
- pandas is easier to use than the "Dataset" in *Mathematica*
- bigger community, hence easier support.

Learning another language is usually beneficial for one's overall understanding of programming. So learning *Wolfram* might be a nice addition. We use the *Wolfram* language for quick prototyping of ideas and often come up with interesting combinations of data or feature engineering with the built-in knowledge of the *Wolfram* language. In addition, a quick Manipulate is fun and can help a lot in understanding the problem and data better.

In *Mathematica,* with just one line, one can deploy our model as an Application Programming Interface (API) or web-app, although only in the *Wolfram* infrastructure, which might not fit inside one's infrastructure or policy. Also, the high level functions Classify and Predict are too much of a black box and even standard scikit learn algorithms outperform them.

Overall, we hope that both languages inspire each other, as the *jupyter notebook* was certainly inspired by *Mathematica*. On the other hand *Wolfram* will have a difficult future if they continue to try to do everything on their own and lock users into their infrastructure. Therefore, a combination of the two languages will be more and more fruitful in the future. For more details see: *Wolf-*

ram Language (*Mathematica*) vs. *Python* for Data Science Projects, 2019, ATSEDA AB (https://atseda.com/en/blog/2019/02/12/mathematica - and - python)

References

Adeli, H., and A. Panakkat (2009), A probabilistic neural network for earthquake magnitude prediction, Neural Networks, 22(7): 1018-1024, doi:https://doi.org/10.1016/j.neunet.2009.05.003.

Al-Kaff A, Martín D, García F, de la Escalera A, Armingol JM (2018) Survey of computer vision algorithms and applications for unmanned aerial vehicles. Expert Systems with Applications 92: 447-463, doi: 10.1016/j.eswa.2017.09.033

Anantrasirichai N, Biggs J, Albino F, Hill P, Bull D (2018) Application of machine learning to classiffication of volcanic deformation in routinely generated insar data, Journal of Geophysical Research: Solid Earth 123 (8) : 6592-6606. doi:10.1029/2018JB015911.

Anantrasirichai N, Biggs J, Albino F, Bull D (2019) A deep learning approach to detecting volcano deformation from satellite imagery using synthetic datasets. Remote Sensing of Environment, doi: 10.1016/j.rse.2019.04.032.

Awange JL and Kiema JBK (2019) Environmental Geoinformatics. Extreme Hydro-Climatic and Food Security Challenges: Exploiting the Big Data. Springer Nature, Heidelberg.

Awange JL (2018) GNSS Environmental Sensing. Revolutionizing Environmental Monitoring. Springer International Publishing AG.

Brown, M., D. Lary, A. Vrieling, D. Stathakis, and H. Mussa (2008), Neural networks as a tool for constructing continuous ndvi time series from AVHRR and MODIS, In ternational Journal of Remote Sensing, 29(24), 7141–7158, doi:https://doi.org/10.1080/01431160802238435.

Chen CH (2016) A handbook of pattern recognition and computer vision. 5th Edition. World Scientific, Dartmouth, USA.

Han J, Shao L, Xu D and Shotton J (2013) Enhanced computer vision with microsoft Kinect sensor: A review. IEEE Transactions on Cybernetics 43:1318-1334, doi: 10.1109/TCYB.2013.2265378.

Hartley R and Zisserman A (2003) Multiple view geometry in computer vision. 2nd Edition, Cambridge University Press, Cambridge, UK.

Jähne B and Haußecker H (2000) Computer Vision and Applications. A guide for students and practitioners. Academic Press. Elsevier.

Jain R, Kasturi R, Schunck BG (1995) Machine vision. McGraw-Hill.

Lary, D. J., A. H. Alavi, A. H. Gandomi, and A. L. Walker (2016), Machine learning in geosciences and remote sensing, Geoscience Frontiers, 7(1), 3 –10, doi: 10.1016/j.gsf.2015.07.003, special Issue: Progress of Machine Learning in Geosciences.

Li, C., J. Wang, L. Wang, L. Hu, and P. Gong (2014), Comparison of classification algorithms and training sample sizes in urban land classification with landsat thematic mapper imagery, Remote Sensing, 6(2), 964–983, doi: 10.3390/rs6020964

Lutz M (2001) Programming Python. 2nd Edition. O'Reilly & Associates.

Maeder RE (1991) Programming in Mathematica. 2nd Edition. Addison-Wesley Longman Publishing Co., Inc. Boston, MA, USA

Moeslund TB and Granum E (2001) A Survey of Computer Vision-Based Human Motion Capture. Computer Vision and Image Understanding 81:231-268, doi: 10.1006/cviu.2000.0897

Nixon M and Aguado A (2012) Feature extraction and image processing for computer vision. 3rd Edition. Academic Press, Elsevier, London.

Oliphant TE (2007) Python for Scientific Computing. Computing in Science & Engineering 9(3): 10-20, doi: 10.1109/MCSE.2007.58

Parker JR (2011) Algorithms for image processing and computer vision. 2nd Edition. Wiley Publishing, Indianapolis, Indiana.

Yilmaz, I. (2010), Comparison of landslide susceptibility mapping methodologies for koyulhisar, turkey: conditional probability, logistic regression, artificial neural networks, and support vector machine, Environmental Earth Sciences, 61(4), 821–836, doi: 10.1007/s12665-009-0394-9.

Chapter 1
Dimension Reduction

In this chapter we discuss some important lossy data reduction methods, which are very important in machine learning as well as digital image processing and visualization.

1.1 Principal Component Analysis

Basic theory

Principal Component Analysis (PCA) is a well-known and widely used technique applicable to a wide variety of applications such as dimensionality reduction, data compression, feature extraction, and visualization (Preisendorfer, 1988). The basic idea is to project a dataset from many correlated coordinates onto fewer uncorrelated coordinates called principal components while still retaining most of the variability present in the data (Baeriswyl and Rebertez, 1997; Haroon and Rasul, 2009) on the one hand. On the other hand, it allows appropriate examinations by simplifying complex sets of interrelationships into two or more new variables (e.g., Awange et al., 2019a). PCA has widely been used in many fields, e.g., climate and water storage change studies (Jolliffe and Cadima 2016; Hu et al. 2017; Anyah et al. 2018; Awange et al., 2019b), hydrometeorology (Awange et al., 2014, 2016), and even in teaching and learning studies (Awange et al. 2017). Variants of PCA exist, e.g., rotated PCA (RPCA; Agutu et al. 2017). Detailed coverage of PCA, its variants as well as its higher version "the Independent Component Analysis (ICA)" are adequately covered, e.g., in Forootan (2016), and specifically for image compresion in Martín-Clemente and Hornillo-Mellado (2006). In what follows, its principle and applications to data compression are presented.

J. Awange et al., *Hybrid Imaging and Visualization*,
https://doi.org/10.1007/978-3-030-26153-5_1

1.1.1 Principal Component

Principal component analysis is a statistical procedure that converts data with possibly correlated variables into a set of linearly uncorrelated variables (e.g., Widman and Schär, 1997).

We seek a linear combination of the columns of a matrix X as

$$\sum_{j=1}^{p} a_j x_j = X\,a$$

with maximum variance,

$$\text{Var}\,(X\,a) = a^{\mathrm{T}} S\,a$$

where S is the *covariance matrix* associated with the dataset consisting of the row vectors of X. For this problem to have a well-defined solution, an additional restriction must be imposed and the most common restriction involves working with unit-norm vectors, i.e. requiring $a^{\mathrm{T}}a = 1$. The problem is equivalent to find the *eigenvectors* of S. Let us consider an X matrix of row vectors with size of two,

```
⇒ X=N[{{1, 2}, {2, 3}, {4, 10}}];X//MatrixForm
```

$$\Leftarrow \begin{pmatrix} 1. & 2. \\ 2. & 3. \\ 4. & 10. \end{pmatrix}$$

First, we define a function, which standardizes the column vectors, (zero mean, and unit variance).

```
⇒ std[data_] := Module[{datatr = Transpose[data]},Transpose[N[
    Table[Standardize[datatr[[i]]], {i, 1, Length[datatr]}]]]];
```

Applying this function, we get the centered column vectors with unit variance

```
⇒ Xstd=std[X];Xstd//MatrixForm
```

$$\Leftarrow \begin{pmatrix} -0.872872 & -0.688247 \\ -0.218218 & -0.458831 \\ 1.091089 & 1.147079 \end{pmatrix}$$

Fig. 1.1 indicates that both coordinates are involved considerably in the 2D representation of the row vectors,

```
⇒ MatrixPlot[Xstd]//Binarize
```

⇐

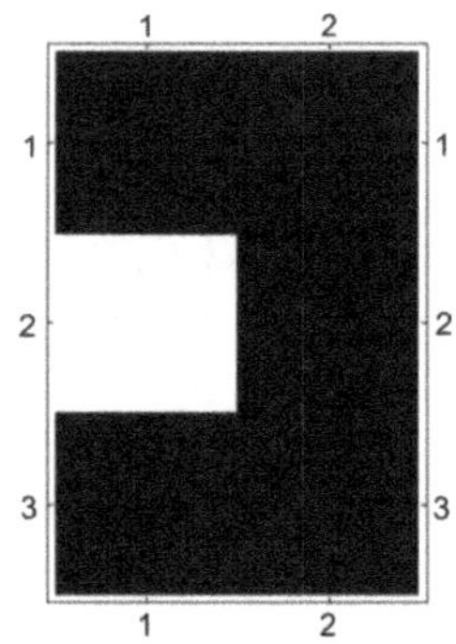

Fig. 1.1 Density plot of the elements of *X*.

We compute the covariance matrix (*n* is the numbers of row vectors)

```
⇒ n=3;
⇒ S=(Xstd//Transpose).Xstd/(n-1);MatrixForm[S]
```

$$\Leftarrow \begin{pmatrix} 1. & 0.976221 \\ 0.976221 & 1. \end{pmatrix}$$

The covariance matrix shows strong correlation between the different coordinates of the 2D space.

Alternatively we may use the built-in function

```
⇒ Covariance[Xstd]
⇐ {{1.,0.976221},{0.976221,1.}}
```

The eigenvalues of the covariance matrix,

```
⇒ Eigenvalues[S]
⇐ {1.97622,0.023779}
```

The eigenvalues are quite different. The eigenvectors are

```
⇒ Eigenvectors[S]
⇐ {{-0.707107,-0.707107},{0.707107,-0.707107}}
```

Then the *principal components* of *X* matrix are

```
⇒ pc=Xstd.Eigenvectors[S];pc//MatrixForm
```

$$\Leftarrow \begin{pmatrix} 0.130549 & 1.103878 \\ -0.170139 & 0.478746 \\ 0.039590 & -1.582624 \end{pmatrix}$$

It means that the principal components are the *X* projection into a space defined by the eigenvectors. Fig. 1.2 shows the dominance of the second coordinates

```
⇒ MatrixPlot[pc]//Binarize
```

⇐

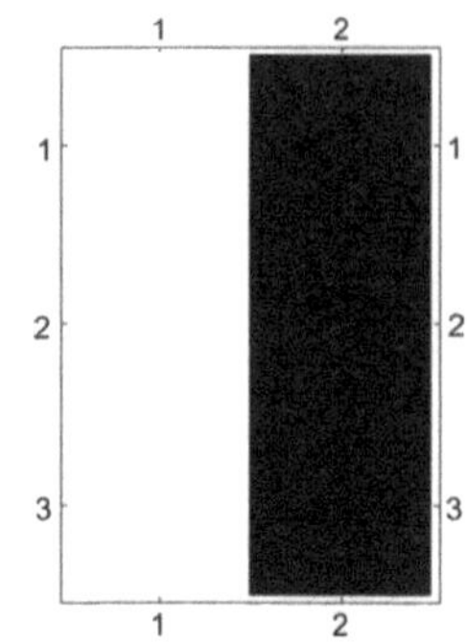

Fig. 1.2 Density plot of the elements of the projection of X into the eigenspace.

The principal components can be computed via built-in function, too

```
⇒ z=PrincipalComponents[Xstd];MatrixForm[z]
```

$$\Leftarrow \begin{pmatrix} 1.103878 & 0.130549 \\ 0.478747 & -0.170139 \\ -1.582624 & 0.039590 \end{pmatrix}$$

One can realize that the order of the column vectors are different.

```
⇒ Map[Reverse[#]&,z]//MatrixForm
```

$$\Leftarrow \begin{pmatrix} 0.130549 & 1.103878 \\ -0.170139 & 0.478746 \\ 0.039590 & -1.582624 \end{pmatrix}$$

In the eigenspace practically there is no correlation between the coordinates

```
⇒ Correlation[z]//MatrixForm
```

$$\Leftarrow \begin{pmatrix} 1. & 1.66533\times 10^{-16} \\ 1.66533\times 10^{-16} & 1. \end{pmatrix}$$

The back projection can be computed as

```
⇒ pc.Inverse[Eigenvectors[S]]==Xstd
⇐ True
```

The principal components can be computed in other ways, too. A very robust technique is the *Singular Value Decomposition.*

1.1.2 Singular Value Decomposition

Singular Value Decomposition (SVD; e.g., Grafarend and Awange 2003), is a computational method often employed to calculate principal components of a dataset. Using SVD to perform PCA is efficient and numerically robust technique. Moreover, the intimate relationship between them can guide our intuition about what PCA actually does and help us gain additional insights into this technique. Using built -in function

```
⇒ {u, s, v} = SingularValueDecomposition[Xstd];
```

where

```
⇒ u//MatrixForm
```

$$\Leftarrow \begin{pmatrix} 0.555249 & 0.598636 & 0.577350 \\ 0.240809 & -0.780178 & 0.577350 \\ -0.796058 & 0.181542 & 0.577350 \end{pmatrix}$$

```
⇒ s//MatrixForm
```

$$\Leftarrow \begin{pmatrix} 1.988075 & 0. \\ 0. & 0.218078 \\ 0. & 0. \end{pmatrix}$$

```
⇒ v//MatrixForm
```

$$\Leftarrow \begin{pmatrix} -0.707107 & -0.707107 \\ -0.707107 & 0.707107 \end{pmatrix}$$

Then the principal components,

```
⇒ u.s//MatrixForm
```

$$\Leftarrow \begin{pmatrix} 1.103878 & 0.130549 \\ 0.478746 & -0.170139 \\ -1.582624 & 0.039590 \end{pmatrix}$$

1.1.3 Karhunen-Loeve Decomposition

Karhunen–Loeve decomposition is typically used to reduce the dimensionality of data and capture the most important variation in the first few components. Now the principle components can be computed as

```
⇒ KL=KarhunenLoeveDecomposition[ Transpose[Xstd],
   Standardized  →  True ] [[1]] // Transpose; MatrixForm[KL]
```

$$\Leftarrow \begin{pmatrix} 1.103878 & 0.130549 \\ 0.478747 & -0.170139 \\ -1.582624 & 0.039590 \end{pmatrix}$$

Let us consider the components of the row vectors of X as the (x, y) coordinates of a 2D point. Then we can visualize the row vectors of the original and the projected X matrix, see Fig. 1.3.

```
⇒ Join[z,First[z]]
⇐ {{1.10388,0.130549},{0.478746,-0.170139},
   {-1.58262,0.0395904},1.10388,0.130549}
```

```
⇒ p0=Show[{ListPlot[Join[z,{First[z]}],
    PlotRange → All,PlotStyle → {PointSize → Large,Red},
    AspectRatio → 0.65],ListPlot[Join[Xstd,{First[Xstd]}],
    PlotRange → All, PlotStyle → {PointSize → Large,Blue},
    AspectRatio → 0.65]},PlotRange → All,Frame → True]
```

⇐

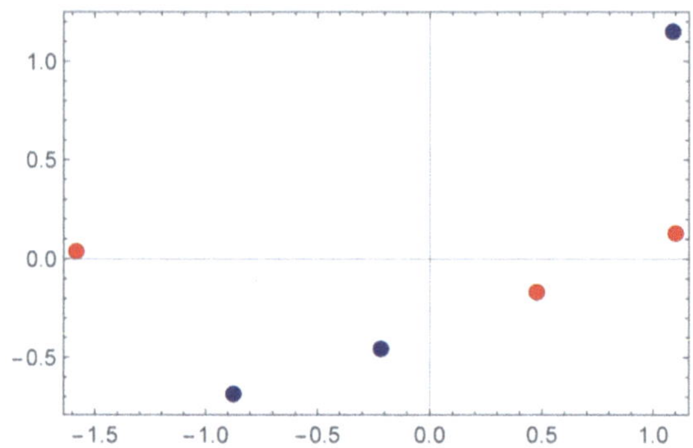

Fig. 1.3 The row vectors of X matrix as 2D points. The blue points stand for the original and the red points for the projected vectors.

1.1.4 PCA and Total Least Square

One can easily realize that the *approximate representation* of the projected points in the eigenspace is possible via fewer coordinates than in the original space. Namely in this example, one can consider only the *first* coordinate of the red points, the x coordinates, since the y coordinates are negligibly small.

Let us consider the original blue points as data points of a linear regression problem. Then the dominant eigenvector of the PCA as a regression line minimizes the error orthogonal (perpendicular) to the model line in the original space, consequently it provides the *Total Least Square* (TLS) solution.

For example if we consider the dominant eigenvector, which is

```
⇒ EigenvectorsX=Transpose[{Transpose[Eigenvectors[S]][[2]],{0,0}}]
⇐ {{-0.707107,0},{-0.707107,0}}
```

and projecting back the object into the original space, we get a line of the TLS solution, see Fig. 1.4

```
⇒ XstdX=z.PseudoInverse[EigenvectorsX]
⇐ {{-0.780559,-0.780559},{-0.338525,-0.338525},{1.11908,1.11908}}
```

```
⇒ p1=Show[{p0,ListPlot[XstdX,Joined → True,PlotStyle → Green]}]
```

⇐

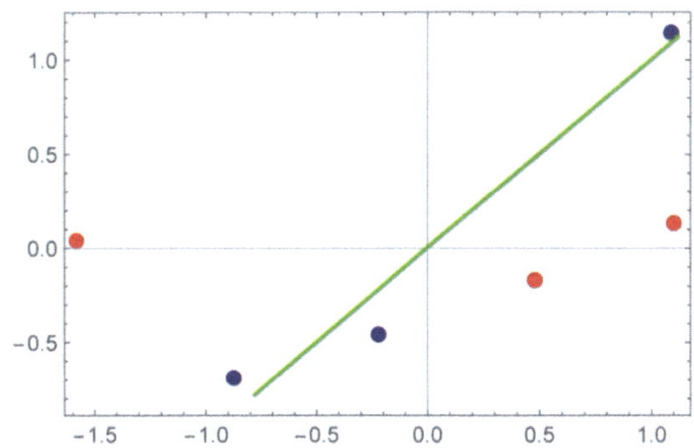

Fig. 1.4 The dominant eigenvector, green line provides the TLS solution for the blue data points.

To illustrate the situation let us compute the TLS per definition. In order to compute the total least squares, let Δx_i and Δy_i stand for the error components of the *i*-th point, then the objective function to be minimized is

$$\sum_{i=1}^{n} \Delta x_i^2 + \Delta y_i^2$$

under the constrains

$$y_i - \Delta y_i = \alpha\left(x_i - \Delta x_i\right) + \beta \qquad i = 1,...,n$$

Since the constrains are linear, eliminating Δy_i,

$$\sum_{i=1}^{n} \Delta x_i^2 + \left(y_i - \alpha\left(x_i - \Delta x_i\right) - \beta\right)^2$$

Considering the coordinates of the original blue points,

```
⇒ {x,y}=Transpose[Xstd];
```

The objective function is,

```
⇒ obj=Apply[Plus,Table[Δxᵢ²+(y[[i]]-α(x[[i]]-Δxᵢ)-β)²,{i,1,n}]];
```

Now the unknown variables are not only the parameters (α, β) but the adjustments Δx_i, too.

```
⇒ vars=Join[Table[{Δxᵢ},{i,1,n}],{α,β}]//Flatten;
```

Let us employ built-in optimization method,

```
⇒ AbsoluteTiming[sol=NMinimize[obj,vars];]
⇐ {0.0592062,Null}
```

Then the solution is

```
⇒ sol
```

```
⇐ {0.0475579, {Δx₁ → -0.0923122, Δx₂ → 0.120307,
   Δx₃ → -0.0279946, α → 1, β → 1.4957 × 10^-16}}
```

Fig. 1.5 shows the fitted line $y(x) = x$

```
⇒ Show[{p1,Plot[x,{x,-1,1.2},PlotStyle → {Dashed,Blue}]}]
```

⇐ 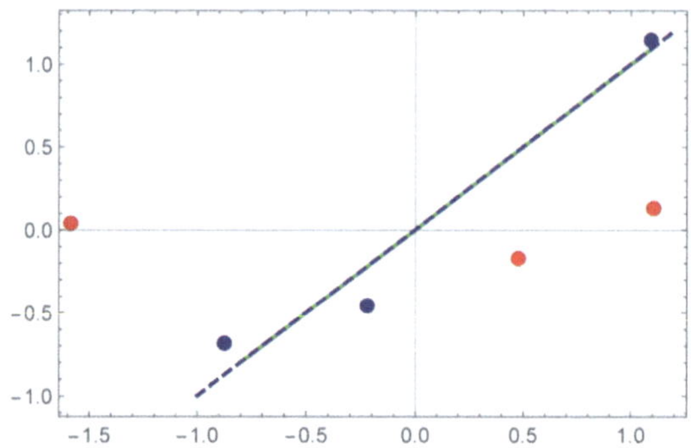

Fig. 1.5 The line of the dominant eigenvector, green line and the fitted TLS dashed blue line.

1.1.5 Image Compression

In case of image data reduction we can do the same. Let us consider the following image, see Fig. 1.6.

```
⇒ image=
```

Fig. 1.6 Density plot of the image to be compressed.

The image data

```
⇒ M=ImageData[image];
⇒ Dimensions[M]
⇐ {180,180}
```

The used storage space is

```
⇒ org=ByteCount[M]
⇐ 259352
```

Let us standardize the column vectors

```
⇒ Xstd=std[M];
```

The covariance matrix is

```
⇒ S=Covariance[Xstd];
```

The eigenvalues are

```
⇒ ES=Eigenvalues[S];
```

The numbers of them,

```
⇒ Dimensions[ES]
⇐ {180}
```

Let us display 20 of these (180) eigenvalues, see Fig. 1.7.

```
⇒ ListPlot[Take[ES,{1,20}],Joined → True,Frame → True,
   PlotStyle → {Red,Thin},PlotRange → All]
```

⇐

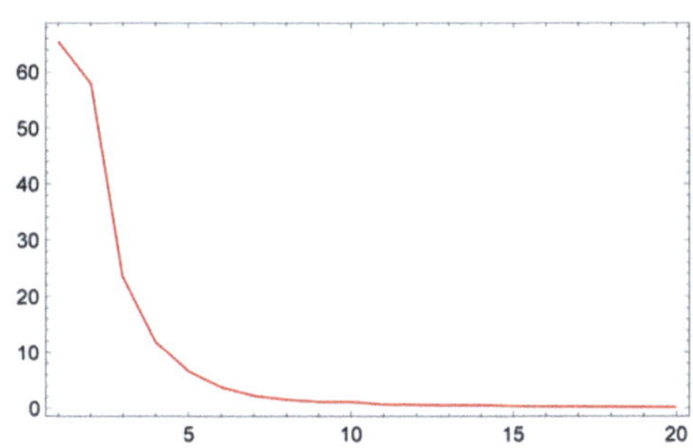

Fig. 1.7 The eigenvalues.

The eigenvectors

```
⇒ EV=Eigenvectors[S];
```

Let us take the nine eigenvectors from the 180, which means 0.05%

```
⇒ n=9;
⇒ EVC=Transpose[Take[EV,{1,n}]];
⇒ Dimensions[EVC]
⇐ {180,9}
```

The image reduction means to project the standardized image matrix into the eigenspace,

```
⇒ MM=Xstd.EVC;
```

Let us project it back into the original space, see Fig. 1.8

```
⇒ Mv=MM.PseudoInverse[EVC];
⇒ imageR=Image[Mv]//ImageAdjust
```

⇐ 

Fig. 1.8 The compressed image, the compression is 95%.

We store the image data in the eigenspace

```
⇒ comp95=ByteCount[MM]
⇐ 13432
```

Consequently the reduction is

```
⇒ 1-comp95/org//N
⇐ 0.948209
```

Alternatively, considering the ratio of the employed eigenvectors,

```
⇒ 1-9/180.
⇐ 0.95
```

A less compressed image (75%) can be seen in Fig. 1.9

```
⇒ n=45;
```

Since

```
⇒ 1-45./180
⇐ 0.75
```

Then

```
⇒ EVC=Transpose[Take[EV,{1,n}]];
```

Image reduction in the eigenspace

```
⇒ MM=Xstd.EVC;
```

Let us project it back into the original space

```
⇒ Mv=MM.PseudoInverse[EVC];
⇒ imageR45=Image[Mv]//ImageAdjust
```

⇐ 

Fig. 1.9 The compressed image, the compression is 75%.

Remark

We can get better quality if we transform back the image matrix from the standardized form.

1.1.6 Color Image Compression

In case of RGB images, one should map all of the three color-channel matrices. However, employing *Mathematica* built in function, we could solve this problem easily, since this function can reduce the vectors of a list. For example, if we have a list of n elements of vectors of m dimensions, then there are $nm = n \times m$ elements. One can partition these nm elements as $r \times q = nm$, and then reduce every vector of length of q to a vector of length of $qR < q$. The number of the reduced elements is therefore $nmR = r \times qR < r \times q = nm$.

Let us consider the following image

```
⇒ img=
```

Fig. 1.10 Colored image to be compressed.

The image data structure is obtained by

```
⇒ data=ImageData[img];data//Dimensions
⇐ {168,250,3}
```

The number of the elements (nm),

```
⇒ vector=Flatten[data];
⇒ Dimensions[vector]
⇐ {126000}
```

Let $q = 30$,

```
⇒ dvector=Partition[vector,30];
```

Now, we should like to reduce the original image to an image which is ten times smaller, $nm = 126000 \to nmR = 12600$. Let us employ the built-in function `DimensionReduce` with `PrincipalComponentsAnalysis` method,

```
⇒ autoencoder=DimensionReduction[dvector,3,
   Method → "PrincipalComponentsAnalysis"]
⇐ DimensionReducerFunction[
     Input type: NumericalVector (Length: 30)
     Output dimension: 3                      ]
```

Every vector of length 30 will be reduced to a vector of length 3 (encoding),

```
⇒ reduced=autoencoder[dvector];
⇒ Dimensions[reduced]
⇐ {4200,3}
```

The reconstruction (decoding)

```
⇒ reconstructed = autoencoder[reduced, "OriginalVectors"];
⇒ Dimensions[reconstructed]
⇐ {4200,30}
```

Organizing the elements in a color image data structure

```
⇒ diti=Partition[Partition[Flatten[reconstructed],3],250];
⇒ diti//Dimensions
⇐ {168,250,3}
```

Then the reconstructed image requiring 10% of the original storage space, can be seen in Fig. 1.11.

```
⇒ img090=Image[diti]
⇐
```

Fig. 1.11 The reconstructed color image from data, after 90%compression.

1.1.7 Image Compression in Python

Let us solve the same problem with Python code.

To use *Python* in *Mathematica,* we start a *Python* session,

```
⇒ session=
   StartExternalSession[<|"System" → "Python",
    "Version" → "3.5.4","Executable" →
     "C:\Users\Ben\AppData\Local\Programs\Python\Python35\
       python.exe"|>]//Quiet
```

```
⇐ ExternalSessionObject[
   SummaryPanel[ System: Python   EvaluationCount: None
                 UUID: 7b76966c2-ebb7-4ee8-bb25-02322d1456a8 ]]
```

Setup

```
# import packages needed for thi section
from sklearn.decomposition import PCA
from sklearn.preprocessing import normalize
import scipy.io as sio
import matplotlib.image as image
import pandas as pd
import matplotlib.pyplot as plt
```

Load image data

```
# Image is stored in MATLAB dataset
X = sio.loadmat('M:\\ex7faces.mat')
X = pd.DataFrame(X['X'])
# Normalize data by subtracting mean and scaling
X_norm = normalize(X)
```

Run PCA

```
# Set pca to find principal components that explain 99%
# of the variation in the data
pca = PCA(.99)
# Run PCA on normalized image data
lower_dimension_data = pca.fit_transform(X_norm)
# Lower dimension data from 5000x1024 to 5000x353
lower_dimension_data.shape
```

```
⇐ {5000,353}
```

Reconstruct images

```
# Project lower dimension data onto original features
approximation = pca.inverse_transform(lower_dimension_data)
# Approximation is 5000x1024
approximation.shape
# Reshape approximation and X_norm to 5000x32x32 to display
# images
approximation = approximation.reshape(-1,32,32)
X_norm = X_norm.reshape(-1,32,32)
```

Display images

The following code displays the original images next to their 99% of variation counterparts. Because of how *matplotlib* displays images, the pictures may be rotated. If you really want to fix this, you can transpose each row of X_norm and approximation using a for loop.

```
for i in range(0,X_norm.shape[0]):
 X_norm[i,] = X_norm[i,].T
 approximation[i,] = approximation[i,].T
fig4, axarr = plt.subplots(3,2,figsize=(8,8))
axarr[0,0].imshow(X_norm[0,],cmap='gray')
axarr[0,0].set_title('Original Image')
axarr[0,0].axis('off')
axarr[0,1].imshow(approximation[0,],cmap='gray')
axarr[0,1].set_title('99% Variation')
axarr[0,1].axis('off')
axarr[1,0].imshow(X_norm[1,],cmap='gray')
axarr[1,0].set_title('Original Image')
axarr[1,0].axis('off')
axarr[1,1].imshow(approximation[1,],cmap='gray')
axarr[1,1].set_title('99% Variation')
axarr[1,1].axis('off')
axarr[2,0].imshow(X_norm[2,],cmap='gray')
axarr[2,0].set_title('Original Image')
axarr[2,0].axis('off')
axarr[2,1].imshow(approximation[2,],cmap='gray')
axarr[2,1].set_title('99% variation')
axarr[2,1].axis('off')
plt.show()
```

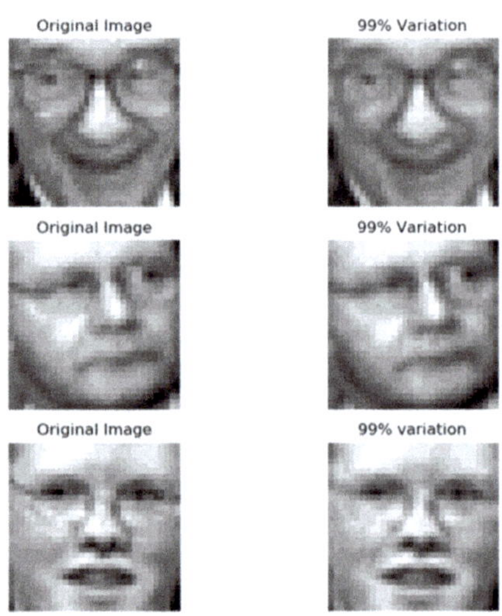

Fig. 1.12 The results of the Python code.

1.2 Independent Component Analysis

Basic theory

Independent Component Analysis (ICA, Forootan et al., 2012) was originally developed to deal with problems that are closely related to the cocktail-party problem. Since the recent increase of interest in ICA, it has become clear that this principle has a lot of other interesting applications as well (see e.g., Agutu et al., 2019; Ndehedehe et al., 2017a, b). Another, very different application of ICA is on feature extraction. A fundamental problem in digital signal processing is to find suitable representations for image, audio or other kind of data for tasks like compression and denoising. Data representations are often based on (discrete) linear transformations, see (Hyvärinen and Oja 2000).

1.2.1 Independent Component Analysis

Let us consider the Independent Component Analysis (ICA) technically as a nonnegative matrix factorization method, which factorizes a nonnegative M matrix of dimension $n{\times}m$ as $S{\times}A$ matrices with dimensions $n{\times}k$ and $k{\times}m$, respectively, so that

$$\text{Norm}\,(M - S \times A) \rightarrow \min$$

for a fixed $k \leq \max(n,m)$. Here k is the number of the independent component vectors. Higher k value provides better fitting.

For factorization one may use the so called *FastICA* algorithm see, (Hyvärinen and Oja, 1997). Let us see an example.

```
⇒ RandomSeed[1234];
⇒ R=RandomReal[{0,1},{6,5}];MatrixForm[R]
```

$$\Leftarrow \begin{pmatrix} 0.841172 & 0.830508 & 0.418279 & 0.874579 & 0.659077 \\ 0.340847 & 0.806480 & 0.502507 & 0.587620 & 0.961298 \\ 0.279034 & 0.189383 & 0.328972 & 0.946375 & 0.194114 \\ 0.488847 & 0.023039 & 0.349878 & 0.013693 & 0.156696 \\ 0.771758 & 0.114210 & 0.617770 & 0.959671 & 0.121220 \\ 0.715040 & 0.040767 & 0.081169 & 0.546480 & 0.264697 \end{pmatrix}$$

Let us employ the *FastICA* algorithm as implemented in *Mathematica* package written by (Antonov 2016)).

```
⇒ Import["https://raw.githubusercontent.com/antononcube/
   MathematicaForPrediction/
   master/IndependentComponentAnalysis.m"]
```

Let $k = 4$

```
⇒ k=4;
⇒ {S,A}=IndependentComponentAnalysis[Transpose[R],k,
    PrecisionGoal → 12];
```

Then

```
⇒ Norm[R-Transpose[S.A]]/Norm[R]
⇐ 0.64741
```

where

```
⇒ Transpose[S.A]//MatrixForm
```

$$\Leftarrow \begin{pmatrix} 0.496913 & 0.486248 & 0.074020 & 0.530320 & 0.314818 \\ -0.095124 & 0.370509 & 0.066536 & 0.151649 & 0.525327 \\ 0.331211 & 0.241559 & 0.381149 & 0.998552 & 0.246291 \\ 0.082743 & -0.383065 & -0.056227 & -0.392412 & -0.249409 \\ 0.452686 & -0.204862 & 0.298698 & 0.640599 & -0.197851 \\ 1.004190 & 0.329912 & 0.370315 & 0.835626 & 0.553843 \end{pmatrix}$$

Let us increase the k value.

```
⇒ k=5;
⇒ {S,A}=IndependentComponentAnalysis[Transpose[R],k,
    PrecisionGoal → 12];
⇒ Norm[R-Transpose[S.A]]/Norm[R]
⇐ 0.585572
```

So we have better fitting. Increasing k further

```
⇒ k=6;
⇒ {S,A}=IndependentComponentAnalysis[Transpose[R],k,
    PrecisionGoal → 12];
⇒ Norm[R-Transpose[S.A]]/Norm[R]
```

$\Leftarrow 1.03105 \times 10^{-15}$

We get quite perfect reconstruction of matrix R.

1.2.2 Image Compression via ICA

Now let us consider the following image,

⇒ imgR= 

Fig. 1.13 Original gray image (Range Image) of 128×128

Mathematica

The image matrix is,

```
⇒ R=ImageData[imgR];
⇒ Dimensions[R]
⇐ {128,128}
```

First we consider $k = 20$, then

```
⇒ {S,A}=IndependentComponentAnalysis[Transpose[R],20,
    PrecisionGoal → 15];
```

The error of the fitting,

```
⇒ Norm[R-Transpose[S.A]]/Norm[R]
⇐ 0.204661
```

where

```
⇒ Dimensions[S]
⇐ {128,20}
```

```
⇒ Dimensions[A]
⇐ {20,128}
```

The reconstructed image is

```
⇒ imdata=Transpose[S.A];
⇒ Image[imdata]
```

⇐ 

Fig. 1.14 The reconstructed image with $k = 20$

Let us increase $k = 50$

```
⇒ {S,A}=IndependentComponentAnalysis[Transpose[R],50,
   PrecisionGoal → 15];
```

Now the error decreased

```
⇒ Norm[R-Transpose[S.A]]/Norm[R]
⇐ 0.117028
```

Then

```
⇒ Dimensions[S]
⇐ {128,50}

⇒ Dimensions[A]
⇐ {50,128}
```

The image is,

```
⇒ imdata=Transpose[S.A];
⇒ imgC=Image[imdata]
```

⇐ 

Fig. 1.15 The reconstructed image with $k = 50$

The quality of the reconstructed image can be improved via convolution,

```
⇒ GaussianFilter[imgC,1.]//ImageAdjust
```

⇐ 

Fig. 1.16 The reconstructed image with $k = 50$ after improvement via convolution

Now let us employ *Python*. Our image is the well known

```
⇒ lena=Import["G:\\lena.png"]
```

⇐ 

Fig. 1.17 The original image

```
⇒ Dimensions[ImageData[lena]]
⇐ {512,512,4}
```

Python

```
⇒ Off[Syntax::stresc]
⇒ session=StartExternalSession[<|"System" → "Python",
   "Version" → "3.5.4","Executable" → "C:\Users\Bela\AppData\
   Local\Programs\Python\Python35\python.exe"|>]//Quiet
⇐ ExternalSessionObject[
   SummaryPanel[ System: Python    EvaluationCount: None
                 UUID: 7b76966c2-ebb7-4ee8-bb25-02322d1456a8 ]]
```

The code developed by Treadway (2018),

```python
# load packages
from sklearn.decomposition import FastICA
from pylab import *
from skimage import data, io, color
```

Let us employ $k = 20$ and $k = 50$,

```
ica = FastICA(n_components = 50)
```

```
emc2_image = io.imread("G:\\lena.png", as_grey = True)
ica.fit(emc2_image)
```

```
⇐ ExternalObject[ Name: FastICA          System:   Python
                  Type: PythonFunction   IsModule: False ]
```

```
emc2_image_ica = ica.fit_transform(emc2_image)
emc2_restored = ica.inverse_transform(emc2_image_ica)
```

```
io.imshow(emc2_restored)
show()
```

```
⇐ $Aborted
```

The results can be seen below.

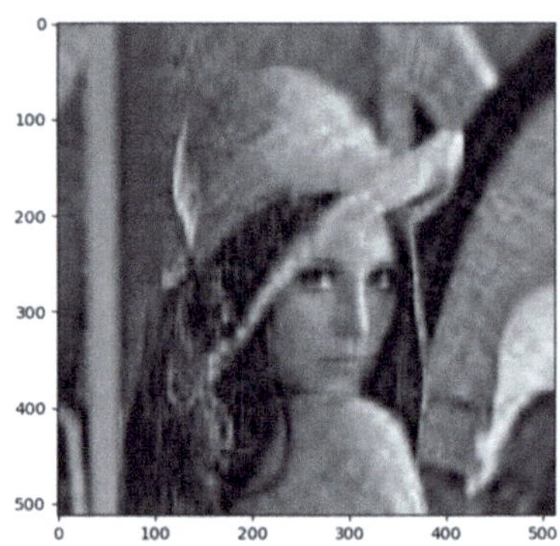

Fig. 1.18 The compressed image, $k = 20$

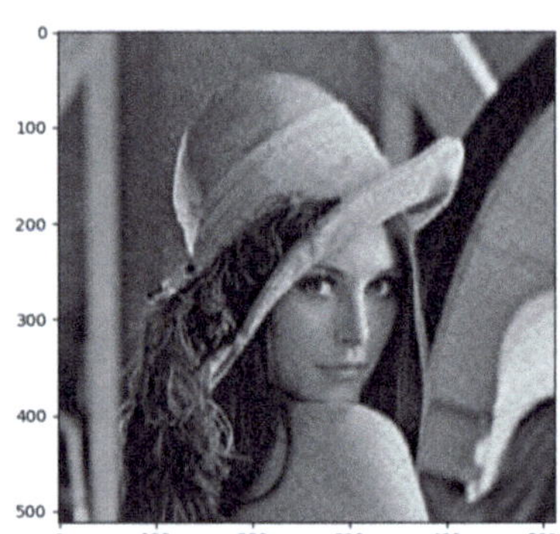

Fig. 1.19 The compressed image, $k = 50$

1.3 Discrete Fourier Transform

Basic theory

The field of digital signal processing relies heavily on operations in the frequency domain (i.e. on the Fourier transform). For example, several lossy image and sound compression methods employ the discrete Fourier transform: the signal is cut into short segments, each is transformed, and then the Fourier coefficients of high frequencies, which are assumed to be unnoticeable, are discarded. The decompressor computes the inverse transform based on this reduced number of Fourier coefficients.

In general *Discrete Fourier Transformation* (DFT) maps a list of real numbers into another list of complex numbers. In our case - image compression - we have real intensity values equally spaced. Anyway the result will be complex.

1.3.1 Data Compression via DFT

First let us consider the basic idea of the *compression in 1D*. We have a list of real numbers

```
⇒ u={1,1,2,2,1,1,0,0};
```

Now, let us carry out its Fourier transform

```
⇒ v=Fourier[{1,1,2,2,1,1,0,0}]
⇐ {2.82843 +0. I,-0.5+1.20711 I,0. +0. I,0.5 -0.207107 I,
   0. +0. I,0.5 +0.207107 I,0. +0. I,-0.5-1.20711 I}
```

The absolute values and the arguments of the complex elements,

```
⇒ vAbs=Abs[v]
⇐ {2.82843,1.30656,0.,0.541196,0.,0.541196,0.,1.30656}
```

```
⇒ vArg=Arg[v]
⇐ {0.,1.9635,0.,-0.392699,0.,0.392699,0.,-1.9635}
```

The transformation is invertable

```
⇒ InverseFourier[vAbs Exp[I vArg]]
⇐ 1.,1.,2.,2.,1.,1.,0.,-1.57009 × 10^-16
```

Replacing elements that are close to zero by 0 digit.

```
⇒ Chop[%]
⇐ {1.,1.,2.,2.,1.,1.,0,0}
```

For data compression, let us eliminate elements (replace by zero) representing higher frequencies, elements with small absolute values. In this case the compressed list contains only three nonzero elements instead of six,

```
⇒ vvAbs=Map[If[#<0.6,0,#]&,vAbs]
⇐ {2.82843,1.30656,0,0,0,0,0,1.30656}
```

Transforming this list back leads to

```
⇒ uu=InverseFourier[vvAbs Exp[I vArg]]//Chop
⇐ {0.646447,1.35355,1.85355,1.85355,1.35355,0.646447,
   0.146447,0.146447}
```

Fig. 1.20 shows the visualization of the two list.

```
⇒ ListPlot[{u,uu},Joined → True,Frame → True]
```

⇐

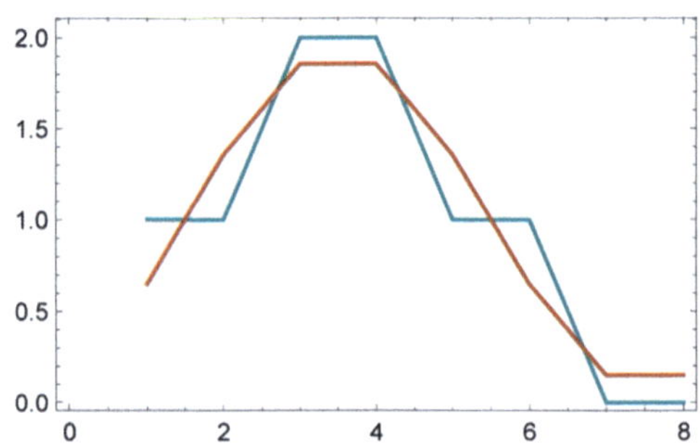

Fig. 1.20 The graph of the original list (blue) and the compressed list (brown).

In case of 2D, the technique is very similar. Let us consider a very simple binary matrix

⇒ Face=

$$
\begin{bmatrix}
1&1&1&1&1&1&1&0&0&0&0&0&0&1&1&1&1&1&1&1\\
1&1&1&1&1&0&0&0&0&0&0&0&0&0&0&1&1&1&1&1\\
1&1&1&0&0&0&0&0&0&0&0&0&0&0&0&0&0&1&1&1\\
1&1&0&0&0&0&0&0&0&0&0&0&0&0&0&0&0&0&1&1\\
1&1&0&0&0&0&0&0&0&0&0&0&0&0&0&0&0&0&1&1\\
1&0&0&0&0&0&1&1&0&0&0&0&1&1&0&0&0&0&0&1\\
1&0&0&0&0&1&1&1&1&0&0&1&1&1&1&0&0&0&0&1\\
0&0&0&0&0&1&1&1&1&0&0&1&1&1&1&0&0&0&0&0\\
0&0&0&0&0&0&1&1&0&0&0&0&1&1&0&0&0&0&0&0\\
0&0&0&0&0&0&0&0&0&0&0&0&0&0&0&0&0&0&0&0\\
0&0&0&0&0&0&0&0&0&0&0&0&0&0&0&0&0&0&0&0\\
0&0&0&0&0&0&0&0&0&0&0&0&0&0&0&0&0&0&0&0\\
0&0&0&0&1&1&1&1&1&1&1&1&1&1&1&1&1&0&0&0\\
1&0&0&0&0&1&1&1&1&1&1&1&1&1&1&1&0&0&0&1\\
1&0&0&0&0&0&0&1&1&1&1&1&1&1&0&0&0&0&0&1\\
1&1&0&0&0&0&0&0&1&1&1&1&0&0&0&0&0&0&1&1\\
1&1&0&0&0&0&0&0&0&0&0&0&0&0&0&0&0&0&1&1\\
1&1&1&0&0&0&0&0&0&0&0&0&0&0&0&0&0&1&1&1\\
1&1&1&1&1&0&0&0&0&0&0&0&0&0&0&1&1&1&1&1\\
1&1&1&1&1&1&1&0&0&0&0&0&0&1&1&1&1&1&1&1
\end{bmatrix}
$$

which is the dataset of the following image

```
⇒ Image[Face]
```

⇐

Fig. 1.21 The image of the binary matrix.

Let us carry out its DFT and visualize the absolute value of the Fourier coefficients, see Fig. 1.22

```
⇒ FaceFu=Fourier[Face];
⇒ AbsFaceFu=Abs[FaceFu];
⇒ ListPlot3D[AbsFaceFu,PlotRange → {0,4}]
```

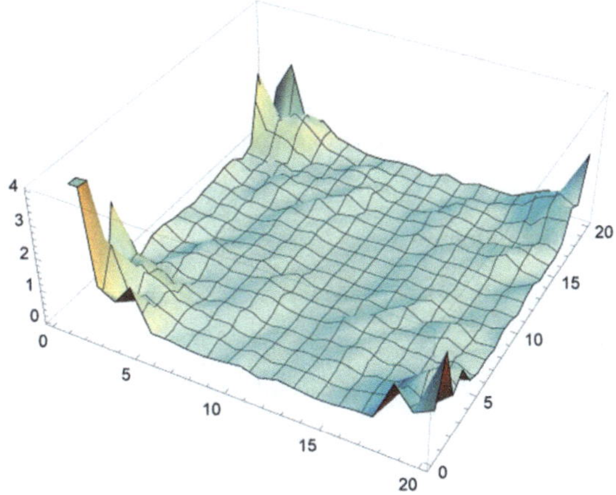

⇐

Fig. 1.22 The absolute values of the DFT coefficients.

Let us see the histogram of these values, see Fig. 1.23

```
⇒ data=Flatten[AbsFaceFu];
⇒ Histogram[data]
```

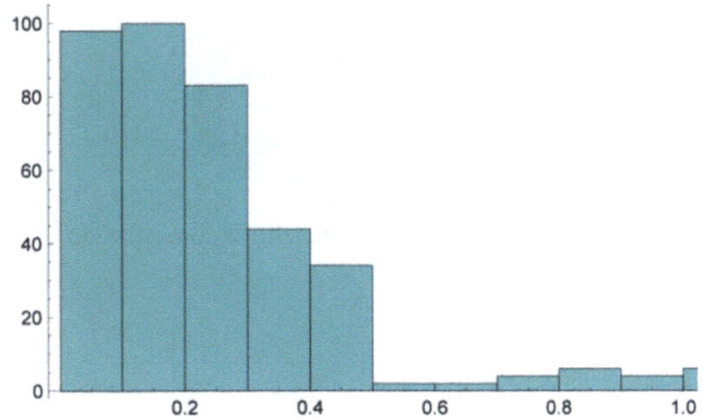

⇐

Fig. 1.23 The histogram of the absolute values of the DTF coefficients.

We can see that there are a lot of small and a few big elements. Let us kick out the big ones, for example those are bigger than 0.5, see Fig. 1.24

```
⇒ M=AbsFaceFu;
⇒ Do[If[AbsFaceFu[[i,j]]>0.5,M[[i,j]]=AbsFaceFu[[i,j]],
    M[[i,j]]=0],{i,1,20},{j,20}]
⇒ ListPlot3D[M,PlotRange → {0,2}]
```

⇐

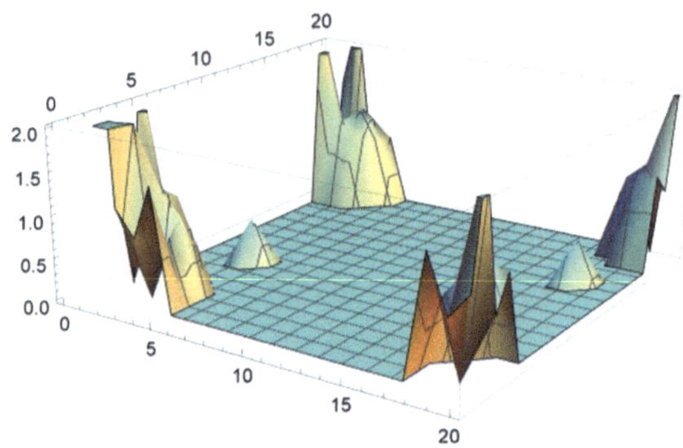

Fig. 1.24 The absolute values of the remained nonzero DFT coefficients.

The number of the zero elements

```
⇒ n0=Select[Flatten[M],#==0&]//Length
⇐ 359
```

This means that roughly 90 % of the total numbers of the 400 coefficients are considered

```
⇒ n0/(20 × 20)
⇐ 0.8975
```

Employing inverse transform, the compressed image can be seen in Fig. 1.25

```
⇒ Chop[InverseFourier[M Exp[I Arg[FaceFu]]]] // Image
```

⇐

Fig. 1.25 The image after 90 % compression.

1.3.2 DFT Image Compression

Now we can employ this technique for compressing a digital image, see Fig. 1.26

⇒ img=

Fig. 1.26 Image to be compressed.

The matrix of the image data

```
⇒ M=ImageData[img];
⇒ M//Dimensions
⇐ {148,223,3}
```

We consider only the first color channel, see Fig. 1.27

```
⇒ MM=Table[First[M[[i,j]]],{i,1,148},{j,1,223}];
⇒ Dimensions[MM]
⇐ {148,223}
```

Then the image is,

```
⇒ imgK=Image[MM];imgK//BrightnessEqualize
```

⇒

Fig. 1.27 Employing a single color channel of the image data.

Let us carry out DFT on the image data

```
⇒ dft=Fourier[MM];
```

The absolute values and the arguments of the coefficients and the arguments

```
⇒ abs=Abs[dft];
⇐ arg=Arg[dft];
```

Fig. 1.28 shows the absolute values of the DFT coefficients

```
⇒ ListPlot3D[abs,PlotRange → {0,1.5}]
```

⇒ 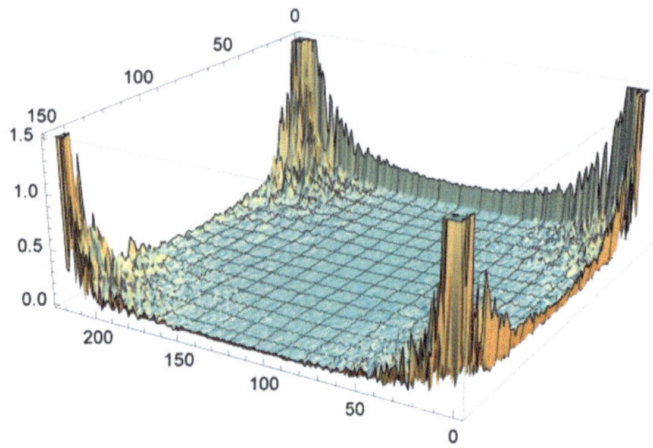

Fig. 1.28 The absolute values of the DFT coefficients.

Let us consider the histogram, see Fig. 1.29

```
⇒ data=Flatten[Abs];
⇒ Histogram[data]
```

⇐ 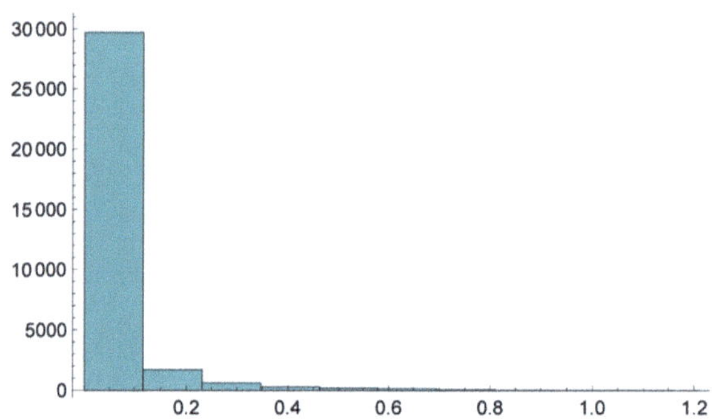

Fig. 1.29 The histogram of the absolute values of the DTF coefficients.

We eliminate – replace them with zero – the coefficients smaller than 0.2, see Fig. 1.30.

```
⇒ M=abs;
⇒ Do[If[abs[[i,j]]>0.2,M[[i,j]]=abs[[i,j]],M[[i,j]]=0],
    {i,1,148},{j,1,223}]
⇒ ListPlot3D[M,PlotRange → {0,1.5}]
```

⇐

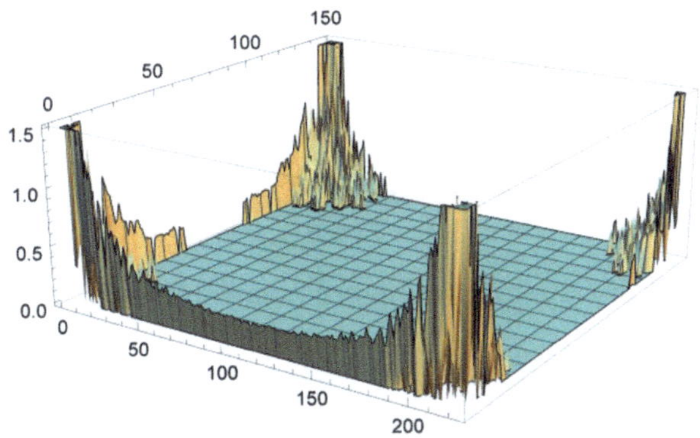

Fig. 1.30 The absolute values of the DFT coefficients bigger than 0.2.

Now, we can compute the number of zeros in the image data matrix

```
⇒ data=Flatten[M];
⇒ n0=Select[data,#==0&]//Length
⇐ 31165
```

This means roughly 95 % compression, since

```
⇒ 1.-n0/(148 223)
⇐ 0.0557205
```

The compressed image is,

```
⇒ Chop[InverseFourier[M Exp[I arg]]] // Image//BrightnessEqualize
```

⇐

Fig. 1.31 The reconstructed image after ~ 95 % compression.

1.4 Discrete Wavelet Transform

Basic theory

The Discrete Wavelet Transform (DWT) projects the information of an image into *orthonormal subspaces* constructed by wavelet bases. The coefficients of these basis functions are representing the *local* details. The very small coefficients can be set to zero without significantly changing the other part of the image. The greater the number of zeros the greater the compression ratio.

1.4.1 Concept of Discrete Wavelet Transform

Let us consider a discrete signal computed from the function

```
⇒ f[x_]:=Sin[20 x] Log[x]²
```

The list of the discrete values is

```
⇒ u=N[Table[f[i],{i,1/32,1,1/32}]]
⇐ {7.02779,7.29508,5.346,2.58784,0.057173,-1.60163,-2.17964,
   -1.84287,-0.98427,-0.044889,0.636128,0.90238,0.781799,0.426935,
   0.0285651,-0.261377,-0.372927,-0.320387,-0.173266,-0.0146507,
   0.0940374,0.130003,0.10599,0.0538185,0.00504998,-0.0222418,
   -0.0265446,-0.017396,-0.00642286,-0.000413988,0.0005056,0.}
```

Let us visualize it, see Fig. 1.32

```
⇒ p0=ListPlot[u,PlotStyle → {Red,Thin},Joined → True,
   Frame → True,PlotRange → All]
```

⇐

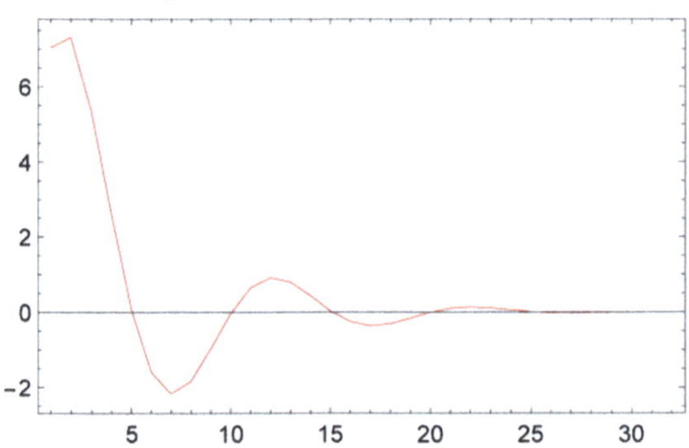

Fig. 1.32 Discrete value signal.

This can be considered as a vector in a V_5 space of dimension of $2^5 = 32$. Using DWT this vector can be projected into a space V_4, which is an orthogonal subspace of V_5. This is done using orthonormal wavelet basis.

There are many different types of orthonormal wavelets, for example classic Haar wavelet, the Daubechies wavelets and Battle-Lemarié wavelets based on B-spline, see Figs.1.33 - 1.35,

```
⇒ Plot[WaveletPsi[HaarWavelet[],x],{x,-1,2},Exclusions → None]
```

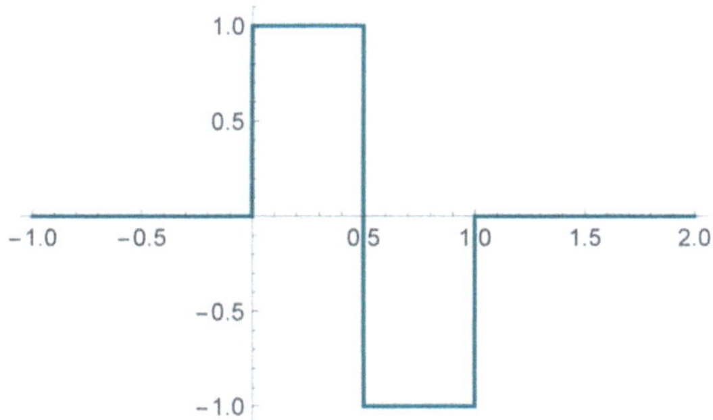

⇐

Fig. 1.33 Classic Haar wavelet.

```
⇒ Plot[WaveletPsi[DaubechiesWavelet[4],x],{x,-3,4},PlotRange → All]
```

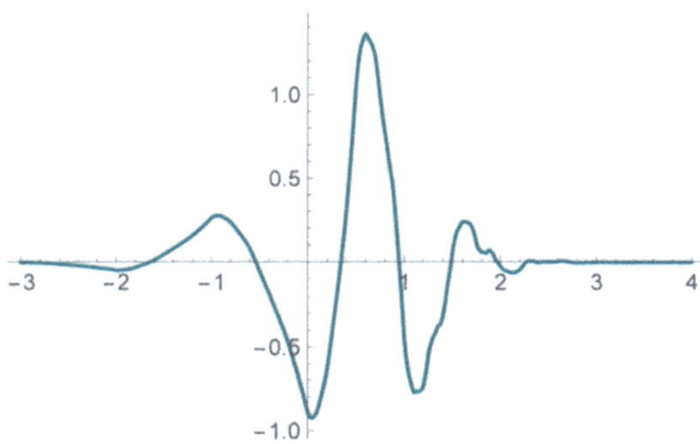

⇐

Fig. 1.34 Daubechies wavelet of order 4.

```
⇒ Plot[WaveletPhi[BattleLemarieWavelet[3,10],x],{x,-10,10},
   PlotRange → All]
```

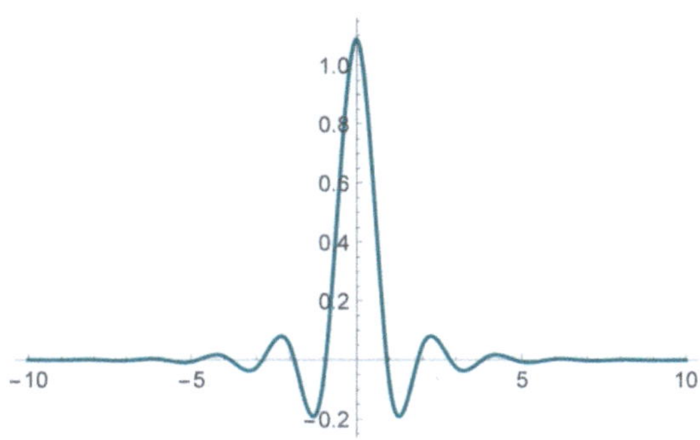

⇐

Fig. 1.35 Battle-Lemarié wavelets based on B-spline of order 3 evaluated on equally spaced interval {-10, 10}.

Here we employ one of the most simple ones, *the Haar wavelet families*. Let n be the level of the decomposition (refinement),

```
⇒ n = 1;
```

The DWT of the signal u employing Haar wavelet basis,

```
⇒ dwd = DiscreteWaveletTransform[u,HaarWavelet[],n]
⇐ DiscreteWaveletData[ Data dimension: {32}
                       Refinements: 1 ]
```

This orthogonal decomposition can be visualized via the wavelet tree, see Fig. 1.36

```
⇒ dwd["TreeView"]
```

⇐

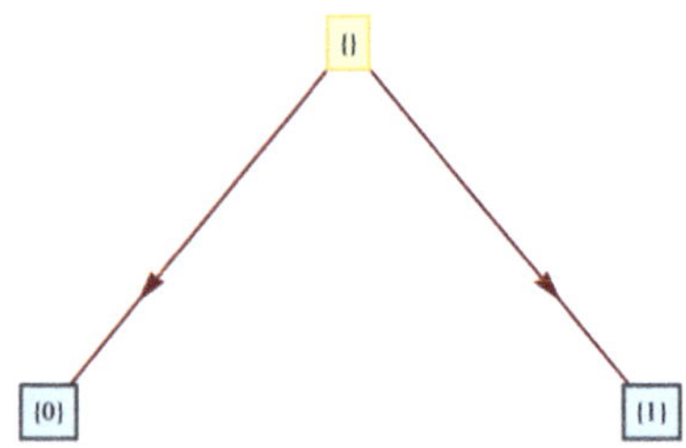

Fig. 1.36 Wavelet tree of refinement level, $n = 1$.

The projected vector in V_4 space of dimension of $2^4 = 16$

```
⇒ p1=Normal[dwd];
⇒ p1[[1]]
⇐ {0} → {10.1278,5.61007,-1.09209,-2.84435,-0.727726,1.08789,
  0.854704,-0.164623,-0.490247,-0.132877,0.158421,0.113001,
  -0.0121564,-0.0310707,-0.00483438,0.000357513}
```

and its residual

```
⇒ p1[[2]]
⇐ {1}->{-0.189,1.95032,1.17295,-0.238134,-0.664243,-0.188269,
  0.250927,0.20502,-0.0371513,-0.112158,-0.0254319,0.0368906,
  0.0192982,-0.00646902,-0.00424891,0.000357513}
```

see Fig. 1.37 which shows the projected signal and its residual in $V_4^{\perp}$, which is the orthonormal space of V_4

```
⇒ GraphicsGrid[
  {{ListPlot[p1[[1]][[2]]/1.5,PlotStyle → {Red},Joined → True,
  Frame → True,PlotRange → {-3,11},AspectRatio → 1.5],
  ListPlot[p1[[2]][[2]]/1.5,PlotStyle → {Blue,Thin},
  Joined → True,Frame → True,PlotRange → {-3,11},
  AspectRatio → 1.5]}}]
```

⇐ 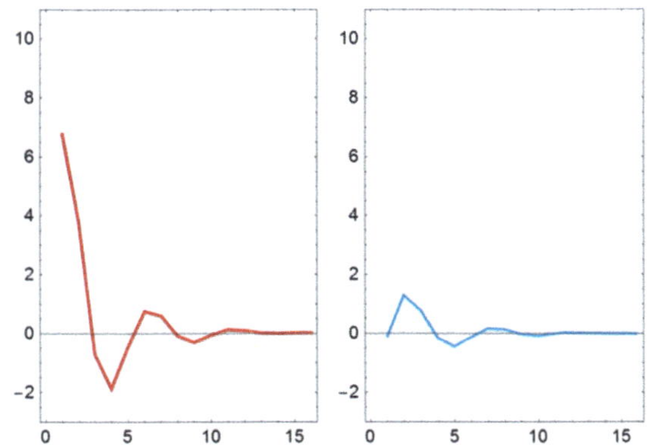

Fig. 1.37 An image box, showing the projections of the signal onto V_4 and $V_4^{\perp}$.

We can see that the residual in $V_4^{\perp}$ is quite big. Therefore we need deeper decomposition (higher level refinement).

```
⇒ n = 2;
⇒ dwd = DiscreteWaveletTransform[u,HaarWavelet[],n]
⇐ DiscreteWaveletData[ Data dimension: {32} Refinements: 2 ]
```

This orthogonal decomposition can be visualize via the wavelet tree,

```
⇒ dwd["TreeView"]
```

⇐ 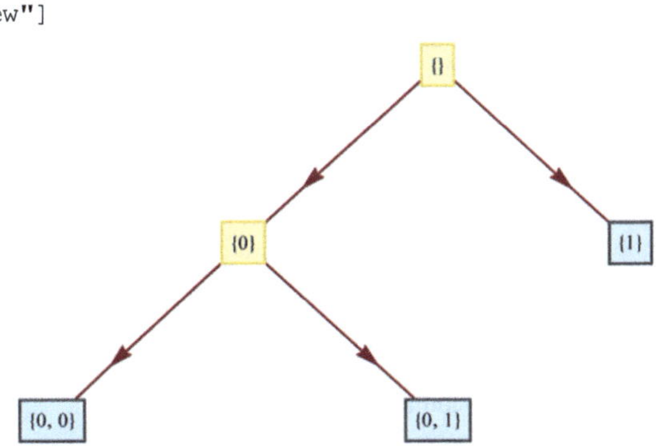

Fig. 1.38 Wavelet tree of refinement level, $n = 2$.

The projected vector will further be project onto V_3 orthogonal subspace, whose space is orthogonal to V_4 space,

```
⇒ p2=Normal[dwd];
⇒ p2[[1]]
```

and its residual is

```
⇒ p2[[2]]
```

```
⇐ {1} → {-0.189,1.95032,1.17295,-0.238134,-0.664243,-0.188269,
    0.250927,0.20502,-0.0371513,-0.112158,-0.0254319,0.0368906,
    0.0192982,-0.00646902,-0.00424891,0.000357513}
```

Then the projected vector will be further project into V_3 orthogonal subspace, whose space is orthogonal to V_4 space,

```
⇒ p2[[3]]
⇐ {{0,0} → {11.1284,-2.78349,0.254674,0.487961,
    -0.440615,0.191924,-0.0305662,-0.00316562}
```

and its residual

```
⇒ p2[[4]]
⇐ {0,1} → {3.19452,1.23903,-1.28383,0.720772,
    -0.252698,0.0321164,0.0133744,-0.00367122}
```

see Fig. 1.39

```
⇒ GraphicsGrid[
    {{ListPlot[p2[[3]][[2]]/1.5,PlotStyle → {Red},Joined → True,
    Frame → True,PlotRange → {-3,12},AspectRatio → 1.5],
    ListPlot[p2[[4]][[2]]/1.5,PlotStyle → {Green,Thin},
    Joined → True,Frame → True,PlotRange → {-3,12},
    AspectRatio → 1.5],ListPlot[p1[[2]][[2]]/1.5,
    PlotStyle → {Blue,Thin},Joined → True,Frame → True,
    PlotRange → {-3,12},AspectRatio → 1.5]}}]
```

⇐

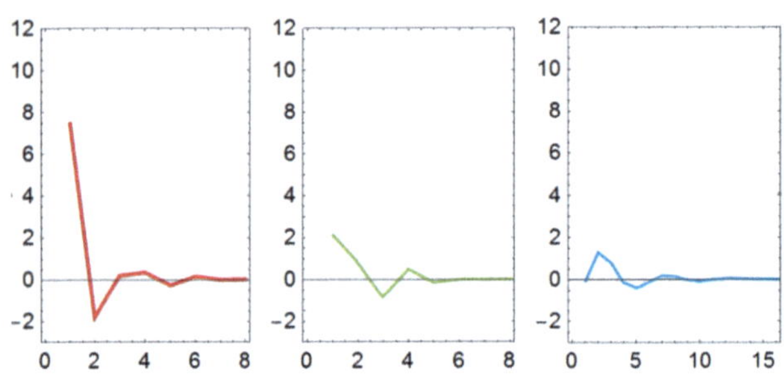

Fig. 1.39 An image box, containing the projections onto V_3, $V_3^{\perp}$ and $V_4^{\perp}$.

1.4.2 2D Discrete Wavelet Transform

We can proceed similarly in case of a 2D discrete signal represented by a matrix. Let us suppose that the signal J is represented by an array of 32×32 elements. The first step is to transform (to project) the 32 rows, as 1D signal. The result will be two arrays of 32×16 elements, (J_h and J_g), where the rows of J_h are the signals projected into the space of V_4, while the rows of J_g are the residuals in $V_4^{\perp}$.

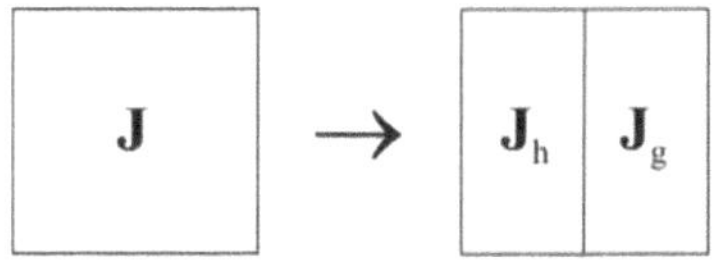

Fig. 1.40 The first step of creating an image box in 2D.

In the following step, we carry out the projections for the columns of the matrices J_h and J_g, see Fig. 1.41

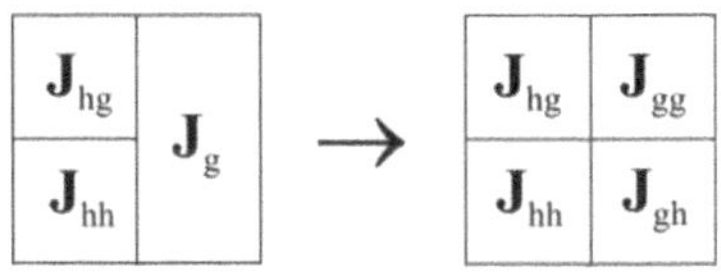

Fig. 1.41 The next step is the projections of the columns.

Let us illustrate this process with an image, see Fig. 1.42

⇒ A=  ;

Fig. 1.42 A tennis player, Martina Hingis.

We define a projection function as

```
⇒ H[s_]:=Module[{sV,dV},sV=Map[Apply[Plus,#]&,Partition[s,2]]/2;
   dV=Map[Apply[Subtract,#]&,Partition[s,2]]/2;Join[sV,dV]]
```

Let us employ this function for the image, but before that, we resize it as a gray image of size 128×128.

```
⇒ AA=ImageResize[A,{128,128}];
⇒ hingis=ColorConvert[AA,"GrayScale"][
```

⇐ 

Fig. 1.43 The resized and "grayed" image.

Checking the dimensions and assigning the image data

```
⇒ ImageData[hingis]//Dimensions
⇐ {128,128}

⇒ Hingis=ImageData[hingis];
```

Now we can carry out the first step, projecting rows

```
⇒ ImageBox1=Partition[Flatten[Map[H,Hingis]],128];
⇐ ImageBox11=Table[If[j>64,If[j==65,63,(63-ImageBox1[[i,j]])/1.5],
    ImageBox1[[i,j]]],{i,1,128},{j,1,128}];
```

The result can be seen in Fig. 1.44

```
⇒ ListDensityPlot[ImageBox1//Reverse,Mesh → False,
    ColorFunction → GrayLevel]
```

⇐ 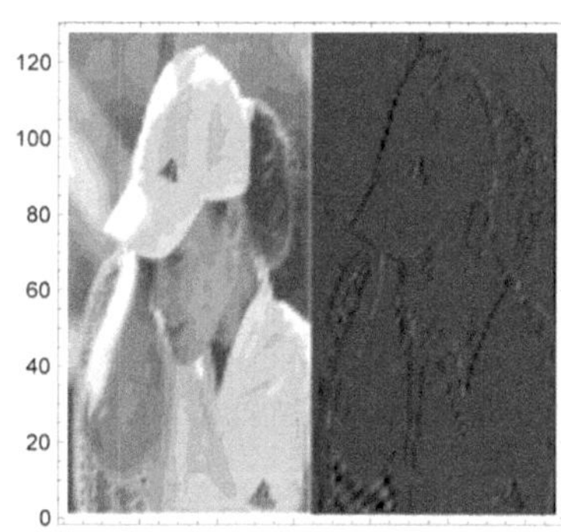

Fig. 1.44 The image after projection of the rows.

Then let us project the columns

```
⇒ ImageBox2=
    Transpose[Partition[Flatten[Map[H,Transpose[ImageBox1]]],128]];
```

```
⇒ ImageBox21=Table[If[i>64||j>64,If[j==65||i==65,63,
   (63-ImageBox2[[i,j]])/1.5],ImageBox2[[i,j]]],{i,1,128},
   {j,1,128}];
```

The result is in Fig. 1.45

```
⇒ p2=ListDensityPlot[
   ImageBox2//Reverse,Mesh → False,ColorFunction → GrayLevel]
```

⇐

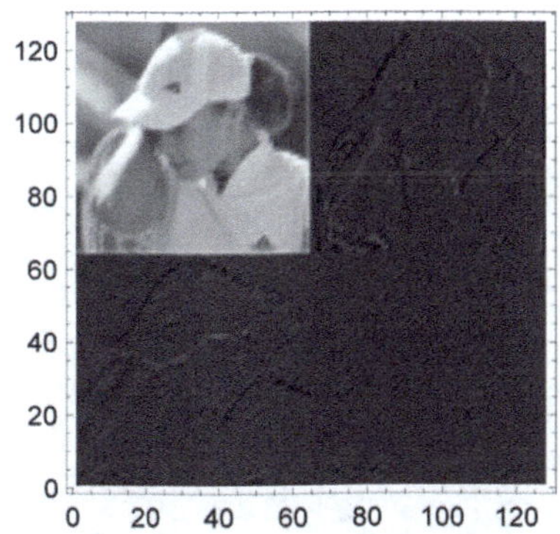

Fig. 1.45 The result of the 2D DWT.

The image size in Fig. 1.45 is 64×64, which means the compressed form of the original image of size 128×128.

1.4.3 DWT Image Compression

Now let us employ this technique for a color image of size 288×492, see in Fig. 1.46.

⇒ img=

Fig. 1.46 A view of Budapest.

Now we can use *DaubechiesWavelet* family of order four, ensuring better quality, since contrary to Haar Wavelet, this family satisfies some smoothness conditions. Let us employ $n = 2$ refinement,

```
⇒ dwd=DiscreteWaveletTransform[img,DaubechiesWavelet[4],2];
```

The image boxes can be seen in Fig. 1.47

```
⇒ z=WaveletImagePlot[dwd]
⇐
```

Fig. 1.47 The image boxes of the DWT.

Let us display the compressed image, see Fig. 1.48

```
⇒ hu=ImageData[z];hu//Dimensions
⇐ {288,492,3}
⇒ Take[hu,{5,72},{5,120}]//Image//ImageAdjust
⇐
```

Fig. 1.48 The compressed image.

The compression ratio

```
⇒ 1 - 1. / 2^4
⇐ 0.9375
```

which means roughly 94%.

In order to get more fancy images of the residual images, we may introduce the following function,

```
⇒ imgFunc[img_,{___,1|2|3}]:=
    Composition[Sharpen[#,0.5]&,ImageAdjust[#,{0,1}]&,ImageAdjust,
    ImageApply[Abs,#1]&][img]
  imgFunc[img_,wind_]:=Composition[ImageAdjust,
    ImageApply[Abs,#1]&][img]
```

Then we get the nice image boxes, see in Fig. 1.49

```
⇒ z=WaveletImagePlot[dwd,Automatic,imgFunc[#1,#2]&,
    BaseStyle → Red,ImageSize → 500]
```

⇐

Fig. 1.49 The colorized image boxes.

1.5 Radial Basis Function

Basic theory

The Radial Basis Method (RBF) is one of the kernel methods, which can be employed for *approximation* of data or functions given by values especially when we do not have regular grid but *scattered data* in *many dimensions*. The RBF method employs the linear combination of the so called radial basis functions to carry out basically local approximation. This method is also applied in machine learning, namely as activation function of artificial neural networks (ANN).

The wide spread successful application of RBF is based on the theoretical fact that the radial basis functions, like algebraic polynomials and sigmoid activation functions of ANN, are so called *universal approximators.*

1.5.1 RBF Approximation

Let $\varphi(x)$ be a nonconstant, bounded, and monotonically increasing continuous function. Let I_m be denoted by an m-dimensional unit hypercube where $x \in I_m$. Considering the space of the continuous functions on this hypercube $C(I_m)$, then for any $g(x) \in C(I_m)$ and $\epsilon > 0$, there exist an integer n with real constants b_i and w_i, such that

$$G(x) = \sum_{i=1}^{n} c_i \varphi\left(w_i^T x + b_i\right)$$

where $G(x)$ is independent of $g(x)$ and

$$|G(x) - g(x)| < \epsilon$$

for all $x \in I_m$. In other words, functions of the form $G(x)$ are dense in $C(I_m)$. The RBF method employs the linear combination of $\varphi(x)$ basis functions,

$$f(x) = \sum_{i=1}^{n} c_i \varphi(\|x - x_i\|)$$

The coefficients c_i can be computed from the known data pairs $f(x_i) = f_i$ via solution of the following linear system,

$$Mc = \begin{pmatrix} \varphi(\|x_1 - x_1\| & \varphi(\|x_1 - x_2\| & \cdots & \varphi(\|x_1 - x_n\| \\ \varphi(\|x_2 - x_1\| & \varphi(\|x_2 - x_2\| & \cdots & \varphi(\|x_2 - x_n\| \\ \vdots & \vdots & \ddots & \vdots \\ \varphi(\|x_n - x_1\| & \varphi(\|x_n - x_2\| & \cdots & \varphi(\|x_n - x_n\| \end{pmatrix} \begin{pmatrix} c_1 \\ c_2 \\ \vdots \\ c_n \end{pmatrix} = \begin{pmatrix} f_1 \\ f_2 \\ \vdots \\ f_n \end{pmatrix} = f$$

Let us consider a 1D problem. The function is,

$$f(x) = x\sin(x)$$

We generate $n = 10$ points $x_i \in [1/2,\ 3]$, i = 1, 2, ...,n

```
⇒ data=Table[{i 0.3,i 0.3 Sin[i 0.3]},{i,1,10}]
⇐ {{0.3,0.0886561},{0.6,0.338785},{0.9,0.704994},{1.2,1.11845},
   {1.5,1.49624},{1.8,1.75293},{2.1,1.81274},{2.4,1.62111},
   {2.7,1.15393},{3.,0.42336}}
⇒ p1=ListPlot[data,Filling → Axis]
```

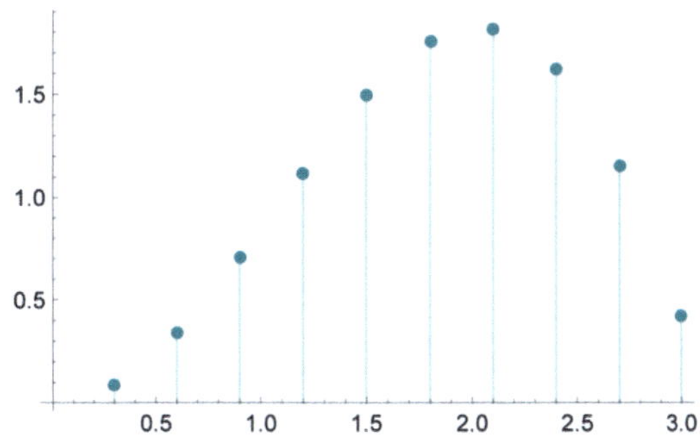

⇐

Fig. 1.50 The generated discrete points.

To approximate the function, we employ a *Thin- Plate Spline* (TPS), which is a type of the radial basis function family, where the basis function,

$$\varphi(r) = r^{2\beta} \log(r)$$

where $r = \| x - x_i \|$.

Fig. 1.51 shows two basic functions with $x_i = 1.5$ and $x_i = 1.8$ locations in case $\beta = 1$,

```
⇒ TPSpline = Function[x,xi,If[x ≠ xi,Abs[x-xi]² Log[Abs[x-xi]],0]]
⇐ Function[x,xi,If[x ≠ xi,Abs[x-xi]² Log[Abs[x-xi]],0]]
⇒ Plot[{TPSpline[x,1.5],TPSpline[x,1.8]},{x,0.3,3},
    PlotRange → {-0.5,0.5}]
```

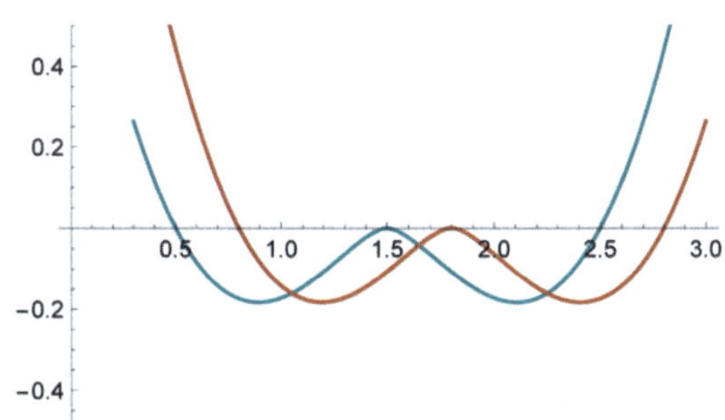

⇐

Fig. 1.51 Two TPS basis functions with $\beta = 1$.

Separating the coordinates data into two arrays,

```
⇒ np=Length[data];datax=Transpose[data][[1]];
  datay=Transpose[data][[2]];
```

the matrix of the linear system is

```
⇒ M=Table[First[TPSpline[datax[[i]],datax[[j]]]],
    {i,1,np},{j,1,np}];
```

This matrix has a relative high condition number

```
⇒ Norm[Inverse[M]]Norm[M]
⇐ 308.843
```

Therefore to compute the coefficents, we use pseudoinverse,

```
⇒ c=PseudoInverse[M].datay
⇐ {1.94971,-2.26345,-0.412961,-0.419176,-0.282941,
   -0.22277,-0.224389,-0.178729,-1.23521,1.00129}
```

The coefficients can be seen in Fig. 1.52

```
⇒ ListPlot[c,Filling → Axis]
⇐
```

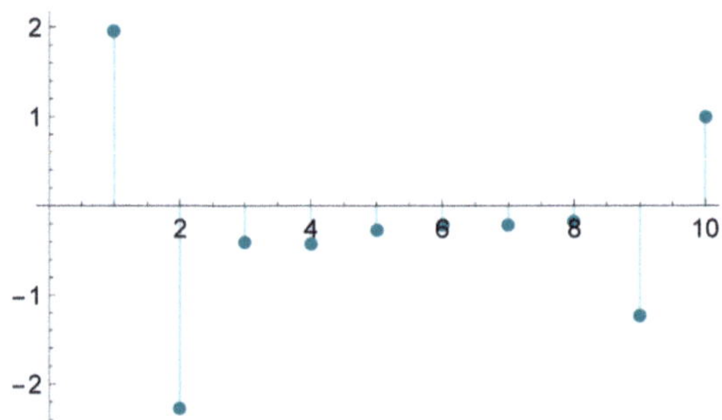

Fig. 1.52 The c_i coefficients.

The approximation function can be computed as an expression,

```
⇒ f=First[c.Map[TPSpline[x,#]&,datax]]
⇐ 1.94971If[x ≠ 0.3,Abs[x-0.3]^2 Log[Abs[x-0.3]],0]
```

or in function form, which provides a very good approximation, see Fig. 1.53

```
⇒ g[x_]:=First[c.Map[TPSpline[x,#]&,datax]]
```

For example

```
⇒ g[1.75]
⇐ 1.72293
⇒ Show[{Plot[x Sin[x],{x,0.5,3},PlotStyle → Red],
   Plot[g[x],{x,0.5,3},PlotRange → {0,3}],p1}]
⇐
```

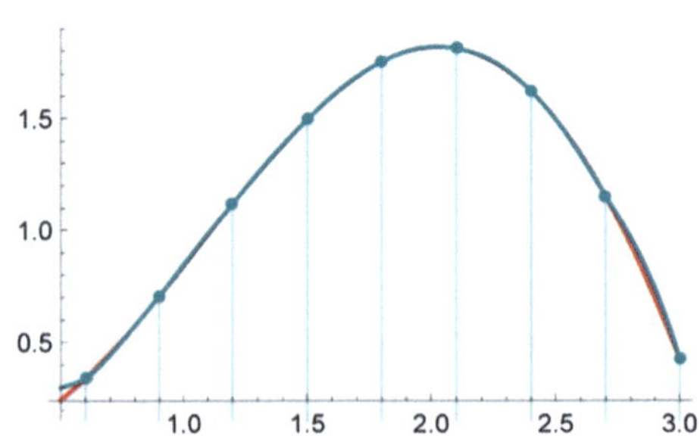

Fig. 1.53 RBF approximation.

1.5.2 RBF Image Compression

The image compression methods tries to reduce the necessary data of an image without loosing the basic information for the proper image reconstruction. Let us consider the following image, Fig. 1.54

⇒ img=

Fig. 1.54 A young squirrel.

The image size

```
⇒ M=ImageData[img];M//Dimensions
⇐ {180,180}]
```

Fig. 1.55 shows the intensity values of the gray image

```
⇒ SListPlot[Flatten[M],PlotStyle → PointSize[Tiny],Frame → True]
```

⇐

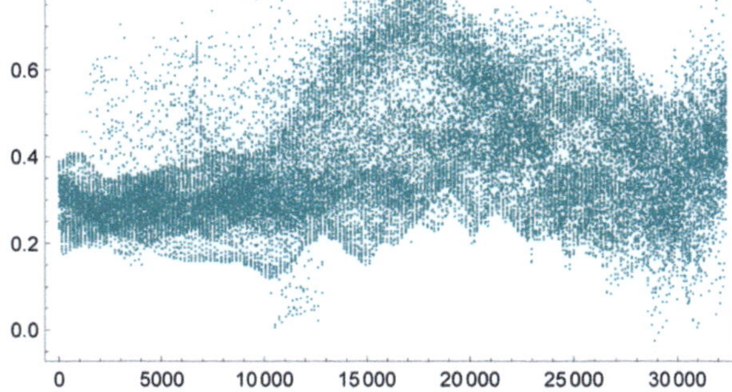

Fig. 1.55 The intensity value of the image pixels.

The size of the image is 180×180, i.e., 32 400 pixels. In order to create compressed image, some non-white pixels will be randomly whitened. Pixel value 1 will be assigned to these pixels. Let us initialize a matrix MR as

```
⇒ MR=M;
⇒ SeedRandom[4567]
```

Then randomly whitening, let us say 60 000 pixels,

```
⇒ Do[MR[[RandomInteger[
    {1,180}],RandomInteger[{1,180}]]]=1.,{i,1,60000}];]
```

Fig, 1.56 shows the image after randomly carrying out 6×10^4 times the whitening process,

```
⇒ Image[MR]
```

⇐

Fig. 1.56 The image after whitening.

Now, the coordinates of the retained black pixels will be normalized, see Fig. 1.57

```
⇒ dataz={};dataxy={};
⇒ Do[If[MR[[i,j]]!=1.,AppendTo[dataxy,{i /180.,j/180.}];
    AppendTo[dataz,MR[[i,j]]]],{i,1,180},{j,1,180}]
⇒ ListPlot[dataxy,Frame → True,AspectRatio → 1,
    PlotStyle → PointSize[Tiny]]
```

⇐

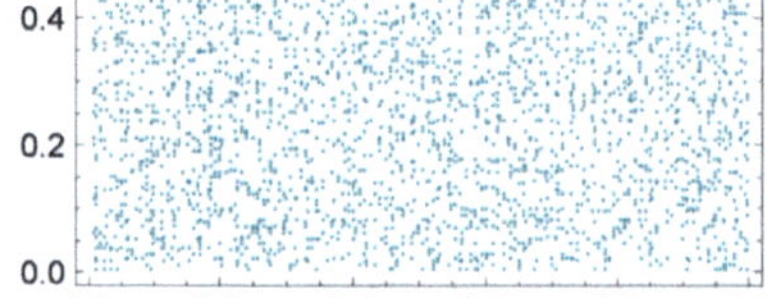

Fig. 1.57 The black pixels with normalized coordinates.

These pixels will be the basic points and their intensity represents the function value to be approximated. Their number is

```
⇒ nj=Length[dataxy]
⇒ 5091
```

which is only the

```
⇒ Length[dataxy]/(32400) 100.
⇒ 15.713
```

percent of the total number of pixels (32 400). This can be considered as roughly 85 % compression. Now, we are going to reconstruct the original image from these pixels. Let us employ the Thin-Plate-Spline (TPS) approximation

$$\varphi(r) = r^{2\beta}\log(r)$$

with $\beta = 2$,

```
⇒ TPSpline2=
    Function[x,xi,If[x ≠ xi,Norm[x-xi,2]^4 Log[Norm[x-xi,2]],0]]
⇐ Function[x,xi,If[x ≠ xi,Norm[x-xi,2]^4 Log[Norm[x-xi,2]],0]]
```

We can employ parallel computation for computing the elements of the coefficient matrix,

```
⇒ DistributeDefinitions[TPSpline2]
⇐ {TPSpline2}
⇒ M=ParallelTable[TPSpline2[dataxy[[i]],dataxy[[j]]],
    {i,1,nj},{j,1,nj}];
```

Using pseudoinverse

```
⇒ PM=PseudoInverse[M];
```

Then we compute the coefficients, see Fig. 1.58.

```
⇒ c=PM.dataz;
⇒ ListPlot[c,Filling → Axis]
```

⇐

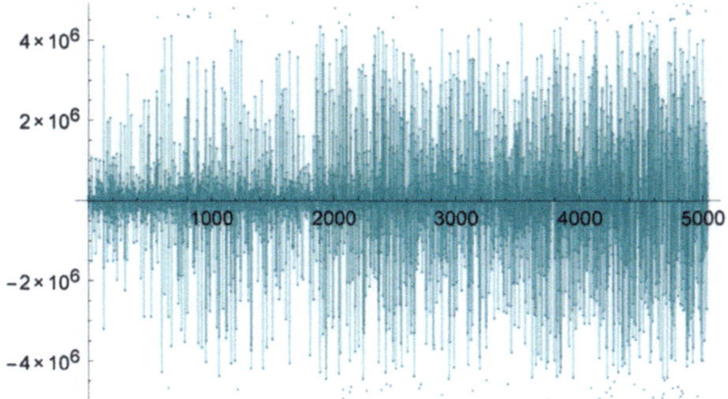

Fig. 1.58 The coefficients of the TPS function.

Let us define the reconstruction (approximation) function as

```
⇒ g[x_,y_]:=c.Map[TPSpline2[{x,y},#]&,dataxy]
```

Now generating 10^4 pixel values in parallel computation, we get the reconstructed image, see Fig. 1.59

```
⇒ DistributeDefinitions[g]
⇐ {g,c,TPSpline,datax}
⇒ RImage=ParallelTable[g[x,y],{x,0,1,0.01},{y,0,1,0.01}];
⇒ Image[RImage]
```

⇐ 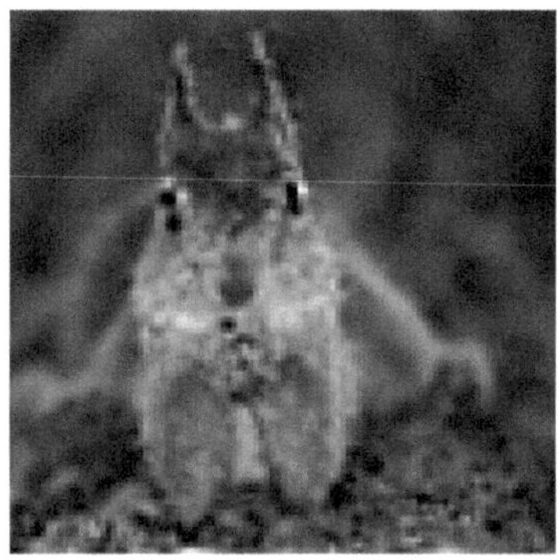

Fig. 1.59 The reconstructed image, which means roughly 85 % compression.

It goes without saying that the reconstructed image is less attractive than the original one, but do not forget that in case of the original image, one needs to store 32 400 intensity values *plus* their (x, y) coordinates, while for the reconstructed image, only the 5091 coefficients are stored, i.e., roughly 85% compression.

1.6 AutoEncoding

Basic theory

Autoencoders are similar to other dimensionality reduction techniques like principal component analysis. They create a space where the essential parts of the data are preserved, while non-essential (or noisy) parts are removed. There are two parts of an autoencoder:

a) Encoder: This is the part of the network that compresses the input into a fewer number of bits. The space represented by these fewer number of bits is called the "latent-space" and the point of maximum compression is called the bottleneck. These compressed bits that represent the original input are together called an "encoding" of the input.

b) Decoder: This is the part of the network that reconstructs the input image using the encoded information.

Let's look at an example to understand the concept better.

1.6.1 Concept of AutoEncoding

Let us suppose we have the following 2D dataset (x_i, y_i)

```
⇒ x=Range[-1,1,0.1 ];
  y=Map[2#+RandomReal[{-0.2,0.2}]&,x];
```

or

```
⇒ vectors=Transpose[{x,y}];
```

See the visualization of these points in Fig. 1.60,

```
⇒ p0=ListPlot[vectors,PlotStyle → Red,PlotMarkers → *,
    AspectRatio → 1.5,AxesLabel → {"x","y"}]
```

⇐

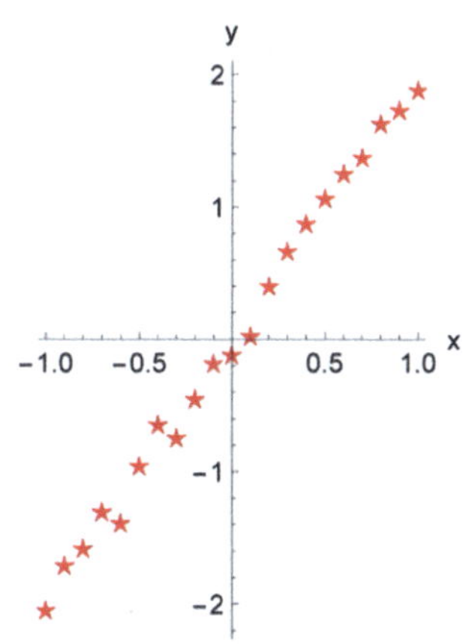

Fig. 1.60 2D dataset.

Now we would like to reduce the 2D set into a 1D using linear mapping represented by the parameters W_1 and b_1 as $(x_i, y_i) \rightarrow z_i$

$$z_i = W_1 \begin{pmatrix} x_i \\ y_i \end{pmatrix} + b_1$$

we call it *encoding*. The *inverse mapping* is, similarly

$$\begin{pmatrix} \tilde{x}_i \\ \tilde{y}_i \end{pmatrix} = W_2 z_i + b_2$$

we call it *decoding*.

The optimal parameters can be computed via minimizing the total error, namely

$$G(W_1, W_2, b_1, b_2) = \sum_{i=1}^{N} \left(\begin{pmatrix} \tilde{x}_i \\ \tilde{y}_i \end{pmatrix} - \begin{pmatrix} x_i \\ y_i \end{pmatrix} \right)^2 = \sum_{i=1}^{N} \left(W_2 z_i + b_2 - \begin{pmatrix} x_i \\ y_i \end{pmatrix} \right)^2 =$$

$$\sum_{i=1}^{N} \left(W_2 \left(W_1 \begin{pmatrix} x_i \\ y_i \end{pmatrix} + b_1 \right) + b_2 - \begin{pmatrix} x_i \\ y_i \end{pmatrix} \right)^2$$

In our case

```
⇒ G=Total[MapThread[((w21(w1#1+w2#2+b1)+b21-#1)^2+
    (w22(w1#1+w2#2+b1)+b22-#2)^2)&,x,y]];
```

Minimizing the error

```
⇒ sol=NMinimize[G,{w1,w2,w21,w22,b1,b2,b21,b22}]
⇐ {0.0367586,
    {w1 → -0.378812,w2 → -0.755743,w21 → -0.530069,w22 → -1.05751,
    b1 → 0.160262,b2 → 0.0952464,b21 → 0.0900884,b22 → 0.166903}}
```

The parameters are

```
⇒ {W1, W2, W21, W22, B1, B2, B21, B22}=
    {w1,w2,w21,w22,b1,b2,b21,b22}/.sol[[2]];
```

Then mapping into 1D

```
⇒ encoding=Map[W1 #[[1]]+W2 #[[2]]+B1&,vectors];
```

and for inverse mapping

```
⇒ decoding=Map[{W21 #+B21,W22 #+B22}&,encoding];
```

Let us visualize the results, see Fig 1.61

```
⇒ p1 = ListPlot[decoding, PlotStyle → Blue,
    PlotMarkers → ●, AspectRatio → 1.5];
⇒ Show[{p0,p1}]
```

⇐

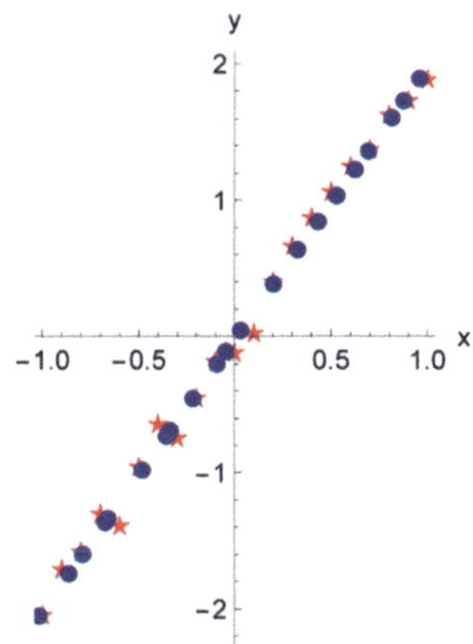

Fig. 1.61 Inverse mapping (blue).

There are two effects using this technique: *we can reduce dimension of the input object* (*encoding*) *and also can eliminate noise from the output* (*decoding*). Let us fit a line to the original data, see Fig. 1.62.

```
⇒ b=Fit[vectors,{1,a},a]
⇐ -0.0128264+1.98551 a
```

Visualizing it, see Fig. 1.62

```
⇒ Show[{p0,p1,Plot[b,{a,-1,1},PlotStyle → {Thin,Red}]}]
```

⇐

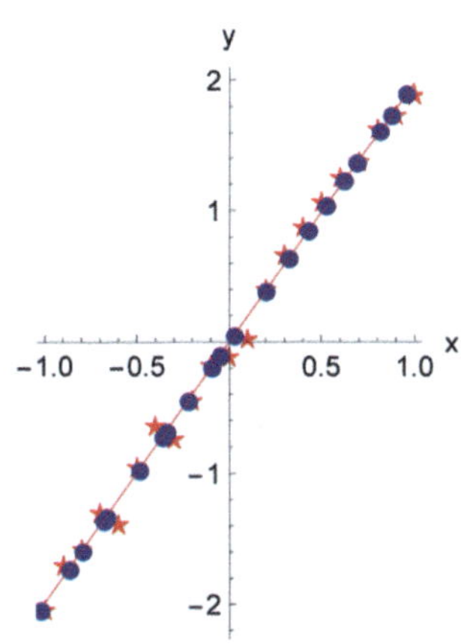

Fig. 1.62 The decoded data points are on the least squares line fitted to the original red points.

The concept can be extended to nonlinear multidimensional mapping using neural networks with activation functions, see Fig. 1.63.

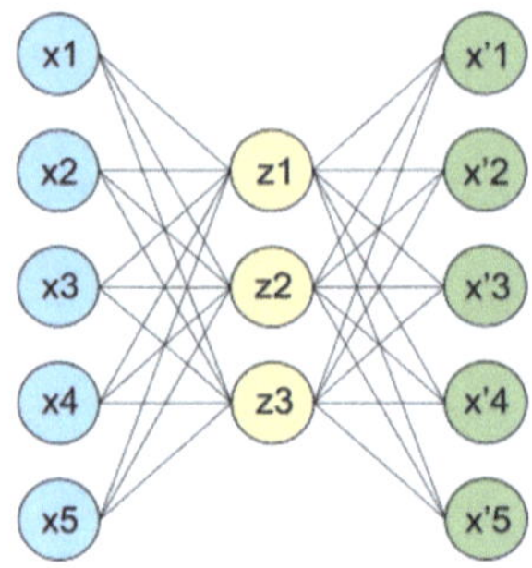

Fig. 1.63 The principle of the AutoEncoder.

The network represents the identity function between the input and output layers, $x_i \to (x')_i$, while the hidden layer represents the reduced dimensional feature of the input, see Fig. 1.64

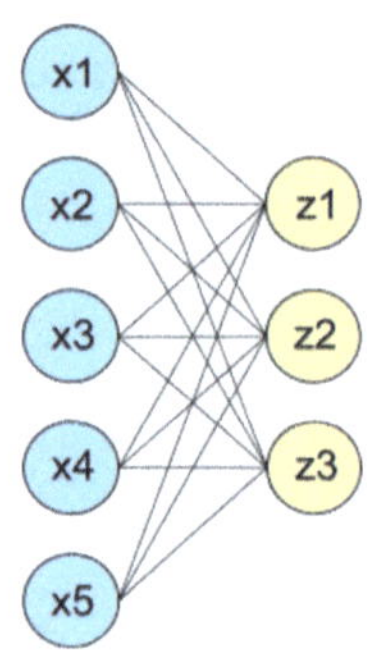

Fig. 1.64 The Encoder part of the AutoEncoder.

1.6.2 Simple Example

Let us solve the problem discussed in the previous section via built in functions of *Mathematica*

```
⇒ vectors
⇐ {{-1.,-1.92412},{-0.9,-1.77127},{-0.8,-1.72194},{-0.7,-1.34765},
   {-0.6,-1.10272},{-0.5,-0.8349},{-0.4,-0.930942},
   {-0.3,-0.516347},{-0.2,-0.333443},{-0.1,-0.159662},
   {0.,-0.0867159},{0.1,0.159127},{0.2,0.327988},{0.3,0.40351},
   {0.4,0.829282},{0.5,0.896315},{0.6,1.00913},{0.7,1.48511},
   {0.8,1.71077},{0.9,1.85131},{1.,2.19269}}
```

Employing built-in function,

```
⇒ autoencoder=DimensionReduction[vectors,1,Method → "AutoEncoder"]
```

```
⇐ DimensionReducerFunction[
   Input type: NumericalVector (Length: 2)
   Output dimension: 1 ]
```

Using it for encoding

```
⇒ reduced=autoencoder[vectors]
⇐ {{2.11631},{1.91152},{1.76693},{1.43337},{1.17505},{0.903404},
   {0.843341},{0.486353},{0.261511},{-0.00218918},{-0.291566},
   {-0.560727},{-0.781182},{-0.940042},{-1.3241},{-1.4775},
   {-1.66034},{-2.07668},{-2.33208},{-2.53274},{-2.86254}}
```

and decoding

```
⇒ reconstructed = autoencoder[reduced, "OriginalVectors"]
⇐ {{-0.986715,-1.95732},{-0.896418,-1.77797},
   {-0.832666,-1.65133},{-0.692416,-1.35956},
   {-0.583796,-1.13359},{-0.469578,-0.895981},
   {-0.444323,-0.843443},{-0.294219,-0.531176},
   {-0.200175,-0.332512},{-0.0995873,-0.160038},
   {0.0156269,-0.115671},{0.120819,0.12227},
   {0.206976,0.317154},{0.26906,0.457587},
   {0.419156,0.797099},{0.479107,0.932707},
   {0.550563,1.09434},{0.713275,1.46239},
   {0.813087,1.68816},{0.891509,1.86555},{1.0204,2.15709}}
```

The visualized original and reconstructed datasets can be seen in Fig. 1.65.

```
⇒ p1 = ListPlot[reconstructed, PlotStyle → Blue,
   PlotMarkers → •, AspectRatio → 1.5];
⇒ Show[{p0,p1}]
```

⇐

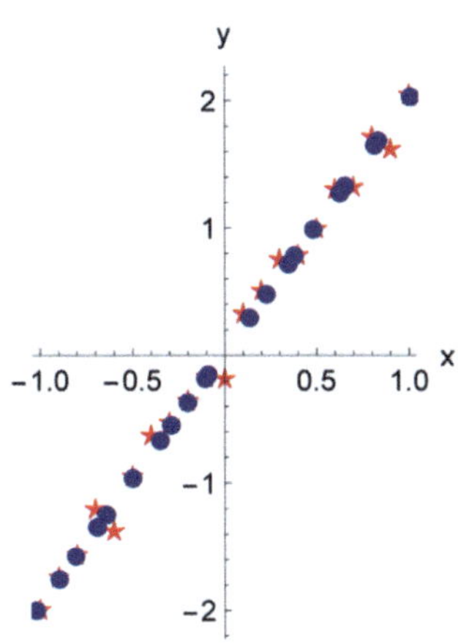

Fig. 1.65 The original data (red stars) and the reconstructed data (blue circles)

1.6.3 Compression of Image

Let us consider an image of a baboon

⇒ img=

Fig. 1.66 The baboon image to be compressed.

The image data structure

```
⇒ data=ImageData[img];data//Dimensions
⇐ {160,160,3}
```

The first element vector (pixel vector),

```
⇒ data[[1,1]]
⇐ {0.392157,0.337255,0.152941}
```

Let $q = 3$,

```
⇒ vector=Flatten[data,1];
```

then

```
⇒ Dimensions[vector]
⇐ {25600,3}
```

and

```
⇒ vector[[1]]
⇐ {0.392157,0.337255,0.152941}
```

Let us employ the built-in function `DimensionReduce` with `AutoEncoder` method, to reduce the vectors of length 3 to vector of length 2, (encoding)

```
⇒ autoencoder=DimensionReduction[vector,2,Method → "AutoEncoder"]
⇐ DimensionReducerFunction[
      Input type: NumericalVector (Length: 3)
      Output dimension: 2                       ]
```

The list of the reduced vectors

```
⇒ reduced=autoencoder[vector];
```

```
⇒ Dimensions[reduced]
⇐ {25600,2}
```

The reconstruction (decoding)

```
⇒ reconstructed = autoencoder[reduced, "OriginalVectors"];
⇒ Dimensions[reconstructed]
⇐ {25600,3}
```

Restructuring the image matrix

```
⇒ diti=Partition[reconstructed,160];
⇒ diti//Dimensions
⇐ {160,160,3}
```

For example, the first pixel vector of the original image data[[1,1]] = {0.392157, 0.337255, 0.152941} can be compared with that of the reconstructed image,

```
⇒ diti[[1,1]]
⇐ {0.399873,0.330421,0.1686}
```

The visualization of the reconstructed image from the (1–2/3) = 1/3 = 33 % compressed data.

```
⇒ img2=Image[diti]
```

⇐ 

Fig. 1.67 The image reconstructed from 33% compressed data.

Let us repeat the process using 1–1/3 = 2/3 = 66% compression,

```
⇒ autoencoder=DimensionReduction[vector,1,Method → "AutoEncoder"]
⇐ DimensionReducerFunction[
      Input type: NumericalVector (Length: 3)
      Output dimension: 1                       ]
```

```
⇒ reduced=autoencoder[vector];
⇒ Dimensions[reduced]
⇐ {25600,1}
```

```
⇒ reconstructed = autoencoder[reduced, "OriginalVectors"];
⇒ diti=Partition[reconstructed,160];
```

```
⇒ img1=Image[diti]
```

⇐ 

Fig. 1.68 The image reconstructed from 66% compressed data.

Finally let us employ 90 % compression,

```
⇒ vector=Flatten[data];
⇒ Dimensions[vector]
⇐ {76800}
⇒ dvector=Partition[vector,30];
⇒ autoencoder=DimensionReduction[dvector,3,Method → "AutoEncoder"]
⇐ DimensionReducerFunction[
     Input type: NumericalVector (Length: 30)
     Output dimension: 3                          ]
⇒ reduced=autoencoder[dvector];
⇒ Dimensions[reduced]
⇐ {2560,3}
⇒ reconstructed = autoencoder[reduced, "OriginalVectors"];
⇒ Dimensions[reconstructed]
⇐ 2560,30}
⇒ diti=Partition[Partition[Flatten[reconstructed],3],160];
⇒ diti//Dimensions
⇐ {160,160,3}
⇒ diti[[1,1]]
⇐ {0.347188,0.304335,0.110974}
⇒ img090=Image[diti]
```

⇐ 

Fig. 1.69 The image reconstructed from 90% compressed data.

Fig. 1.70 shows the original image and the images reconstructed from differently compressed data,

```
⇒ GraphicsGrid[{{img,img2,img1,img090}}]
```

⇐ 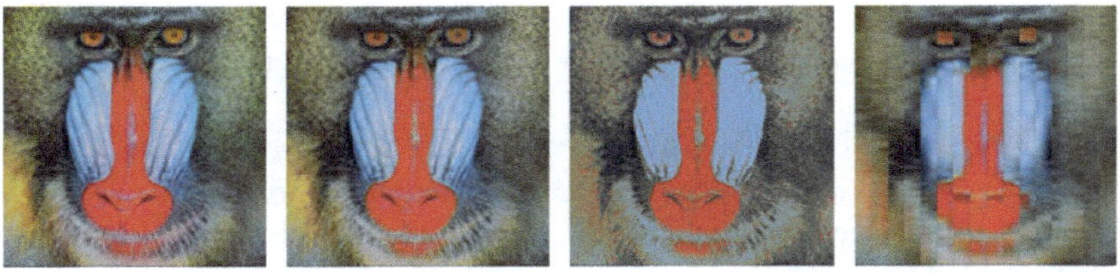

Fig. 1.70 The original image and the images reconstructed from 33%, 66% and 90% compressed data.

1.7 Fractal Compression

Basic theory

Employing a collection of geometric transformations one may reconstruct an image knowing the parameters of these transformations. The compression ratio of the process can be measured by the ratio of the number of these parameters to the number of the pixel information of the image.

This procedure is based on two theoretical concepts:

a) Theory of the contractive transformation,
b) Contractive mapping fixed point theorem

The contractive transformation always bring points closer together (by some factor less than 1) and the contractive mapping theorem says something that is intuitively obvious: if a transformation is contractive then when applied repeatedly starting with any initial point, we converge to a unique fixed point.

1.7.1 Concept of Fractal Compression

The algorithm consists of two steps. The first step is the *encoding*. The image to be compressed (Range Image) should be downsampled into a smaller one (Domain Image), see Fig. 1.71.

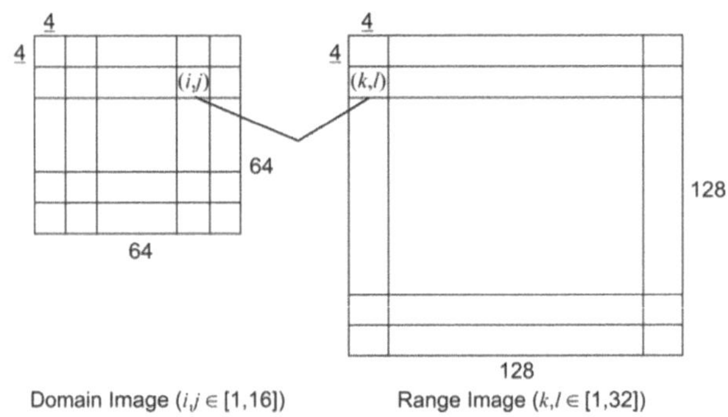

Fig. 1.71 The original Range Image and the downsampled Domain Image.

Let us refer to the elements (usually 4×4 blocks) of these two images as $R_{k,l}$ and $D_{i,j}$. We are looking for the best transformation *mapping* for every single elements of the Range Image from the collection of the elements of the Domain Image, which provides

$$\left\| R_{k,l} - \left(\alpha D_{i,j} + t \right) \right\| \to \min_{D}$$

under the conditions $0 \le \alpha \le 1$ and $-255 \le t \le 255$ (grayscale images).

It means we should find the optimal parameters of (α, t) for all $R_{k,l}$ from the collection of $D_{i,j} \in D$. This is a very hard computational task. The result is a *Fractal Code Book* (FCB) containing the optimal $(\alpha, t)_{k,l}$ pairs.

The next step is the reconstruction of the Range Image, *decoding*, using the Fractal Code Book. Since the mappings in the FCB are contractive and their fixed points are the blocks of the Range Image, the following mapping series results into the approximation of the Range Image:

1) Start a Range Image whose pixel are randomly generated, R_0,
2) Employing down sampling to create a corresponding Domain Image, D_0
3) Apply the mappings of the FCB to D_0, creating a new Range Image, R_1
4) Repeat steps 2) and 3) creating a series R_0 , R_1 , R_2 ,..., $R_n \to R$

Usually n = 7 - 8 iteration is satisfactory.

1.7.2 Illustrative Example

Let us consider an example to illustrate how *Mathematica* can achieve this algorithm. Since the encoding is computationally a very demanding task, we use a simple snippet image.

Our image is, see Fig. 1.72

```
⇒ imgR=
```

Fig. 1.72 The original binary Range Image of 24×24 pixels.

Mathematica

This will be the Range Image. We sample it down to an image of 12×12. This will be the Domain Image, see Fig. 1.73

```
⇒ imgD=ImageResize[imgR,{12,12}]//Round
⇐
```

Fig. 1.73 The Domain Image of 12×12 pixels.

Now we consider 2×2 blocks and divide the two images 12×12= 144 and 6×6= 36 blocks, respectively.

```
⇒ imgRP=ImagePartition[imgR, 2];
⇒ Grid[imgRP, Spacings → {0.8, 0.1}]
⇐
```

Fig. 1.74 The 2×2 image blocks of the Range Image.

```
⇒ imgDP=ImagePartition[imgD, 2];
⇒ Grid[imgDP, Spacings → {0.8, 0.1}]
```

⇐

Fig. 1.75 The 2×2 image blocks of the Domain Image.

We put these blocks in two lists, RP and DP

```
⇒ RP=Flatten[Table[ImageData[imgRP[[i,j]]],{i,1,12},{j,1,12}],1];
⇒ Length[RP]
⇐ 144
⇒ DP=Flatten[Table[ImageData[imgDP[[i,j]]],{i,1,6},{j,1,6}],1];
⇒ Length[DP]
⇐ 36
```

Here are functions for computing Range Block (RP) and the Domain Block (DB),

```
⇒ RangeBlocks[imgR_]:=Module[{imgRP,RP},
    imgRP=ImagePartition[imgR, 2];
    RP=Flatten[Table[ImageData[imgRP[[i,j]]],{i,1,12},
     {j,1,12}],1]];
⇒ RangeBlocks[imgR][[105]]//MatrixForm
```

$$\Leftarrow \begin{pmatrix} 1. & 1. \\ 1. & 0. \end{pmatrix}$$

```
⇒ DomainBlocks[imgR_]:=Module[{imgD,imgDP,DP},
    imgD=ImageResize[imgR,{12,12}]//Round;
    imgDP=ImagePartition[imgD, 2];
    DP=Flatten[Table[ImageData[imgDP[[i,j]]],{i,1,6},
     {j,1,6}],1]];
⇒ DomainBlocks[imgR][[10]]//MatrixForm
```

$$\Leftarrow \begin{pmatrix} 0. & 0. \\ 0. & 1. \end{pmatrix}$$

The next module is a function for computing the best DB for a given RB. The input is a given block of the Range Image (Ri), and the output is a block from the Domain Image blocks, which can be fitted in the best way to the given input block. We should find the best (optimal parameters α and t using constrained minimization. Although the problem is linear, here local nonlinear method is employed, since it requires less programming effort, however at the same time the more computation power, too.

```
⇒ BestDomain[Ri_,DP_]:=Module[{sD,sG,j,minD,s1,smin,sindex,best},
    sD={};
   Do[sG=Norm[\[Alpha] DP[[j]]+t IdentityMatrix[2]-Ri];
   minD=FindMinimum[{sG,0. ≤ α ≤ 1.,-255. ≤ t ≤ 255.},
    {{α,0.5},{t,0.}},MaxIterations → 15]//Quiet;
   AppendTo[sD,minD],{j,1,36}];
   s1=Map[First[#]&,sD];
   smin=Min[s1];
   sindex=Position[s1,n_ /; n==smin]//Flatten;
   best=sD[[First[sindex]]];
   {First[sindex],{α,t}/.best[[2]]}]
```

For example

```
⇒ AbsoluteTiming[BestDomain[RP[[35]],DP]]
⇐ {0.72433,{3,{0.5,0.}}}
```

This computation took 0.7 sec. The three output are the number of the optimal block from the Domain Image, and 0.5 and 0 are the optimal parameters that ensure the best fitting to the block 35 form the Range Image blocks.

Computing the Fractal Code Book means finding the optimal mate block from the Domain Image blocks for every blocks of the Range Image.

Let us employ parallel computing.

```
⇒ FractalCodeBook[RP_, DP_, n1_, n2_] :=
   ParallelMap[BestDomain[#, DP] &, RP[[n1 ;; n2]]]
⇒ DistributedDefinitions[BestDomain, RP0,DP0]
```

This process takes less than 30 sec for the 144 Range Image blocks.

```
⇒ s=AbsoluteTiming[FractalCodeBook[RP0,DP0,1,144]]
⇐ {28.4507,{{1,{0.996886,0.00313009}},{1,{0.996886,0.00313009}},
   {1,{0.996886,0.00313009}},{27,{0.999875,0.0000625009}},
   {3,{0.5,0.}},{3,{0.5,0.}},{3,{0.5,0.}},{3,{0.5,0.}},
   {27,{0.999875,0.0000625009}},{1,{0.996886,0.00313009}},
   {1,{0.996886,0.00313009}},{1,{0.996886,0.00313009}},
   .....
   {21,{0.999875,0.0000625009}},{1,{0.996886,0.00313009}},
   {1,{0.996886,0.00313009}},{1,{0.996886,0.00313009}}}}
```

Having the mates, we have the FCB for the Range Image

```
⇒ S=Map[Flatten[#]&,s[[2]]];
```

Now we can reconstruct the Range Image using these collections of the optimal parameter pairs. Let us start with an image of randomly generated pixels, see Fig. 1.76

```
⇒ SeedRandom[1234]
⇒ imgRN=Image[Table[RandomReal[],{i,1,24},{j,1,24}]]
```

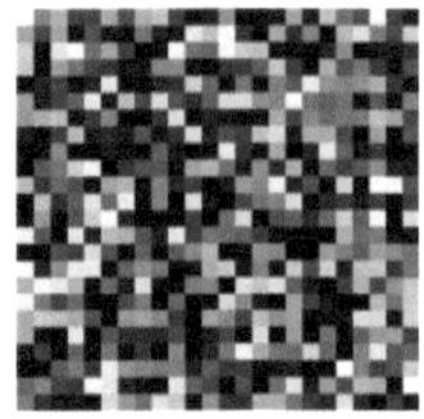

⇐

Fig. 1.76 The starting Range Image of 24×24 pixels.

The next modul can realizes the iterated function system,

```
⇒ FractalDecoding[img0_,FCB_,m_]:=Module[
    {S,u,imgR,imgDP0,DP0,RP1,ImageRP1,k,i1,j1,j},
   S={img0};
   Do[u=Last[S];
   imgR=ImageResize[u,{12,12}]//Round;
   imgDP0=ImagePartition[imgR, 2];
   DP0=Flatten[Table[ImageData[imgDP0[[i1,j1]]],{i1,1,6},
    {j1,1,6}],1];
   RP1={};
   Do[AppendTo[RP1,FCB[[j,2]] DP0[[FCB[[j,1]]]]+FCB[[j,3]]
    IdentityMatrix[2]],{j,1,144}];
   ImageRP1=ImageAssemble[Partition[Map[Image[#]&,RP1],12]];
   AppendTo[S,ImageRP1],{k,1,m}];
   S]
```

Let us employ it,

```
⇒ z=FractalDecoding[imgRN,S,5];
```

The results, the elements of the iteration series, can be seen in Fig. 1.77

```
⇒ g=GraphicsGrid[Partition[z,3]]
```

⇐

Fig. 1.77 The series of images after 5 iteration steps.

The last image can be improved a bit via binarizing pixels that do not have binary values, see Fig. 1.78

⇒
```
gc=Closing[z//Last,DiskMatrix[0.9]]//Round
```

⇐

Fig. 1.78 The reconstruction result.

1.7.3 Image Compression with Python

Now let us employ this algorithm in *Python*. This code is considerably faster than the *Mathematica* code, consequently normal size of images can be handled

```
import matplotlib.pyplot as plt
import matplotlib.image as mpimg
from scipy import ndimage
from scipy import optimize
import numpy as np
import math
```

Manipulate channels,

```
def get_greyscale_image(img):
 return np.mean(img[:,:,:2], 2)
```

```
def extract_rgb(img):
 return img[:,:,0], img[:,:,1], img[:,:,2]
```

```
def assemble_rbg(img_r, img_g, img_b):
 shape = (img_r.shape[0], img_r.shape[1], 1)
 return np.concatenate((np.reshape(img_r, shape),
  np.reshape(img_g, shape),
  np.reshape(img_b, shape)), axis=2)
```

Transformations

```
def reduce(img, factor):
 result = np.zeros((img.shape[0] // factor,
  img.shape[1] // factor))
 for i in range(result.shape[0]):
  for j in range(result.shape[1]):
   result[i,j] = np.mean(img[i*factor:(i+1)*factor,j
    *factor:(j+1)*factor])
 return result
```

```
def rotate(img, angle):
 return ndimage.rotate(img, angle, reshape=False)
```

```
def flip(img, direction):
 return img[::direction,:]
```

```
def apply_transformation(img, direction, angle, contrast=1.0,
  brightness=0.0):
 return contrast*rotate(flip(img,direction),angle)+brightness
```

Contrast and brightness

```
def find_contrast_and_brightness1(D, S):
 # Fix the contrast and only fit the brightness
 contrast = 0.75
 brightness = (np.sum(D - contrast*S)) / D.size
 return contrast, brightness
```

```
def find_contrast_and_brightness2(D, S):
 # Fit the contrast and the brightness
 A = np.concatenate((np.ones((S.size, 1)), np.reshape(S,
 (S.size, 1))), axis=1)
 b = np.reshape(D, (D.size,))
 x, _, _, _ = np.linalg.lstsq(A, b)
 #x = optimize.lsq_linear(A, b, [(-np.inf, -2.0),
 (np.inf, 2.0)]).x
 return x[1], x[0]
```

Compression for greyscale images

```python
def generate_all_transformed_blocks(img, source_size,
  destination_size,step):
 factor = source_size // destination_size
 transformed_blocks = []
  for k in range((img.shape[0]-source_size)//step+1):
   for l in range((img.shape[1]-source_size)//step+1):
    # Extract the source block and reduce it to the
    sape of a destination block
    S = reduce(img[k*step:k*step+source_size,
     l*step:l*step+source_size], factor)
    # Generate all possible transformed blocks
     for direction, angle in candidates:
      transformed_blocks.append((k,l,direction,angle,
       apply_transformation(S, direction, angle)))
 return transformed_blocks
```

```python
def compress(img, source_size, destination_size, step):
 transformations = []
 transformed_blocks = generate_all_transformed_blocks(img,
  source_size, destination_size, step)
 for i in range(img.shape[0] // destination_size):
   transformations.append([])
  for j in range(img.shape[1] // destination_size):
   print(i, j)
   transformations[i].append(None)
   min_d = float('inf')
   # Extract the destination block
   D = img[i*destination_size:(i+1)*destination_size,
     j*destination_size:(j+1)*destination_size]
   # Test all possible transformations and take the best one
    for k,l,direction,angle,S in transformed_blocks:
     contrast,brightness = find_contrast_and_brightness2(D,S)
     S = contrast*S + brightness
     d = np.sum(np.square(D - S))
     if d < min_d:
       min_d = d
       transformations[i][j]=(k,l,direction,angle,contrast,
        brightness)
 return transformations
```

```python
def decompress(transformations, source_size, destination_size,
  step, nb_iter=8):
 factor = source_size // destination_size
 height = len(transformations) * destination_size
 width = len(transformations[0]) * destination_size
 iterations = [np.random.randint(0, 256, (height, width))]
 cur_img = np.zeros((height, width))
 for i_iter in range(nb_iter):
  print(i_iter)
  for i in range(len(transformations)):
   for j in range(len(transformations[i])):
    # Apply transform
    k,l,flip,angle,contrast,brightness = transformations[i][j]
    S = reduce(iterations[-1][k*step:k*step+source_size,
     l*step:l*step+source_size], factor)
    D = apply_transformation(S,flip,angle,contrast,brightness)
    cur_img[i*destination_size:(i+1)*destination_size,
     j*destination_size:(j+1)*destination_size] = D
   iterations.append(cur_img)
   cur_img = np.zeros((height, width))
 return iterations
```

Compression for color images

```
def reduce_rgb(img, factor):
 img_r, img_g, img_b = extract_rgb(img)
 img_r = reduce(img_r, factor)
 img_g = reduce(img_g, factor)
 img_b = reduce(img_b, factor)
 return assemble_rbg(img_r, img_g, img_b)
```

```
def compress_rgb(img, source_size, destination_size, step):
 img_r, img_g, img_b = extract_rgb(img)
 return [compress(img_r,source_size,destination_size,step), \
  compress(img_g, source_size, destination_size, step), \
  compress(img_b, source_size, destination_size, step)]
```

```
def decompress_rgb(transformations, source_size,
  destination_size, step, nb_iter=8):
 img_r = decompress(transformations[0], source_size,
  destination_size, step, nb_iter)[-1]
 img_g = decompress(transformations[1], source_size,
  destination_size, step, nb_iter)[-1]
 img_b = decompress(transformations[2], source_size,
  destination_size, step, nb_iter)[-1]
 return assemble_rbg(img_r, img_g, img_b)
```

Plot the results

```
def plot_iterations(iterations, target=None):
 # Configure plot
 plt.figure()
 nb_row = math.ceil(np.sqrt(len(iterations)))
 nb_cols = nb_row
 # Plot
 for i, img in enumerate(iterations):
  plt.subplot(nb_row, nb_cols, i+1)
  plt.imshow(img, cmap='gray', vmin=0, vmax=255,
   interpolation='none')
  if target is None:
   plt.title(str(i))
  else:
   # Display the RMSE
   plt.title(str(i) + ' (' + '{0:.2f}'.format
    (np.sqrt(np.mean(np.square(target - img)))) + ')')
  frame = plt.gca()
  frame.axes.get_xaxis().set_visible(False)
  frame.axes.get_yaxis().set_visible(False)
 plt.tight_layout()
```

Parameters

```
directions = [1, -1]
angles = [0, 90, 180, 270]
candidates = list(zip(directions, angles))
```

Test for grayscale image

```
⇒ Import["M:\\me.jpg"]
```

⇐ 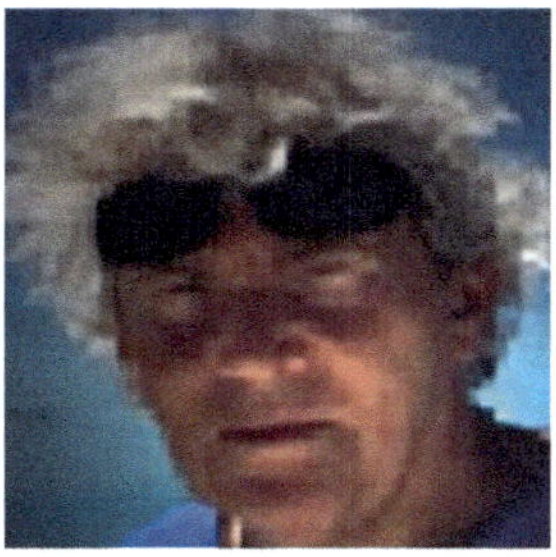

Fig. 1.79 The image to be compressed.

```
⇒ ImageData[%]//Dimensions
⇐ {256,256,3}
```

Range Image Blocks: 8×8, Domain Image Blocks: 4×4 and number of iterations: 8

```
def test_greyscale():
 img = mpimg.imread('M:\\me.jpg')
 img = get_greyscale_image(img)
 img = reduce(img, 4)
 plt.figure()
 plt.imshow(img, cmap='gray', interpolation='none')
 transformations = compress(img, 8, 4, 8)
 iterations = decompress(transformations, 8, 4, 8)
 plot_iterations(iterations, img)
 plt.show()
```

```
test_greyscale()
```

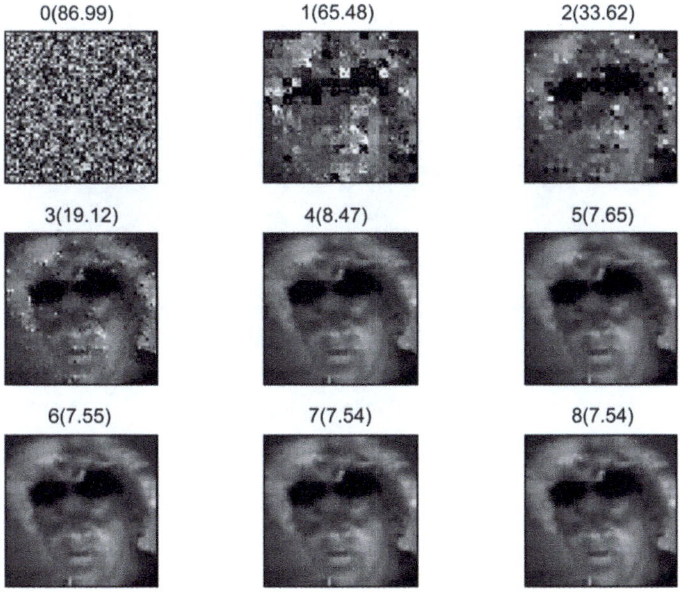

Fig. 1.80 The reconstructed grayscale images.

Test for RGB image

```
def test_rgb():
 img = mpimg.imread('M:\\me.jpg')
 img = reduce_rgb(img, 4)
 transformations = compress_rgb(img, 8, 4, 8)
 retrieved_img = decompress_rgb(transformations, 8, 4, 8)
 plt.figure()
 plt.subplot(121)
 plt.imshow(np.array(img).astype(np.uint8),
  interpolation='none')
 plt.subplot(122)
 plt.imshow(retrieved_img.astype(np.uint8),
  interpolation='none')
 plt.show()
```

```
test_rgb()
```

⇒ cimg=

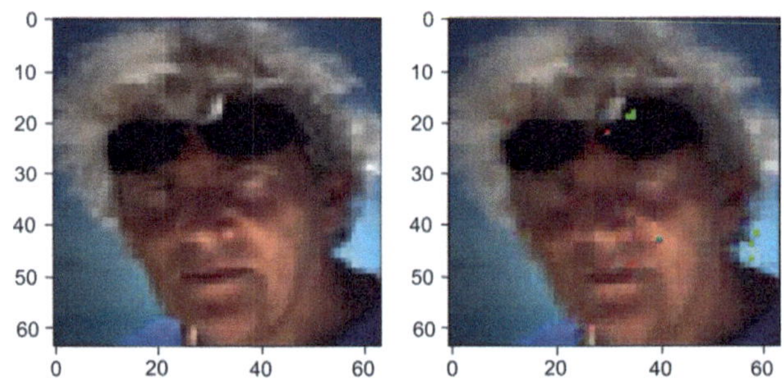

Fig. 1.81 The original and the reconstructed rgb images.

This result somewhat may be improved using lowpass filter

⇒ LowpassFilter[cimg,0.99]

⇐

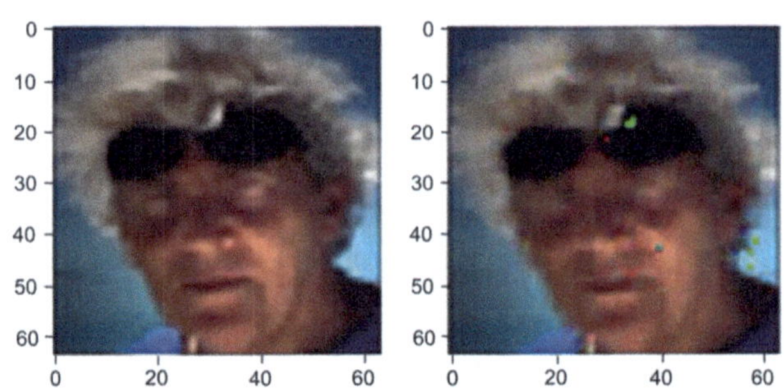

Fig. 1.82 Employing lowpass filter.

1.7.4 Accelerating Fractal Code Book Computation

We may improve the computation of the Fractal Code Book if we use only a *few representant elements in the Domain Blocks* instead of all of the original ones. A possible solution is the *clustering* of the original Domain Blocks, and considering only the center elements of the clusters. It goes without saying that the quality will not be the same as it would be with the original Domain Blocks. A trade off is necessary between the computation time and the decoding quality. In this case even the decoding algorithm should be somewhat modified, too.

Another possibility is to employ *unsupervised clustering via neural network*, and using the codebook vectors as elements in the new Domain Blocks.

All in all the computation can be speed up sometimes with two magnitudes when these techniques are applied.

As an *illustrative example,* a gray image of 128×128 was compressed.

⇒ img=

Fig. 1.83 Original gray image (Range Image) of 128×128.

```
⇒ R=ImageData[imgR];
⇒ Dimensions[R]
⇐ {128,128}
```

The size of the image blocks is 4×4 and therefore the number of the Range Blocks is 32×32 =1024, see Fig. 1.84.

```
⇒ imgRP=ImagePartition[imgR,4];
⇒ Grid[imgRP, Spacings → {0.8, 0.1}]
```

⇐ 

Fig. 1.84 The Range Blocks.

```
⇒ RP=Flatten[Table[ImageData[imgRP[[i,j]]],{i,1,32},{j,1,32}],1];
⇒ Length[RP]
⇐ 1024
```

The domain image size was 64×64, see Fig. 1.85, and the number of the blocks in the Domain Blocks was 16×16 = 256, see Fig. 1.86.

```
⇒ imgD=ImageResize[imgR,{64,64}]
```

⇐ 

Fig. 1.85 The Domain Image size of 64×64.

```
⇒ imgDP=ImagePartition[imgD,4];
⇒ Grid[imgDP, Spacings → {0.8, 0.1}]
```

⇐ 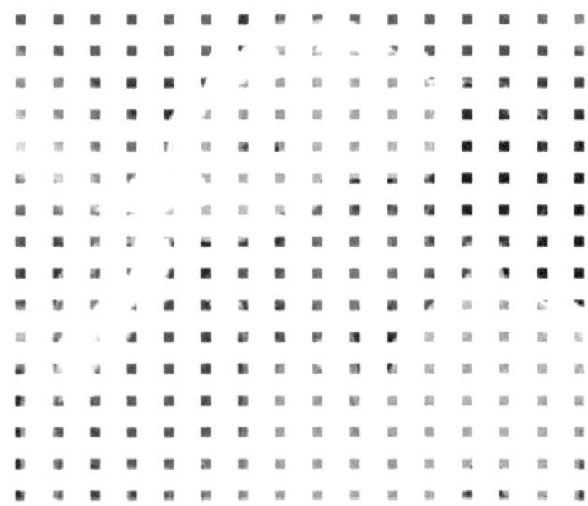

Fig. 1.86 The original Domain Blocks.

```
⇒ DP=Flatten[Table[ImageData[imgDP[[i,j]]],{i,1,16},{j,1,16}],1];
⇒ Length[DP]
⇐ 256
```

In order to reduce the number of the domain blocks from 256 to 5, *clustering* was employed, and the centers of the clusters were considered as the elements of the modified Domain Blocks. It means that the originally 1024×256 = 262 144 computation cycles in the coding process, could be reduced to 1024×5 = 5120 cycles.

```
⇒ clusters=FindClusters[DP,5,PerformanceGoal → "Quality"];
⇒ Length[clusters]
⇐ 5
```

The image matrices of the five elements of the reduced Domain Blocks,

```
⇒ centers=Map[Mean[#]&,clusters];
⇒ Map[TableForm[#]&,centers]
⇐ { 0.422460    0.412478    0.413119    0.413155
    0.431943    0.412299    0.404777    0.429162
    0.423922    0.418004    0.418966    0.439929
    0.445954    0.451266    0.462888    0.473939 ,

    0.148366    0.155065    0.138235    0.251144
    0.143954    0.136928    0.131046    0.236765
    0.133660    0.127124    0.142157    0.226797
    0.160294    0.140686    0.132353    0.210784 ,

    0.681618    0.692927    0.706583    0.716352
    0.687535    0.713725    0.720483    0.727556
    0.692787    0.713796    0.718943    0.720483
    0.706583    0.723284    0.719083    0.713235 ,
```

```
0.774292    0.736383    0.682353    0.606536
0.748148    0.667102    0.593028    0.525054
0.492810    0.382135    0.373856    0.335512
0.367756    0.280610    0.302397    0.281917 ,

0.074510    0.058823    0.027451    0.427451
0.776471    0.384314    0.019608    0.392157
0.984314    0.988235    0.764706    0.674510
0.952941    0.964706    0.980392    0.980392 }
```

Now, the modules should also be modified a bit, namely

```
⇒ RangeBlockes[imgR_]:=Module[{imgRP,RP},
   imgRP=ImagePartition[imgR, 4];
   RP=Flatten[Table[ImageData[imgRP[[i,j]]],{i,1,32},{j,1,32}],1]];

⇒ DomainBlockes[imgR_]:=Module[{imgD,imgDP,DP},
      imgD=ImageResize[imgR,{64,64}];
   imgDP=ImagePartition[imgD, 4];
   DP=Flatten[Table[ImageData[imgDP[[i,j]]],{i,1,16},{j,1,16}],1]];

⇒ BestDomain[Ri_,DP_]:=Module[{sD,sG,j,minD,s1,smin,sindex,best},
      sD={};
    Do[sG=Norm[α DP[[j]]+t IdentityMatrix[4]-Ri];
      minD=FindMinimum[{sG,0. ≤ α ≤ 1.,-255. ≤ t ≤ 255.},
        {{α,0.5},{t,0.}},MaxIterations → 15]//Quiet;
      AppendTo[sD,minD],{j,1,5}];
    s1=Map[First[#]&,sD];
    smin=Min[s1];
    sindex=Position[s1,n_ /; n==smin]//Flatten;
    best=sD[[First[sindex]]];
    {First[sindex],{α,t}/.best[[2]]}]
```

Even in this way, the computation of the Fractal Code Book took a long time, which could have been reduced via parallel computing,

```
⇒ RP0=RP;DP0=centers;
⇒ FractalCodeBook[RP_, DP_, n1_, n2_] :=
   ParallelMap[BestDomain[#, DP] &, RP[[n1 ;; n2]]]
⇒ DistributeDefinitions[BestDomain,RP0,DP0]
⇒ AbsoluteTiming[s=FractalCodeBook[RP0,DP0,1,1024];]
⇐ {1332.06,Null}
```

Having the mates, the optimal pairs, the Fractal Code Book for the Range Image,

```
⇒ S=Map[Flatten[#]&,s];
```

We can reconstruct the Range Image using these collections of the optimal parameter pairs. Let us start with an image of randomly generated pixels.

```
⇒ SeedRandom[1234];
⇒ imgRN=Image[Table[RandomReal[],{i,1,128},{j,1,128}]];
```

However, the decoding module must be modified, too. Theoretically, we should carry out clustering in every step of the decoding cycle to find the actual centers of the clusters for the actually iterated image. To avoid this, we used the momentarily small blocks (4×4) nearest to the fixed centers for transformation. It goes without saying, that this action reduces the quality of the encoded image further. To compensate for this negative effect, we employed a global, nonlinear filter in every iteration step.

```
⇒ FractalDecoding[img0_,FCB_,m_]:=
   Module[{S,u,imgR,imgDP0,DP0,RP1,ImageRP1,k,i1,j1,j,z},
    S={img0};
    Do[u=Last[S];
     imgR=ImageResize[u,{64,64}];
     imgDP0=ImagePartition[imgR,4];
     DP0=Flatten[Table[ImageData[imgDP0[[i1,j1]]],
       {i1,1,16},{j1,1,16}],1];

     RP1={};
     Do[AppendTo[RP1,
       FCB[[j,2]]Flatten[Nearest[DP0,centers[[FCB[[j,
         1]]]]]],1]+FCB[[j,3]]IdentityMatrix[4]],{j,1,1024}];
     z=ImageAssemble[Partition[Map[Image[#]&,RP1],32]];
     ImageRP1=NonlocalMeansFilter[z,1.5,0.07,4];
     AppendTo[S,ImageRP1],{k,1,m}];
   S]
```

Let us employ the new decoding module, the result can be seen in Fig. 1.87.

```
⇒ z=FractalDecoding[imgRN,S,6];
⇒ g=GraphicsGrid[Partition[z,3]]
```

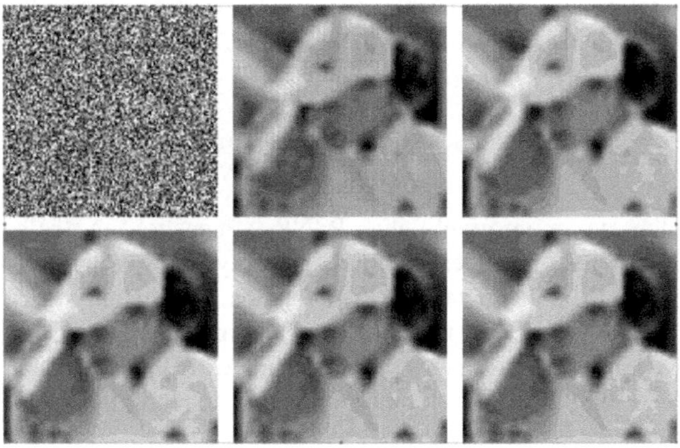

⇐

Fig. 1.87 The iterated encoded images of size 128×128.

The quality is far from perfect, but the computation effort could be reduced from quarter of million cycles down to five thousands. In addition, the image can be sharpened a bit, which provides somewhat better quality.

To compare the original and the decoded image, see Fig. 1.88.

Fig. 1.88 The original and the reconstructed images.

1.8 Comparison of Dimension Reduction Methods

1.8.1 Measure of Image Quality

Comparing restoration results requires a measure of image quality. Two commonly used measures are the Mean-Square-Error (MSE) and Peak Signal-to-Noise Ratio. The mean-square-error (MSE) between two images $g(x,y)$ and $f(x,y)$ is:

$$e_{MSE} = \frac{1}{MN} \sum_{n=1}^{N} \sum_{m=1}^{M} \left(f(x,y) - g(x,y) \right)^2 .$$

One problem with MSE is that it depends strongly on the image intensity scaling. Peak Signal-to-Noise Ratio (PSNR) avoids this problem by scaling the MSE according to the image range:

$$\mathrm{PSNR} = -10 \log_{10} \frac{e_{MSE}}{S^2} .$$

where S is the maximum pixel value. PSNR is measured in decibels (dB). The PSNR measure is also not ideal, but is commonly used. Its main undoing is that the signal strength is estimated as S^2, rather than the actual signal strength for the image. PSNR is a good measure for comparing restoration results for the same image, but between-images, comparisons of PSNR are meaningless. One image with 20 dB PSNR may look much better than another image with 30 dB PSNR.

Fig. 1.89 illustrates the *PSNR* of different images'qualities,

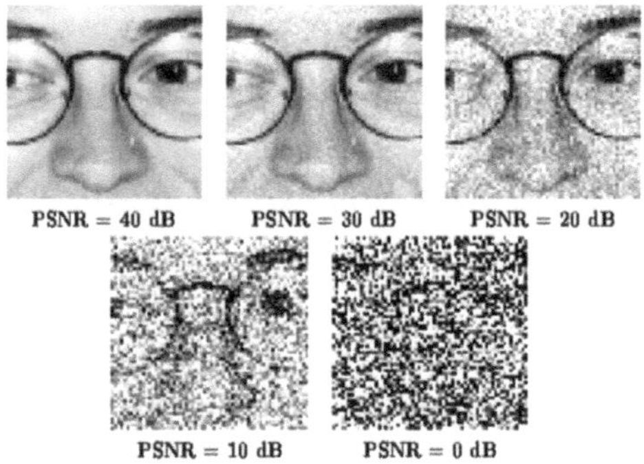

Fig. 1.89 Illustration of PSNR.

MSE and PSNR were calculated after quantization (i.e. after converting floating-point pixel values to integer), but before clipping of the intensity range.

In case of RGB images, we consider the norm of the RGB intensity vector of the pixels.

1.8.2 Comparing Different Images

As mentioned above, PSNR is good for comparing restoration results, but meaningless for comparing different images. To ameliorate this situation, we employ a built-in function of *Mathematica*, which computes distance between two images.

It is possible to use different distance measures. Here, we employ normalized entropy of between different images using *256-bin* histogram.

This measure was basically developed for gray-scale images. One intuitive approach is to consider the image as a bag of pixels and compute

$$H = -\sum_k p_k \log_2 (p_k).$$

where $k \in K$ is the number of gray levels and p_k is the probability associated with gray level k.

⇒ `im1=`  `;` `im2=` 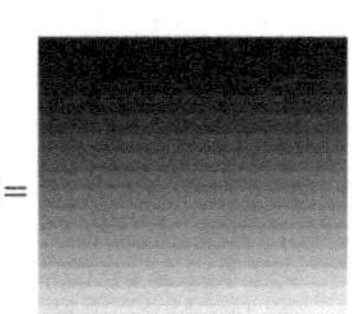 `;`

Fig. 1.90 Comparing two images.

```
⇒ ImageDistance[
    im1,im2,DistanceFunction → "DifferenceNormalizedEntropy"]
⇐ 0.966133
```

1.8.3 Compression of Mandala: Comparison between PCA and DWT methods

We shall compare the two methods, namely the Principal Component Analysis (PCA) and the Discrete Wavelet Transform (DWT) compression methods. The quality of the image will be considered at different compression ratios, since the computation time is negligible.

The image considered is a mandala, see Fig. 1.91.

```
⇒ img=
```

```
;
```

Fig. 1.91 Mandala.

PCA method

The size of the image

```
⇒ data=ImageData[img];data//Dimensions
⇐ {338,338,3}
```

The number of the elements,

```
⇒ vector=Flatten[data];Dimensions[vector]
⇐ {342732}
```

Let us partition

```
⇒ dvector=Partition[vector,338];
```

The 338 vector of length will be reduced to length of size 85, which corresponds ~75% compression since 1 – 85/338 ~0.75. Let us employ the built-in function,

```
⇒ decoder=DimensionReduction[dvector,85,
    Method → "PrincipalComponentsAnalysis",TargetDevice → "GPU"];
```

The application of the decoder function,

```
⇒ reduced=decoder[dvector];
```

we get the reduced data set,

```
⇒ Dimensions[reduced]
⇐ {1014,85}
```

The reconstruction (decoding) can be done as

```
⇒ reconstructed = decoder[reduced, "OriginalVectors"];
```

The size of the reconstructed data set,

```
⇒ Dimensions[reconstructed]
⇐ {1014,338}
```

Organizing the elements in a color image data structure

```
⇒ datarec=Partition[Partition[Flatten[reconstructed],3],338];
⇒ datarec//Dimensions
⇐ {338,338,3}
```

Then the reconstructed image, see Fig. 1.92.

```
⇒ img75=Image[datarec]
```

⇐

Fig. 1.92 Reconstructed Mandala using PCA compression ~75%.

Let us compute the PSNR value. The function for computing PSNR, see section 1.7.2,

```
⇒ PSNR[data100_,datacomp_,n_,m_]:=Module[{d1,d2,eMSE,S,i,j},
    d1=Flatten[Table[Norm[data100[[i,j]]],{i,1,n},{j,1,m}]];
    d2=Flatten[Table[Norm[datacomp[[i,j]]],{i,1,n},{j,1,m}]];
    eMSE=Total[MapThread[(#1-#2)^2&,d1,d2]]/n/m;
    S=Max[Join[d1]];
    -10Log[10,eMSE/S^2]];
⇒ PSNR[data,datarec,338,338]
⇐ 33.0012
```

We may compare the original image with the reconstructed one via built in function computing image distance, see section 1.7.2,

```
⇒ ImageDistance[img,img75,
    DistanceFunction → "DifferenceNormalizedEntropy"]
⇐ 0.643046
```

Similar computation can be done for ~94% compression, too.

Discrete Wavelet Transform

The corresponding wavelet compression in case of 75% compression,

```
⇒ dwd = DiscreteWaveletTransform[img, DaubechiesWavelet[4], 1];
```

In order to get fancy images of the residual images, we may introduce the following function,

```
⇒ imgFunc[img_,{___,1|2|3}]:=Composition[Sharpen[#,0.5]&,
    ImageAdjust[#,{0,1}]&,ImageAdjust,ImageApply[Abs,#1]&][img]
    imgFunc[img_,wind_]:=
     Composition[ImageAdjust,ImageApply[Abs,#1]&][img]
```

Then image boxes, see in Fig. 1.93.

```
⇒ WaveletImagePlot[
    dwd,Automatic,imgFunc[#1,#2]&,BaseStyle → Red,ImageSize → 338]
```

⇐

Fig. 1.93 The image boxes of the DWT.

The reconstructed image,

```
⇒ img75W=dwd[{{0}},
    {"Rules", "Inverse", "Image"}][[1,2]]//RemoveBackground
```

⇐

Fig. 1.94 The reconstructed image of DWT.

Let us compute the image quality measures,

```
⇒ PSNR[data,ImageData[img75W],338,338]
⇐ 16.054
```

and

```
⇒ ImageDistance[
    img,img75W,DistanceFunction → "DifferenceNormalizedEntropy"]
⇐ 0.484494
```

Similar computation can be done for ~94% compression, too.

Table 1.1 shows the images resulting from different compression methods at different compression ratios. This table indicates that PCA method performs better than DWT.

Table 1.1 Comparing the compressed images

Method	Compression 75%	Compression 94%
PCA		
DWT		

This conclusion can be verified by the PSNR values – higher values indicates better quality, however the values of the image distances contradict this statement. This is not very surprising, since entropy is not a perfect measure for color (multichannel) images.

Table 1.2 Comparing the quality of compressed images

Method	Compression	PSNR	Image Distance
PCA	75	33	0.64
DWT	75	16	0.48
PCA	94	22	0.71
DWT	94	15	0.52

References

Agutu NO, Awange JL, Zerihun A, Ndehedehe CE, Kuhn M, Fukuda Y (2017) Assessing Multi-satellite Remote Sensing, Reanalysis, and Land Surface Models' Products in Characterizing Agricultural Drought in East Africa. Remote Sensing of Environment 194: 287-302, doi: 10.1016/j.rse.2017.03.041

Agutu NO, Awange JL, Ndehedehe C, Kirimi F, Kuhn M (2019) GRACE-derived Groundwater Changes over Greater Horn of Africa: Temporal Variability and the Potential for Irrigated Agriculture. Science of the Total Environment, doi: 10.1016/j.scitotenv.2019.07.273.

Antonov A (2016), Independent Component Analysis Mathematica package, source code Mathematica for Prediction at GitHub package IndependentComponentAnalysis.m. https://raw.githubusercontent.com/antononcube/MathematicaForPrediction/master/IndependentComponentAnalysis.m

Anyah RO, Forootan E, Awange JL, Khaki M (2018) Understanding linkages between global climate indices and terrestrial water storage changes over Africa using GRACE products. Science of the Total Environment 635: 1405–1416, doi: 10.1016/j.scitotenv.2018.04.159

Awange JL, Anwar AHMF, Forootan E, Nikraz H, Khandu, Walker J (2017) Enhancing civil engineering surveying learning through workshops. Journal of Surveying Engineering, doi: 10.1061/(ASCE)SU.1943-5428.0000211#sthash.HFRzHTSX.dpuf.

Awange JL, Forootan E, Kuhn M, Kusche J, Heck B (2014) Water storage changes and climate variability within the Nile Basin between 2002 and 2011. Advances in Water Resources 73:1-25, doi: 10.1016/j.advwatres.2014.06.010

Awange JL, Khandu, Forootan E., Schumacher M., Heck B (2016) Exploring hydro-meteorological drought patterns over the Greater Horn of Africa (1979-2014) using remote sensing and reanalysis products. Advances in Water Resources 94: 45-59, doi:10.1016/j.advwatres.2016.04.005.

Awange JL, Saleem A, Sukhadiya R, Ouma YO, Hu K (2019a) Physical dynamics of Lake Victoria over the past 34 years (1984 -2018): Is the lake dying? Science of the Total Environment

Awange JL, Hu K, Khaki M (2019b). The newly merged satellite remotely sensed, gauge and reanalysis-based Multi-Source Weighted-Ensemble Precipitation: Evaluation over Australia and Africa (1981–2016). Science of the Total Environment, 670, 448-465, doi: https://doi.org/10.1016/j.scitotenv.2019.03.148

Baeriswyl PA, Rebetez M (1997) Regionalization of precipitation in Switzerland by means of principal component analysis. Theoretical and Applied Climatology, 58(1-2), 31-41

Fisher Y. (ed.) (1994) Fractal Image Compression, Springer, Heidelberg, Berlin

Forootan E, Awange JL, Kusche J, Heck B, Eicker A (2012) Independent patterns of water mass anomalies over Australia from satellite data and models. Remote Sensing of Environment. 124: 427 – 443, doi: 10.1016/j.rse.2012.05.023.

Forootan E (2016) Statistical signal decomposition techniques for analyzing time-variable satellite gravity data. Dissertation, DGK Reihe C Heft No 763. Munchen

Grafarend EW, Awange JL (2003) Nonlinear analysis of the three-dimensional datum transformation (conformal group C7). Journal of Geodesy 76: 66-76, doi: 10.1007/s00190-002-0299-9

Haroon MA, Rasul G. (2009) Principal component analysis of summer rainfall and outgoing long-wave radiation over Pakistan. Pakistan Journal of Meteorology, 5(10)

Hyvärinen, A. and Oja, E. (1997) A fast fixed-point algorithm for independent component analysis. Neural Computation, 9(7):1483-1492.

Hyvärinen A, Oja E (2000) Independent Component Analysis: Algorithms and Applications, Neural Networks, 13(4-5): 411-430 URL: https://www.cs.helsinki.fi/u/ahyvarin/papers/NN00new.pdf

Hu K, Awange JL, Forootan E, Goncalves RM, Fleming K. (2017) Hydrogeological characterization of groundwater over Brazil using remotely sensed and model products. Science of the Total Environment, 599, 372-386, doi:10.1016/j.scitotenv.2017.04.188

Jolliffe TI, Cadima J (2016) Principal component analysis: a review and recent developments. Philos Trans A Math Phys Eng Sci., 374(2065), doi:10.1098/rsta.2015.0202

Martín-Clemente R. and Hornillo-Mellado S. (2006) Image processing using ICA: a new perspective, IEEE MELECON 2006, May 16-19, Benalmádena (Málaga), Spain, pp 502-505, https://people.math.carleton.ca/~smills/2014-15/STAT5703-inter%202015/Pdf%20Notes/01653148-ICAonImage.pdf

Ndehedehe CE, Awange JL, Kuhn M, Agutu NO, Fukuda Y (2017a) Climate teleconnections influence on West Africa's terrestrial water storage. Hydrological Processes 31(18): 3206-3224, doi: 10.1002/hyp.11237.

Ndehedehe CE, Awange JL, Kuhn M, Agutu NO, Fukuda Y (2017b) Analysis of hydrological variability over the Volta river basin using in-situ data and satellite observations. Journal of Hydrology: Regional Studies 12: 88-110, doi: 10.1016/j.ejrh.2017.04.005.

Preisendorfer RW (1988) Introduction. In C. Mobley (Ed.), Principal Component Analysis in Meteorology and Oceanography (pp. 1-9). Elsevier, Amsterdam

Treadway A (2018) What is Independent Component Analysis (ICA)? in R-bloggers, https://www.r-bloggers.com/ica-on-images-with-python/

Vigier P (2018) Fractal Image Compression, pvigier's blog https://pvigier.github.io/2018/05/14/fractal-image-compression.html

Wayner P (2000) Compression Algorithms for Real Programmers, Academic Press, San Diego

Widmann M, Schaer C (1997) A principal component and long-term trend analysis of daily precipitation in Switzerland. International Journal of Climatology, 17(12), 1333-1356, doi:10.1002/(SICI)1097-0088(199710)17:12<1333::AID-JOC108>3.0.CO2-Q

Chapter 2
Classification

2.1 KNearest Neighbors Classification

Basic theory

KNearest Neighbors Classification is the simplest, purely data-driven algorithm that can be used either for classification or regression tasks. In geosciences literature, it is known as the Voronoi polygons, while in numerical simulations, it is known as Dirichlet cells (Müller and Guido 2017).

To employ it for classification, an existing set of example data is labeled as the training set where for all the data, the class from which each piece of the data belong to is known. Whenever a new piece of data without a label is given, it is compared to the labeled existing piece of data and the most similar (closest according to a given measure) pieces of data (the nearest neighbors) is then taken and their labels considered. We look at the top k most similar pieces of data from our known dataset; this is where the k comes from (k is an integer usually less than 20). Lastly, a majority vote is taken from the k most similar pieces of data, where this majority becomes the new class assigned to the data we were asked to classify, see Fig. 2.1.

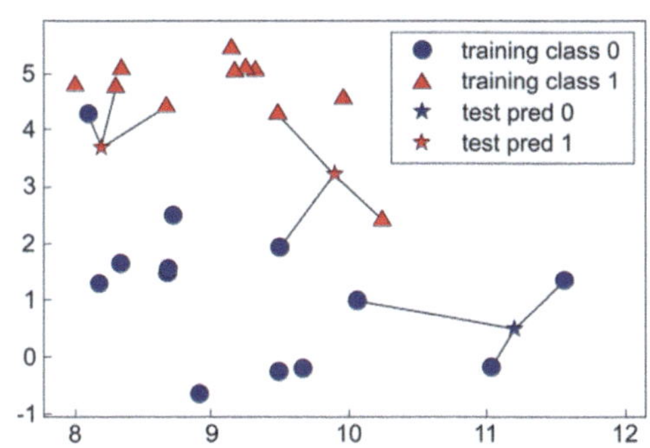

Fig. 2.1 Principle of KNearest Neighbors Classification in case of two classes and employing $k = 3$ neighbors, adopted from Müller and Guido (2017)

J. Awange et al., *Hybrid Imaging and Visualization*,
https://doi.org/10.1007/978-3-030-26153-5_2

Generally *k,* which depends on data and on the presence of structures in the data, is a hyper-parameter that should be tuned adaptively. A common approach to find the optimal value of *k* is to use the *cross-validation procedure*. In *n*-fold *cross-validation,* the original training data set is partitioned into *n*- subsets. In a special case (*leave - one - out*) a single observation is selected from the original data set as validation point. One from the *n*-subset is used as the validation data for testing the model, and the remaining *n*-1 subsets are used as training data. The validation error is calculated with the process repeated *n* times and the averaged validation errors used as a single cross validation error estimation for a specified parameter *k*. The procedure is repeated for different tuning parameter values of *k*, with the model that provides the lowest cross-validation error chosen as the optimal (*k*) one.

The advantages of the algorithm are the high accuracy, insensitivity to outliers and no special assumption about the data. However it is computationally expensive and requires a lot of memory.

2.1.1 Small Data Set

As a first example, we consider a small data set from the `mglearn` package of *Python*. For *Mathematica* we may write the data into an ASCII file (Navlani, 2018)

To use *Python* in *Mathematica,* we start a *Python* session,

```
⇒ session=
   StartExternalSession[<|"System" → "Python",
    "Version" → "3.5.4","Executable" →
     "C:\Users\Ben\AppData\Local\Programs\Python\Python35\
       python.exe"|>]//Quiet
```

```
⇐ ExternalSessionObject[
   SummaryPanel[ System: Python   EvaluationCount: None
                 UUID: 7b76966c2-ebb7-4ee8-bb25-02322d1456a8 ]]
```

```
import mglearn
import numpy as np
X, y = mglearn.datasets.make_forge()
np.savetxt('G:\\dataX.txt',X,fmt='%.5e')
```

The data elements to be classified are featured by 2D vectors,

```
⇒ X=Import["G:\\dataX.txt","Table"];
```

However if the file is short, we do not write the data into a file. The labels of the elements,

```
> y
⇐ {1,0,1,0,0,1,1,0,1,1,1,1,0,0,1,1,1,0,0,1,0,0,0,0,1,0}
⇒ y=%;
```

It means we have two classes. Let us visualize the data, see Fig. 2.2.

```
⇒ class1=Pick[X,y,0];
⇒ class2=Pick[X,y,1];
⇒ p0=ListPlot[{class1,class2},PlotStyle → {Green,Red},Frame → True,
   Axes → None,PlotMarkers → {Automatic,Medium},AspectRatio → 1]
⇐
```

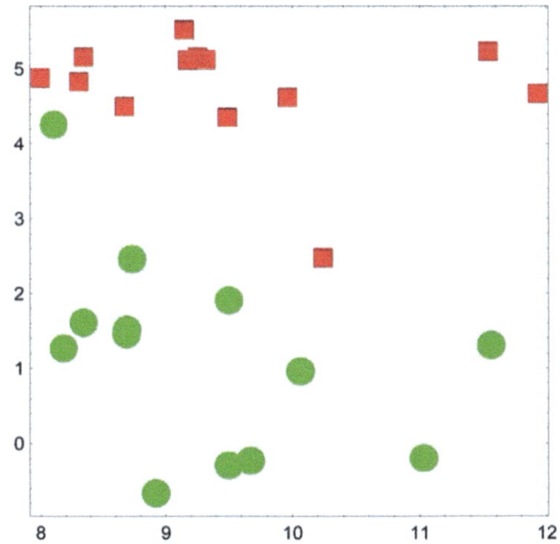

Fig. 2.2 Small dataset

Let us carry out the classification first with *Mathematica*.

Mathematica

We prepare the data for *Mathematica* as *Element* → *Label,*

```
⇒ trainingData=Thread[X → y];
```

First we employ $k = 1$ neighbour with exhaustive search technique on the entire dataset ("Scan")

```
⇒ c1=Classify[trainingData,Method → {"NearestNeighbors",
   "NeighborsNumber" → 1,"NearestMethod" → "Scan"}];
```

The result can be seen on Fig. 2.3.

```
⇒ p1=Show[{DensityPlot[c1[{u,v}],{u,7.5,12},{v,-1,6},
   ColorFunction → "CMYKColors"],p0}]
```

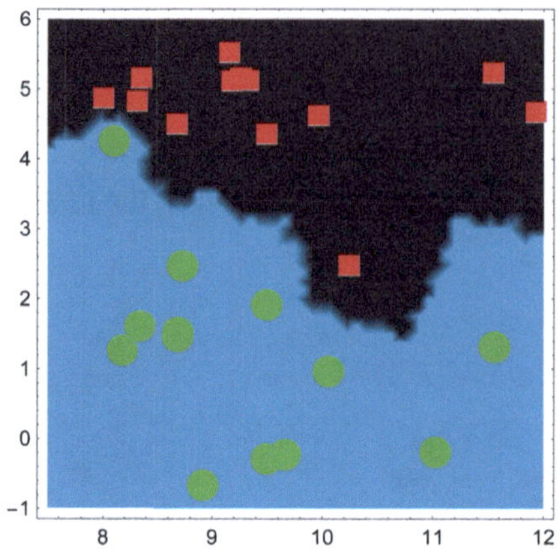

⇐

Fig. 2.3 Classification with $k = 1$

Let us investigate the effect of the number of neighbours on the quality of the classification. We shall consider $k = 3$ and $k = 9$ neighbours.

```
⇒ c3=Classify[trainingData,Method → {"NearestNeighbors",
   "NeighborsNumber" → 3,"NearestMethod" → "Scan"}];
```

```
⇒ c9=Classify[trainingData,Method → {"NearestNeighbors",
   "NeighborsNumber" → 9,"NearestMethod" → "Scan"}];
```

The results can be seen on Fig. 2.4.

```
⇒ p3=Show[{DensityPlot[c3[{u,v}],{u,7.5,12},{v,-1,6},
   ColorFunction → "CMYKColors"],p0}];
```

```
⇒ p9=Show[{DensityPlot[c9[{u,v}],{u,7.5,12},{v,-1,6},
   ColorFunction → "CMYKColors"],p0}];
```

```
⇒ GraphicsGrid[{{p3,p9}}]
```

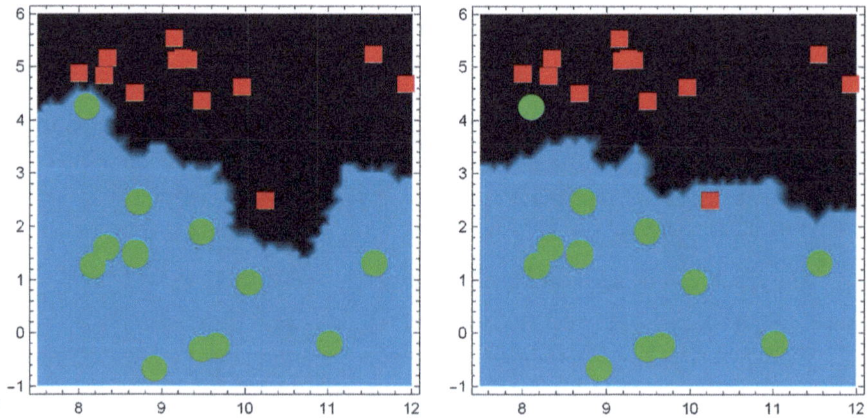

⇐

Fig. 2.4 Classification via Nearest Neighbor $k = 3$ and $k = 9$

We can compare the labels of the data elements with the labels provided by the method using different number of neighbours. The best result is given for $k = 1$.

```
⇒ y1=Map[c1[#]&,X]
⇐ {1,0,1,0,0,1,1,0,1,1,1,1,0,0,1,1,1,0,0,1,0,0,0,0,1,0}
⇒ y3=Map[c3[#]&,X]
⇐ {1,0,1,0,1,1,1,0,1,1,1,0,0,0,1,1,1,0,0,1,0,0,0,0,1,0}
⇒ y9=Map[c9[#]&,X]
⇐ {1,0,1,0,1,1,1,0,1,1,1,0,0,0,1,1,1,0,0,1,0,0,0,0,1,0}
```

The norm of the errors are

```
⇒ Norm[y-y1]
⇐ 0
⇒ Norm[y-y3]
⇐ √2
⇒ Norm[y-y9]
⇐ √2
```

This result indicates, that the $k = 1$ is the best solution. However, the third one ($k = 9$) seems to be more robust, since the different data elements are far from the boundary of the two classes, therefore this configuration is not so sensitive to the measurement errors of the feature vectors.

Python

Let us employ *Python* for $k = 1$,

```
# from sklearn.neighbors import KNeighborsClassifier
clf=KNeighborsClassifier(n_neighbors=1).fit(X,y)
prediction=clf.predict(X)
prediction
```

```
⇐ {1,0,1,0,0,1,1,0,1,1,1,1,0,0,1,1,1,0,0,1,0,0,0,0,1,0}
⇒ yP1=%;
```

Then computing the error vectors,

```
⇒ Norm[y-yP1]
⇐ 0
```

We have got the same result.

In this example only training set was employed. Now let us employ a test set, too, in order to check the generalization ability of our classifier. It can be done easily in *Python.*

```
from sklearn.model_selection import train_test_split
X, y = mglearn.datasets.make_forge()
X_train, X_test, y_train, y_test=
 train_test_split(X, y, random_state=0)
```

The total data set has been randomly split.

```
X_train
```

```
⇐ {{8.9223,-0.639932},{8.73371,2.49162},{9.32298,5.09841},
  {7.99815,4.85251},{11.033,-0.168167},{9.17748,5.09283},
  {11.564,1.33894},{9.15072,5.49832},{8.3481,5.13416},
  {11.9303,4.64866},{8.10623,4.28696},{8.67495,4.47573},
  {9.67285,-0.202832},{9.50169,1.93825},{8.69289,1.54322},
  {9.96347,4.59677},{9.50049,-0.264303},{9.25694,5.13285},
  {8.68937,1.4871}}
⇒ Xtrain=%;
```

```
X_test
```

```
⇐ {{11.5416,5.21116},{10.0639,0.990781},{9.49123,4.33225},
  {8.18378,1.29564},{8.30989,4.80624},{10.2403,2.45544},
  {8.34469,1.63824}}
⇒ Xtest=%;
```

```
y_train
```

```
⇐ {0,0,1,1,0,1,0,1,1,1,0,1,0,0,0,1,0,1,0}
⇒ ytrain=%;
```

```
y_test
```

```
⇐ {1,0,1,0,1,1,0}
⇒ ytest=%;
```

So we have now 19 elements in the training set and 7 elements in the test set.

Mathematica

Now using the training set, we create a classifier with 3 neighbors.

```
⇒ trainingData=MapThread[#2 → #1&,{ytrain,Xtrain}];
⇒ ctrained=Classify[trainingData,Method → {"NearestNeighbors",
  "NeighborsNumber" → 3,"NearestMethod" → "Scan"}];
```

Testing the classifier on the training data

```
⇒ trainingData=MapThread[#2 → #1&,{ytrain,Xtrain}];
⇒ ctraining=ClassifierMeasurements[ctrained,trainingData]
⇐ ClassifierMeasurementsObject[ Classifier: NearestNeighbors
                                Number of test examples: 19 ]
```

The accuracy of the classifier on the training set,

```
⇒ ctraining["Accuracy"]
⇐ 0.947368
```

The confusion matrix shows the number of the misclassified elements as off-diagonal elements, see Fig. 2.5.

```
⇒ ctraining["ConfusionMatrixPlot"]
```

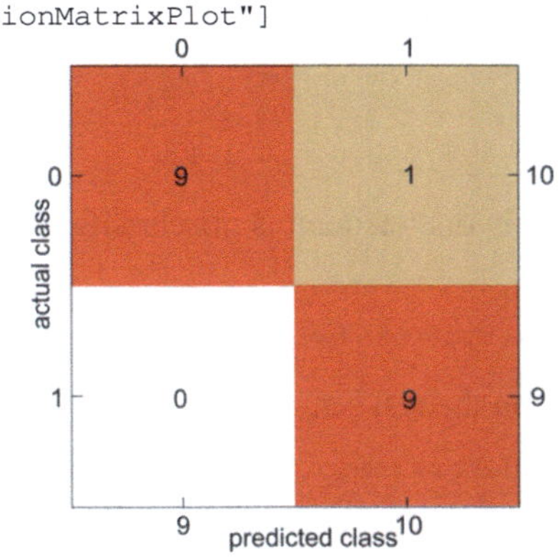

⇐

Fig. 2.5 Confusion matrix for the training set.

Now let us see the same statistics for the test set, which was not involved in the training process.

```
⇒ testingData=MapThread[#2 → #1&,{ytest,Xtest}];
⇒ ctesting=ClassifierMeasurements[ctrained,testingData]
⇐ ClassifierMeasurementsObject[ Classifier: NearestNeighbors
                                Number of test examples: 7 ]
```

The test set accuracy,

```
⇒ ctesting["Accuracy"]
⇒ 0.857143
⇐ ctesting["ConfusionMatrixPlot"]
```

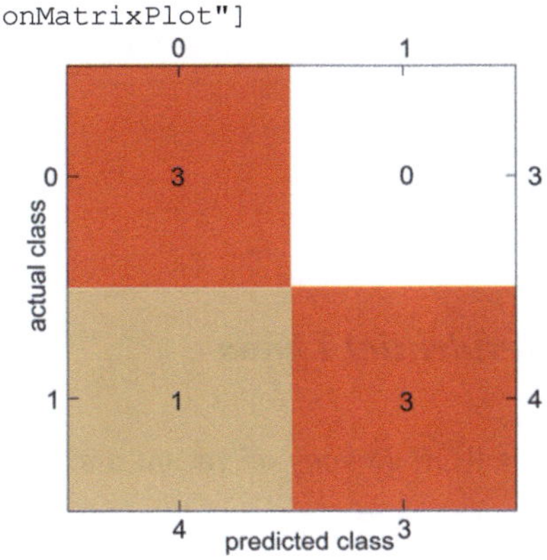

⇐

Fig. 2.6 Confusion matrix for the test set.

Python

Training the classifier (Groswami, 2018)

```
from sklearn.neighbors import KNeighborsClassifier
clf=KNeighborsClassifier(n_neighbors=3).fit(X_train,y_train)
```

Result of the prediction on the training set,

```
prediction=clf.predict(X_train)
prediction)
```

⇐ {0,0,1,1,0,1,0,1,1,1,1,1,0,0,0,1,0,1,0}

and it can be seen that only one element is misclassified.

```
⇒ ytrain
⇐ {0,0,1,1,0,1,0,1,1,1,0,1,0,0,0,1,0,1,0}
```

Result of the prediction on the test set,

```
from sklearn.neighbors import KNeighborsClassifier
clf=KNeighborsClassifier(n_neighbors=3).fit(X_train,y_train)
prediction=clf.predict(X_test)
prediction
```

```
⇐ {1,0,1,0,1,0,0}
⇒ ytest
⇐ {1,0,1,0,1,1,0}
```

Again, only one element is misclassified. The accuracy can be also computed similarly in *Mathematica.*

The accuracy on the training set,

```
print("Test set score: {:.2f}".format(clf.score(X_train,
 y_train)))
```

```
Test set score: 0.95
```

The accuracy on the test set,

```
print("Test set score: {:.2f}".format(clf.score(X_test,
 y_test)))
```

```
Test set score: 0.86
```

Both codes provided the same results.

2.1.2 Vacant and Residential Lands

There are two classes of RGB images of vacant and residential land, see Figs, 2.7 and 2.8.

Vacant Land

```
⇒ vacant=
```

Fig. 2.7 Vacant land areas.

Residential Areas

```
⇒ residental=
```

Fig. 2.8 Residental land areas.

We would like to design a KNearest Neighbors classifier for classifying images into these two classes, namely vacant or residential areas. As can be seen from the images, this classification problem is not an easy task. The images are RGB images with different sizes. It is reasonable to convert these images as nominal data, into a feature vector as numerical data. One may use one of dimension reduction methods discussed in the *Chapter* 1. We have 12 images in the training set, and 4 images in the test set for the both categories (vacant and residential lands). Autoencoding is employed to transform images into a feature vectors of size of 64 elements.

Let us create the training data

```
⇒ lands=Join[vacant,residential];aining["Accuracy"]
```

Since, they have different original sizes, we unify their size to 256×256,

```
⇒ landsReduced=Map[ImageResize[#,{256,256}]&,lands];
```

Now, we can reduce the data size. Here, one should make a trade-off between size and saving characteristic features. The feature extraction method we use here is the *t-SNE* (*t distributed Stochastic Neighbor Embedding*) method, which requires a big amount of computation, therefore we use of GPU (Graphic Processing Unit), instead of CPU (Central Processing Unit),

```
⇒ reduced=DimensionReduce[landsReduced,64,Method → "TSNE",
   TargetDevice → "GPU"];
```

So we have 16 -16 vectors of size 64,

```
⇒ Dimensions[reduced]
⇐ {32,64}
```

The vectors of vacant land images are

```
⇒ V=Take[reduced,{1,16}];
```

similarly those of the residential images are,

```
⇒ R=Take[reduced,{17,32}];
```

Let us label the vacant lands with label 0 and the residential lands with 1,

```
⇒ yV=Table[0,{i,1,16}];
⇒ yR=Table[1,{i,1,16}];
```

For training we randomly select 12 feature vectors from both classes,

```
⇒ SeedRandom[1035];
⇒ VTR=RandomSample[V,12];
⇒ RTR=RandomSample[R,12];
```

The remaining vectors represent the testing set,

```
⇒ VTE=Complement[V,VTR];
⇒ RTE=Complement[R,RTR];
```

Then the training set,

```
⇒ Xtrain=Join[VTR,RTR];
```

The corresponding labels,

```
⇒ ytrain=Join[Table[0,{i,1,12}],Table[1,{i,1,12}]]
⇐ {0,0,0,0,0,0,0,0,0,0,0,0,1,1,1,1,1,1,1,1,1,1,1,1}
```

Similarly, the test set,

```
⇒ Xtest=Join[VTE,RTE];
```

The corresponding labels,

```
⇒ ytest=Join[Table[0,{i,1,4}],Table[1,{i,1,4}]]
⇐ {0,0,0,0,1,1,1,1}
```

Let us save these sets in files for *Python* as *mtx* file,

```
⇒ Export["G:\\Xtrain.mtx",Xtrain];
⇒ Export["G:\\Xtest.mtx",Xtest];
⇒ Export["G:\\ytrain.mtx",{ytrain}];
⇒ Export["G:\\ytest.mtx",{ytest}];
```

Now after the data preparation let start the classification.

Mathematica

Training process with $k = 1$ neighbors,

```
⇒ trainingData=MapThread[#1 → #2&,{Xtrain,ytrain}];
⇒ AbsoluteTiming[c=Classify[trainingData,
    Method → {"NearestNeighbors","NeighborsNumber" → 1}];]
⇐ {0.768199,Null}
```

Checking the accuracy on the training set

```
⇒ ctraining=ClassifierMeasurements[c,trainingData]
⇐ ClassifierMeasurementsObject[ Classifier: NearestNeighbors
                                Number of test examples: 24 ]
⇒ ctraining["Accuracy"]
⇐ 1
⇒ ctraining["ConfusionMatrixPlot"]
```

⇐ 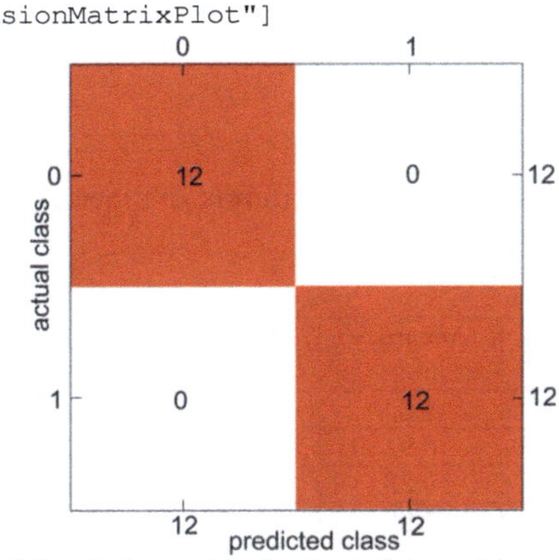

Fig. 2.9 Confusion matrix for the training set in case $k = 1$.

The error on the training set,

```
⇒ yTR=Map[c[#]&,Xtrain]
⇐ {0,0,0,0,0,0,0,0,0,0,0,0,1,1,1,1,1,1,1,1,1,1,1,1}
⇒ Norm[yTR-ytrain]
⇐ 0
```

Let us check the classifier on the test set,

```
⇒ testingData=MapThread[#1 → #2&,{Xtest,ytest}];
⇒ ctesting=ClassifierMeasurements[c,testingData]
⇐ ClassifierMeasurementsObject[ Classifier: NearestNeighbors
                                Number of test examples: 8 ]
⇒ ctraining["Accuracy"]
⇐ 1
⇒ ctraining["ConfusionMatrixPlot"]
```

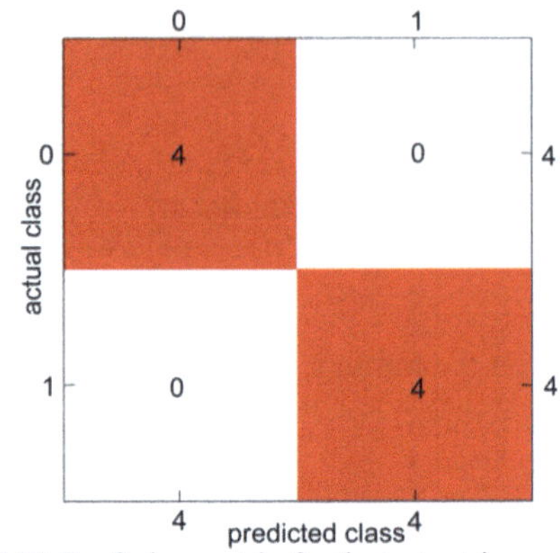

⇐

Fig. 2.10 Confusion matrix for the test set in case $k = 3$.

```
⇒ yTE=Map[c[#]&,Xtest]
⇐ {0,0,0,0,1,1,1,1}
⇒ Norm[yTE-ytest]
⇐ 0
```

Table 2.1 shows the result in case of different number of neighbors.

Table 2.1 Results with *Mathematica*

Neighbors	Time [sec]	Accuracy training set	Accuracy test set
1	0.78	1.00	1.00
2	0.80	0.92	0.88
3	0.76	0.88	0.88
4	0.86	0.83	0.75

Python

```
import numpy as np
from numpy import array, matrix
from scipy.io import mmread, mmwrite
```

Input data for the training

```
Xtrain=mmread('G:\\Xtrain.mtx')
y=mmread('G:\\ytrain.mtx')
ytrain=y[0]
```

Training process

```
from sklearn.neighbors import KNeighborsClassifier
clf=KNeighborsClassifier(n_neighbors=1).fit(Xtrain,ytrain)
```

Prediction accuracy on the training

```
prediction=clf.predict(Xtrain)
prediction
```

```
⇐ {0,0,0,0,0,0,0,0,0,0,0,0,1,1,1,1,1,1,1,1,1,1,1,1}
```

⇒ yPTR=%;

The error on the training set

⇒ Norm[ytrain-yPTR]

⇐ 0

The accuracy on the training set

```
print("Training set score: {:.2f}".format(clf.score
  (Xtrain, ytrain)))
```

```
Training set score: 1.00
```

The error on the test set

```
Xtest=mmread('G:\\Xtest.mtx')
y=mmread('G:\\ytest.mtx')
ytest=y[0]
```

```
prediction=clf.predict(Xtest)
prediction
```

⇐ {0,0,0,0,1,1,1,1}

⇒ yPTE=%;

⇒ Norm[yPTE-ytest]

⇐ 0

The accuracy on the test set

```
print("Test set score:{:.2f}".format(clf.score(Xtest,ytest)))
```

```
Training set score: 1.00
```

Remark

Mathematica and *Python* give the same result. One should keep in mind that since the elements of the training and test sets are selected randomly, the result of a new run can be different.

2.2 Logistic Regression

Basic theory

Let us consider an element featured by the vector $z_i = \{x_1, x_2, \ldots, x_m\}$ and labeled by y_i. Logistic regression can be considered as a special shallow neural network, where the activation function is a sigmoid function, see Fig. 2.11 (Brownlee, 2016).

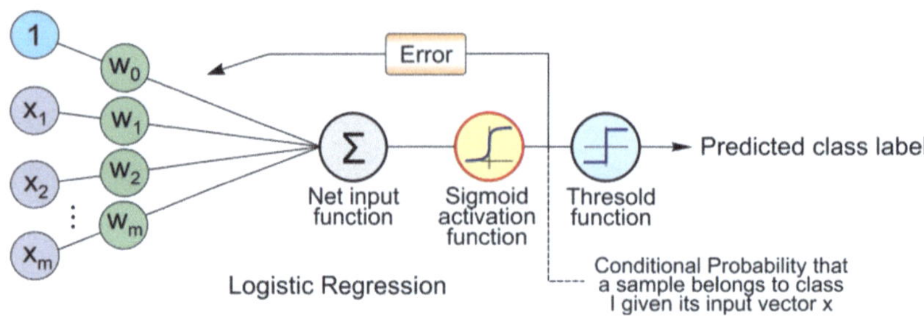

Fig. 2.11 Principle of Logistic Regression.

The output of the sigmoid function can be interpreted as the probability of a particular sample belonging to class 1,

$$\phi(z_i) = \frac{1}{1+\text{Exp}(-z)} = P(y=1|z:w),$$

given its feature vector z parametrized by weights w. Since this method is basically a binary classification technique, sometimes a threshold function (unitstep or Heaviside) is added. The cost function is the log - likelihood function to be maximized for getting optimal weights,

$$\mathcal{L}(w) = \ln L(w) = \sum_{i=1}^{n}\left[y_i \ln(\phi(z_i)) + (1-y_i)\ln(1-\phi(z_i))\right].$$

In *Mathematica* and in *Python* the objective function can be extended by L1 and L2 regularization terms,

$$\mathcal{L}_{\mathcal{R}}(w) = \ln L(w) + \lambda_1 \sum_{i=1}^{n} |w_i| + \frac{\lambda_2}{2}\sum_{i=1}^{n} w_i^2,$$

where λ_i are regularization parameters (Prince, 2012).

Remark

The basic idea is good for binary classification only. However there is a possibility to generalize the binary technique for example in many successive steps. The technique is called *one againts many*: Assuming k classes, we start with two classes with labels 1 and the other with (2, 3, ... k), then again two classes with elements labeled 2 and others (3,..k).

The advantages of this algorithm are that it is computationally inexpensive, easy to implement and knowledge representation easy to interpret. However it is prone to have low accuracy.

As a first example, let us consider the *Iris Classification* problem.

2.2.1 Iris Data Set

Now, we have 150 samples, two features: petal length and petal width, and three target sets: Iris-setosa (0), Iris-versicolor (1) and Iris-virginica (2). Let us start with *Python*.

Python

Let us read the elements to be classified (X) and their labels (y),

```
from sklearn import datasets
iris=datasets.load_iris()
```

```
X=iris.data[:,[2,3]]
y=iris.target
```

Save X in file for *Mathematica*,

```
import numpy as np
```

```
np.savetxt('M:\\dataX.txt',X,fmt='%.2e')
```

Training *Python* regressor,

```
from sklearn.linear_model  import LogisticRegression
lr=LogisticRegression(C=100.0,random_state=1).fit(X,y)
```

The result of *Python*

```
lr.predict(X[:150,:])
```

```
⇐ {0,0,0,0,0,0,0,0,0,0,0,0,0,0,0,0,0,0,0,0,0,0,0,0,0,
   0,0,0,0,0,0,0,0,0,0,0,0,0,0,0,0,0,0,0,0,0,0,0,0,0,
   1,1,1,1,1,1,1,1,1,1,1,1,1,1,1,1,1,1,1,1,2,1,1,1,1,
   1,1,2,1,1,1,1,1,2,1,1,1,1,1,1,1,1,1,1,1,1,1,1,1,1,
   2,2,2,2,2,2,2,2,2,2,2,2,2,2,2,2,2,2,2,1,2,2,2,2,2,
   2,2,2,2,2,2,2,2,1,1,2,2,2,2,2,2,2,2,2,2,2,2,2,2,2}
```

```
⇒ yP=%;
```

The original training labels,

```
y
```

```
⇐ {0,0,0,0,0,0,0,0,0,0,0,0,0,0,0,0,0,0,0,0,0,0,0,0,0,
   0,0,0,0,0,0,0,0,0,0,0,0,0,0,0,0,0,0,0,0,0,0,0,0,0,
   1,1,1,1,1,1,1,1,1,1,1,1,1,1,1,1,1,1,1,1,1,1,1,1,1,
   1,1,1,1,1,1,1,1,1,1,1,1,1,1,1,1,1,1,1,1,1,1,1,1,1,
   2,2,2,2,2,2,2,2,2,2,2,2,2,2,2,2,2,2,2,1,2,2,2,2,2,
   2,2,2,2,2,2,2,2,1,1,2,2,2,2,2,2,2,2,2,2,2,2,2,2,2}
```

```
⇒ yTR=%;
```

The error is

```
⇒ Norm[yP-yTR]
⇐ √6
```

The accuracy on the training set,

```
> print("Training set score: {:.2f}".format(lr.score(X, y)))
  Training set score: 0.96
```

Mathematica

Reading the data

```
⇒ Xtrain=Import["M:\\dataX.txt","Table"];
```

The total training set

```
⇒ dataT=MapThread[Join[#1,{#2}]&,{Xtrain,yTR}];
```

In order to visualize the three data sets, we divide the training set into the three classes,

```
⇒ data0=Map[{#[[1]],#[[2]]}&,Select[dataT,#[[3]]==0&]];
  data1=Map[{#[[1]],#[[2]]}&,Select[dataT,#[[3]]==1&]];
  data2=Map[{#[[1]],#[[2]]}&,Select[dataT,#[[3]]==2&]];
```

Then Fig. 2.12 shows the elements to be classified,

```
⇒ p0=ListPlot[{data0,data1,data2},PlotStyle → {Green,Blue,Red},
   Frame → True,Axes → None,PlotMarkers → {Automatic},
   AspectRatio → 0.9,FrameLabel → {"petallength","petal width"},
   Frame → True]
```

⇐

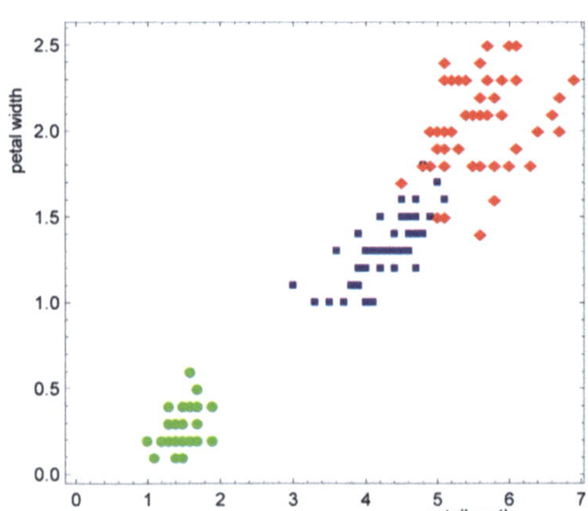

Fig. 2.12 The three classes of the Iris-data set.

The training set,

```
⇒ dataTrain=Map[{#[[1]],#[[2]]} → #[[3]]&,dataT];
```

The training process,

```
⇒ c=Classify[dataTrain,Method → "LogisticRegression",
    PerformanceGoal → "Quality"];
```

Then the accuracy of the classifier on the training set

```
⇐ ClassifierMeasurementsObject[ Classifier: LogisticRegression
                                 Number of test examples: 150 ]
```

and

```
⇒ ctraining["Accuracy"]
⇐ 0.96
```

Fig.2.13 shows the confusion matrix,

```
⇒ ctraining["ConfusionMatrixPlot"]
```

⇐ 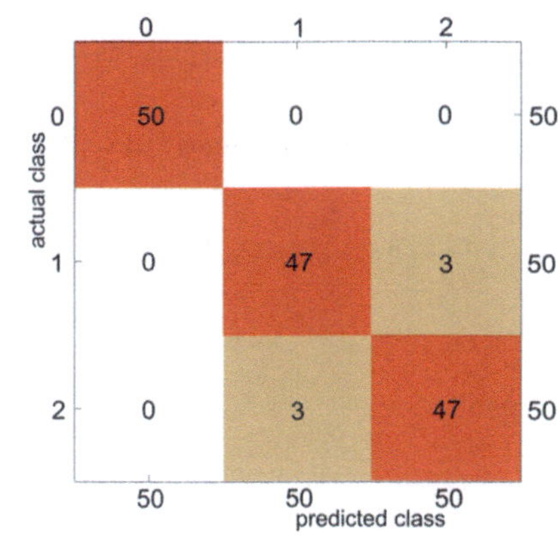

Fig. 2.13 The confusion matrix of the training set.

Let us visualize the three classes,

```
⇒ Show[{DensityPlot[c[{u,v}],{u,0,7},{v,-1,4.5},
    ColorFunction → "CMYKColors",PlotPoints->50],p0},
    AspectRatio->1.2]
```

⇐ 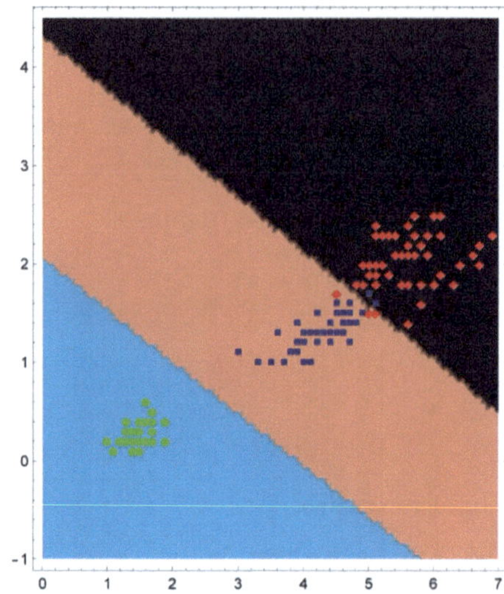

Fig. 2.14 The three classes of the Iris-data set.

The error of the classification can be characterized with the norm of the misclassified elements

```
⇒ yP=Map[c[#]&,Join[{data0,data1,data2}]]//Flatten;
⇒ Norm[yTR-yP]
```
⇐ $\sqrt{6}$

We have got the same result.

2.2.2 Digit Recognition

The digits are represented as grey images of size 26×16, with pixel values $\in [0,1]$. For example let us consider digit 2,

⇒ `image=`  `;`

```
⇒ dataD=ImageData[image];
⇒ Dimensions[dataD]
⇐ {20,16}
```

Let us visualize the image of the digit 2,

```
⇒ MatrixPlot[dataD]
```

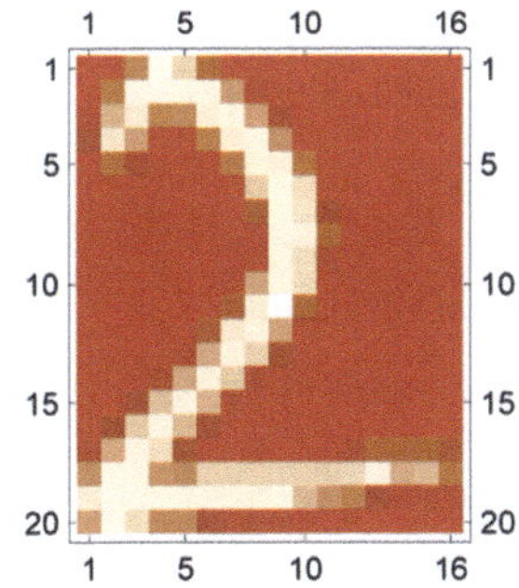

⇐

Fig. 2.15 The image of a digit 2.

Our training set is,

```
⇒ digitset=
```

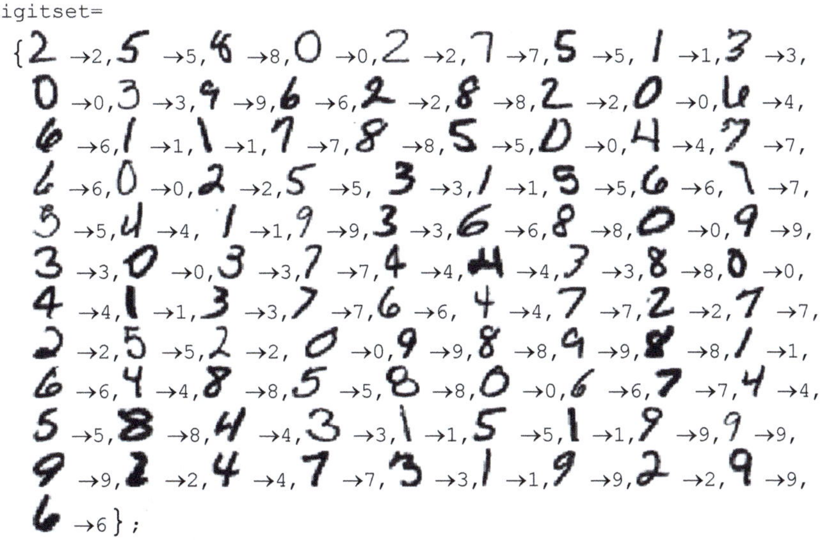

which contains 100 elements.

Now in order to improve the generalization ability, we introduce a *validation set* besides the test set. The validation set will be included in the training process however its error is just monitored and does not influence the parameter change of the classifier. This technique can hinder over-fitting, similarly to the regularization. We randomly select these three sets.

```
⇒ RandomSeed[1234];
⇒ trainingset=RandomSample[digitset,70];
⇒ validationset=RandomSample[digitset,15];
⇒ testset=RandomSample[digitset,15];
```

Mathematica

Now, the training with the validation set results,

```
⇒ logistic=Classify[trainingset,ValidationSet → validationset,
    Method → "LogisticRegression",PerformanceGoal → "Quality"]
⇐ ClassifierFunction[ Input type: Image
                      Number of classes: 10 ]
```

Testing the classifier on the training data

```
⇒ ctraining=ClassifierMeasurements[logistic,trainingset]
⇐ ClassifierMeasurementsObject[ Classifier: LogisticRegression
                                Number of test examples 70 ]
```

```
⇒ ctraining["Accuracy"]
⇐ 1.
```

The confusion matrix is on Fig.2.16

```
⇒ ctraining["ConfusionMatrixPlot"]
⇐
```

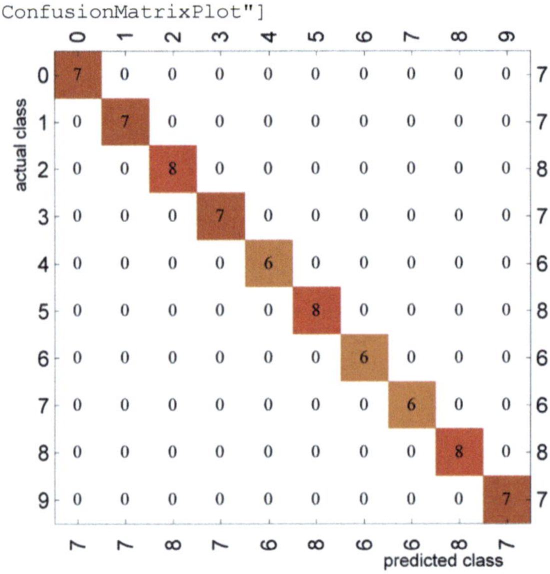

Fig. 2.16 The confusion matrix on the training set.

Testing the classifier on the test data

```
⇒ ctesting=ClassifierMeasurements[logistic,testset]
⇐ ClassifierMeasurementsObject[ Classifier: LogisticRegression
                                Number of test examples 15 ]
```

```
⇒ ctesting["Accuracy"]
⇐ 0.866667
```

The confusion matrix,

```
⇒ ctesting["ConfusionMatrixPlot"]
```

⇐

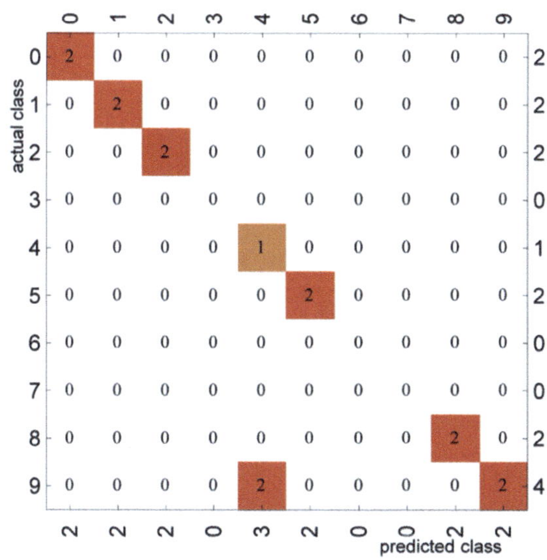

Fig. 2.17 The confusion matrix on the test set.

Here we used images of the digits as nominal input, no numerical encoding was employed.

Python

Now we employ 75 elements in the training set and 25 elements in the test set. There is no validation set.

```
⇒ Xtrain=RandomSample[digitset,75];
⇒ Xtest=Complement[digitset,Xtrain];
```

The size of the images of the digits is somewhat different, that is why we should use resizing process. The resized grey images are of size 16×20. Therefore we use a feature vector of size 320 for each digits.

The elements and labels for the training set (numerical encoding).

```
⇒ Xtra=
   Map[Flatten[ImageData[ImageResize[#[[1]],{16,20}]],1]&,Xtrain];
⇒ ytra=Map[#[[2]]&,Xtrain];
```

Similarly for the test set,

```
⇒ Xtes=
   Map[Flatten[ImageData[ImageResize[#[[1]],{16,20}]],1]&,Xtest];
⇒ ytes=Map[#[[2]]&,Xtest];
```

Saving these data for *Python*,

```
⇒ Export["Xtest.mtx",Xtes];
⇒ yy={ytes};
```

```
⇒ Export["ytest.mtx",yy];
⇒ Export["Xtrain.mtx",Xtra];
⇒ yy={ytra};
⇒ Export["ytrain.mtx",yy];
```

Feature vectors of size 320 representing a digit as an image,

```
⇒ Xtra//Dimensions
⇐ {75,320}
```

Preparing for reading *.mtx files,

```
import numpy as np
from numpy import array, matrix
from scipy.io import mmread, mmwrite
```

Reading data,

```
X=mmread('Xtrain.mtx')
y=mmread('ytrain.mtx')
yy=y[0]
```

Training process,

```
from sklearn.linear_model  import LogisticRegression
lr=LogisticRegression(C=100.,random_state=1).fit(X,yy)
```

The prediction of *Python*

```
ytr=lr.predict(X)
```

```
ytr
```

```
⇐ {0,9,7,7,7,4,1,2,5,1,1,5,0,0,1,3,7,0,1,1,8,9,8,9,4,
   0,6,5,5,5,7,3,8,6,9,8,7,9,1,9,5,6,8,6,4,1,7,2,6,6,
   0,2,2,0,5,9,1,3,3,7,4,2,4,2,5,3,7,2,8,4,0,3,5,8,2}
⇒ yP=%;
```

The original training labels

```
⇒ yy
⇐ {0,9,7,7,7,4,1,2,5,1,1,5,0,0,1,3,7,0,1,1,8,9,8,9,4,
   0,6,5,5,5,7,3,8,6,9,8,7,9,1,9,5,6,8,6,4,1,7,2,6,6,
   0,2,2,0,5,9,1,3,3,7,4,2,4,2,5,3,7,2,8,4,0,3,5,8,2}
⇒ yT=%;
```

The error is

```
⇒ Norm[yP-yT]
⇐ 0
```

The accuracy on the training set

```
print("Training set score: {:.2f}".format(lr.score(X, yy)))
```

```
Training set score: 1.00
```

The *Python* result on the test set,

```
Xt=mmread('Xtest.mtx')
yt=mmread('ytest.mtx')
yt
```

⇐ {4,1,9,4,9,6,6,0,7,8,3,9,6,3,3,4,6,8,2,2,3,4,8,5,0}

⇒ yT=First[%];

```
yte=lr.predict(Xt)
yte
```

⇐ {4,1,9,1,9,6,8,0,7,8,3,9,6,3,5,4,1,8,3,2,3,4,8,5,4}

⇒ yP=%;

The error on the test set

⇒ Total[MapThread[If[Abs[#1-#2]>0,1,0]&,{yT,yP}]]

⇐ 6

In percentage

⇒ 100 -6/25 100

⇐ 76

The accuracy on the test set using built in function

```
print("Test set score: {:.2f}".format(lr.score(Xt, yt[0])))
```

```
Training set score: 0.76
```

Mathematica provides somewhat better result because of using validation set, however *Python* parameter *C*= 100 indicating regularization, also gives certain improvement on the test set (Fortuner, 2019).

2.3 Tree Based Methods

Basic theory

The *Decision Tree* is one of the most commonly used non-parametric classification technique. In case of non parametric model, it has no parameter to be estimated and the applied algorithm dominates. Based on features in data, decision tree models learn a series of questions to infer the class labels of samples. During this process the data will be subsequently split in order to decrease the impurity of the data measured by entropy of data. The more the data is mixed, the higher is the entropy. Information gained is the expected reduction in entropy caused by partitioning the examples according to a given attribute. The idea is to start with mixed classes and to continue partitioning until each node reaches its observations of purest class (Dangeti, 2017).

The advantages of this algorithm are that they are computationally inexpensive, easy to implement and knowledge representation is easy to interpret. However, it is prone to have low accuracy.

Further features of the algorithm include the model's shape not being predefined, the model fitting in the best possible classification based on the data, and providing the best results when most of the variables are categorical in nature on the one hand, and when outliers and missing values are dealt with graciously in decision trees on the other hand.

One may improve the performance of the Decision Tree Classifier using weights or employing an *ensemble of tree models* known as *Random Forest Method.*

This can be considered as an ensemble of decision trees and the algorithm is basically the following:

1) Randomly chose sample size n from the training set with replacement,
2) Carry out decision tree computation,
3) Repeat steps 1 and 2 k times,
4) Aggregate the prediction by each tree to assign the class labelled by majority vote.

Advantages: low variance, robust model, less sensitive to overfitting. The most important parameter is the number of trees (Skeritt, 2018).

2.3.1 Playing Tennis Today?

Explanatory Example

In the following example, the response variable has only two classes; whether to play tennis or not. Will John play tennis or not, when the outlook of the weather is rainy, the humidity is high and the wind is weak? Table 2.2 has been compiled based on various conditions recorded on various days

The following information is available,

Table 2.2 The training set of the problem

Day	Outlook	Humidity	Wind	Play
D1	Sunny	High	Weak	No
D2	Sunny	High	Strong	No
D3	Overcast	High	Weak	Yes
D4	Rain	High	Weak	Yes
D5	Rain	Normal	Weak	Yes
D6	Rain	Normal	Strong	No
D7	Overcast	Normal	Strong	Yes
D8	Sunny	High	Week	No
D9	Sunny	Normal	Weak	Yes
D10	Rain	Normal	Weak	Yes
D11	Sunny	Normal	Strong	Yes
D12	Overcast	High	Strong	Yes
D13	Overcast	Normal	Week	Yes
D14	Rain	High	Strong	No

Features → Outlook: Values → Sunny, Rain, Overcast,
Humidity: Values → High, Normal,
Wind: Values → Weak, Strong,
Class → Play: Values → Yes, No.

The training example indicate 9 Yes vs. 5 No

Our training set has 14 elements. An input element is characterized by a non-numeric feature vector with 3 elements (Outlook, Humidity and Wind). The output is binary (Yes, No) saying that the input belongs to the set Play or does not.

The prediction problem is: Does the input element (Rain, High, Weak) belong to set Play or not? Let us separate the elements of the training set according to the *first feature* Outlook. We get three subsets (values) Sunny, Overcast and Rain. The set *Sunny* can be classified according to the second feature *Humidity* while the set *Rain* can be classified according to the third feature *Wind.* In this way we get five sets: the first two are *pure in the feature Humidity*, the third in the *Outlook* and last two in the *Wind.* See Fig. 2.18.

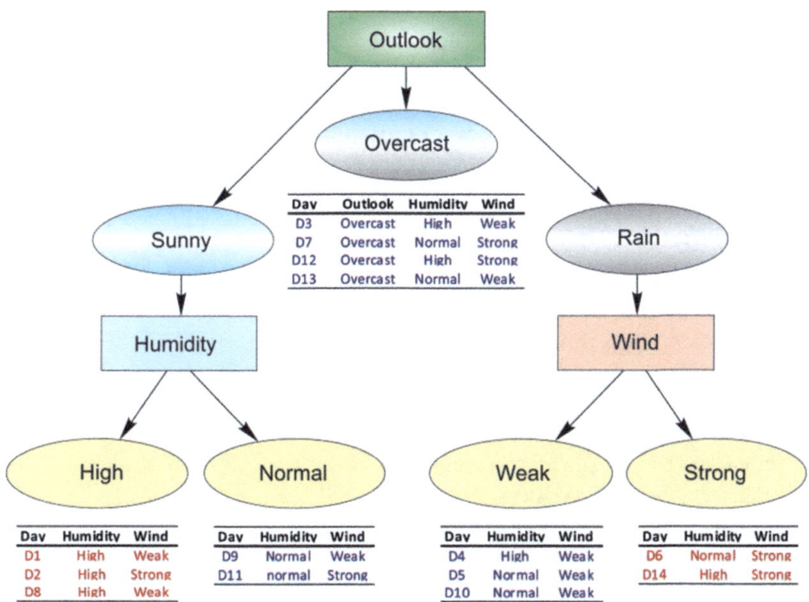

⇐

Fig. 2.18 Pure subsets of the decision tree.

The evaluation of the tree can be seen in Fig.2.19, where according to the results, the prediction is *yes*.

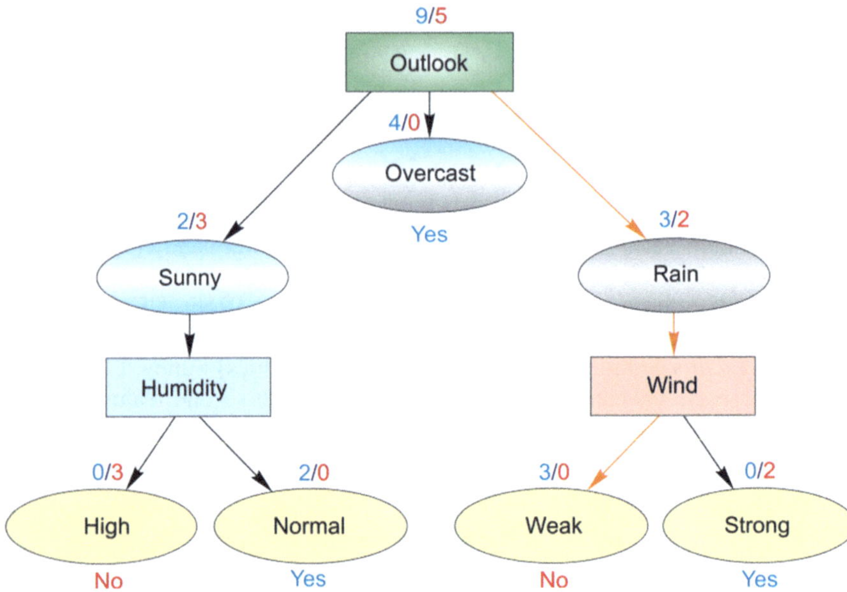

⇐

Fig. 2.19 Evaluation of the decision tree. In our case: (Rain, High, Weak) → Yes.

Let us solve the problem using *Mathematica*. The input-out pairs according to Table 2.2 are.

Mathematica

```
⇒ dataTraining={{"Sunny","High","Weak"} → "No",
   {"Sunny","High","Strong"} → "No",
   {"Overcast","High","Weak"} → "Yes",
   {"Rain","High","Weak"} → "Yes",
   {"Rain","Normal","Weak"} → "Yes",
   {"Rain","Normal","Strong"} → "No",
   {"Overcast","Normal","Strong"} → "Yes",
   {"Sunny","High","Weak"} → "No",
   {"Sunny","Normal","Weak"} → "Yes",
   {"Rain","Normal","Weak"} → "Yes",
   {"Sunny","Normal","Strong"} → "Yes",
   {"Overcast","High","Strong"} → "Yes",
   {"Overcast","Normal","Weak"} → "Yes",
   {"Rain","High","Strong"} → "No"};
⇒ TableForm[dataTraining,TableAlignments → {Right,Center}]
```

```
⇐ {Sunny,High,Weak} → No
   {Sunny,High,Strong} → No
   {Overcast,High,Weak} → Yes
   {Rain,High,Weak} → Yes
   {Rain,Normal,Weak} → Yes
   {Rain,Normal,Strong} → No
   {Overcast,Normal,Strong} → Yes
   {Sunny,High,Weak} → No
   {Sunny,Normal,Weak} → Yes
   {Rain,Normal,Weak} → Yes
   {Sunny,Normal,Strong} → Yes
   {Overcast,High,Strong} → Yes
   {Overcast,Normal,Weak} → Yes
   {Rain,High,Strong} → No
```

Let us create a classification function,

```
⇒ c=Classify[dataTraining,Method → "DecisionTree",
   PerformanceGoal → "Quality"]
⇐ ClassifierFunction[ Input type: {Nominal, Nominal, Nominal}
                       Classes: No, Yes ]
```

Give our data as input for the classifier,

```
⇒ c[{"Rain","High","Weak"}]
⇐ Yes
```

Out of Yes or No, we can get the probabilities, too

```
⇒ c[{"Rain","High","Weak"},"Probabilities"]
⇐ <|No → 0.142857,Yes → 0.857143|>
```

One may get more information about the classification process as well as about the classifier,

```
⇒ ClassifierInformation[c]
```

⇐

Classifier Information

Input time	{Nominal, Nominal, Nominal}
Casses	No, Yes
Method	decision Tree
Accuracy	45.6% ± 6.6%
Loss	0.556 ± 0.059
Single evaluation time	1.59 ms/example
Batch evaluation speed	149. example/s
Classifier memory	105. kB
Training examples used	14 examples
Training time	1.19 s

The accuracy of the classification process as well as the confusion matrix can also be easily computed,

```
⇒ cm=ClassifierMeasurements[c,dataTraining]
⇐ ClassifierMeasurementsObject[ Classifier: DecisionTree
                                Number of test examples 14 ]
```

```
⇒ cm["Accuracy"]
⇐ 1.
⇒ cm["ConfusionMatrixPlot"]
```

⇐ 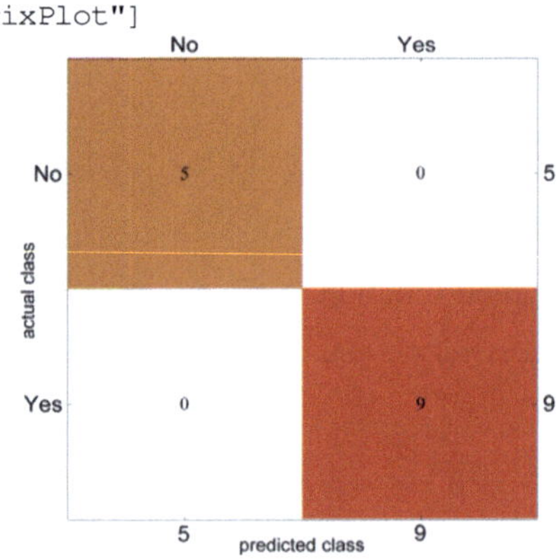

Fig. 2.20 The confusion matrix.

Now let us employ *Python* to solve the problem.

In order to solve the problem with *Python*, we transform the dataset from nominal values into numerical ones, like

```
⇒ dataX=Map[#[[1]]&,dataTraining]/.{"Sunny" → 0.1,
   "Overcast" → 0.2,"Rain" → 0.3,"High" → 1.,
   "Normal" → .1,"Weak" → 2.,"Strong" → 2.1}
⇐ {{0.1,1.,2.},{0.1,1.,2.1},{0.2,1.,2.},{0.3,1.,2.},{0.3,1.1,2.},
   {0.3,1.1,2.1},{0.2,1.1,2.1},{0.1,1.,2.},{0.1,1.1,2.},
   {0.3,1.1,2.},{0.1,1.1,2.1},{0.2,1.,2.1},{0.2,1.1,2.},
   {0.3,1.,2.1}}
```

Similarly, the output is binary value,

```
⇒ datay=Map[#[[2]]&,dataTraining]/.{"Yes" → 1,"No" → 0}
⇐ {0,0,1,1,1,0,1,0,1,1,1,1,1,0}
```

Let us save these data for the *Python* system,

```
⇒ Export["G:\\Xdata.txt",dataX];
⇒ Export["G:\\ydata.txt",datay];
```

Read data from txt files,

```
X=[i for i in range(14)]
f=open('G:\\Xdata.txt','r')
for i in range(0,14):
 X[i]=list(ast.literal_eval(f.readline()))
```

```
X
```

⇐ {{0.1,1.,2.},{0.1,1.,2.1},{0.2,1.,2.},{0.3,1.,2.},{0.3,1.1,2.},
{0.3,1.1,2.1},{0.2,1.1,2.1},{0.1,1.,2.},{0.1,1.1,2.},
{0.3,1.1,2.},{0.1,1.1,2.1},{0.2,1.,2.1},{0.2,1.1,2.},
{0.3,1.,2.1}}

```
y=[i for i in range(14)]
f=open('G:\\ydata.txt','r')
for i in range(0,14):
 y[i]=int(f.readline())
```

```
y
```

⇐ {0,0,1,1,1,0,1,0,1,1,1,1,1,0}

Training the classifier,

```
from sklearn.tree  import DecisionTreeClassifier
tree=DecisionTreeClassifier(criterion='gini',max_depth=4,
 random_state=1).fit(X,y)
```

In our case the input feature vector is {Rain,High,Weak}→{0.3,1.,2.}. Then the prediction of the classifier is,

```
import numpy as np
X_new=np.array([[0.3,1.,2.]])
prediction=tree.predict(X_new)
prediction
```

⇐ {1}

Which means: Play!

2.3.2 Snowmen and Dice

We now like to classify images into two different categories, snowman and dice, see Fig. 2.21 and Fig.2.22.

Fig. 2.21 Representatives of snowman category.

Fig. 2.22 Representatives of dice category.

The feature extraction of these pictures represented by image matrices of 64×64 pixels, was carried out by wavelet transform using second order Daubechies filter, then employing averaging technique for seven nonoverlapping bands of the spectrum. For example let us consider the grayscale version of an image,

Fig. 2.23 Image.

Do create the list form of the image matrix,

```
⇒ dims = Flatten[ImageData[im]]
```

The DFT using 6 levels of refinement,

```
⇒ wtr=DiscreteWaveletTransform[dims,DaubechiesWavelet[2],6]
⇐ DiscreteWaveletData[ Data dimension: {4096}
                       Refinements: 6 ]
```

Plotting the wavelet transform coefficients in the different refinement levels,

```
⇒ WaveletListPlot[wtr,Ticks → Full]
```

⇐

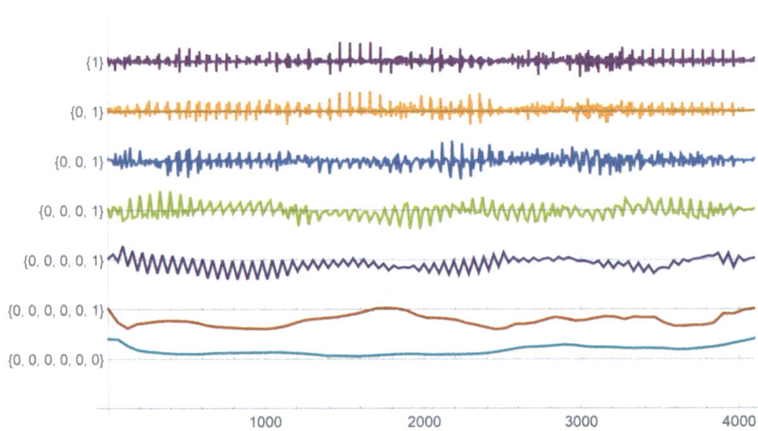

Fig. 2.24 DFT coefficients at different refinement levels.

The energy content of the coefficients $W_{i,j}$, $j = 1,2,3, ...n_i$ at the i-th level may be expressed as,

$$\left|\log\left[\frac{1}{n_i}\sum_{j=1}^{n_i} W_{i,j}{}^2\right]\right|.$$

This can represent the i-th component of the feature vector of the image.

For example considering that the first level

```
⇒ Normal[wtr][[2]][[1]]
⇐ {1}
```

and the corresponding coefficients are,

```
⇒ ListPlot[Normal[wtr][[2]][[2]],Joined → True,
   PlotStyle → Purple,PlotRange → All,AspectRatio → 0.2]
```

⇐ 

Fig. 2.25 DFT coefficients at first refinement level.

The energy content of this level, the first element of the feature vector can be computed as

```
⇒ Log[Apply[Plus,Normal[wtr][[2]][[2]]^2] / Length[Normal[wtr][[2]][[2]]]]//Abs
⇐ 5.97594
```

Since

```
⇒ Map[#[[1]]&,Normal[wtr]]
⇐ {{0},{1},{0,0},{0,1},{0,0,0},{0,0,1},{0,0,0,0},{0,0,0,1},
   {0,0,0,0,0},{0,0,0,0,1},{0,0,0,0,0,0},{0,0,0,0,0,1}}
```

Then the feature vector of the image is, see Fig. 2.24,

```
⇒ Map[Abs[Log[Apply[Plus,Normal[wtr][[2]][[2]]^2] / Length[Normal[wtr][[2]][[2]]]]]
   &,{2,4,6,8,10,11,12}]
⇐ {5.97594,3.93134,2.74834,1.71002,0.515124,2.58898,0.746729}
```

Consequently the dimension of these feature vectors is seven, $n = 7$. The feature vectors of snowmen,

```
⇒ Snowman=Import["M:\\Snowman.dat"];
⇒ Dimensions[Snowman]
⇐ {10,7}
```

and those of the dices,

```
⇒ DiCes=Import["G:\\Dice.dat"];
```

The output value for the snowman class is 1 and for the dice class is 0.

```
⇒ dataSetSnowman=Map[# → 1&,SnowMan];
⇒ dataSetDice=Map[# → 0&,DiCes];
```

From the 10 - 10 images, the first 7 - 7 ones are considered as elements of the training set, and the last 3 - 3 represent the test set.

```
⇒ trainingData=Join[RandomSample[dataSetSnowman,7],
   RandomSample[dataSetDice,7]];
```

First we solve the problem using *Mathematica.*

Mathematica

Let us create a classification function, employing Random Forest method,

```
⇒ c=Classify[trainingData,Method → "RandomForest",
   PerformanceGoal → "Quality"]
⇐ ClassifierFunction[ Input type: NumericalVector (Length: 7)
                      Classes: 0, 1 ]
```

The accuracy of the classification process as well as the confusion matrix can also be easily computed on the training data, see Fig. 2.26

```
⇒ cm=ClassifierMeasurements[c,trainingData]
⇐ ClassifierMeasurementsObjec[ Classifier: RandomForest
                               Number of test examples 14 ]

⇒ cm["Accuracy"]
⇐ 1.

⇒ cm["ConfusionMatrixPlot"]
⇐
```

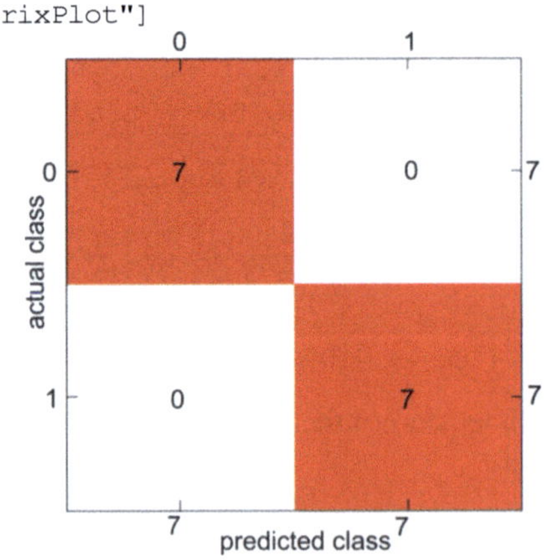

Fig. 2.26 The confusion matrix for the training set.

The same information for the test data, see Fig. 2.27

The test set

```
⇒ dataTest=Complement[Join[dataSetSnowman,dataSetDice],
    trainingData];
⇒ cm=ClassifierMeasurements[c,dataTest]
⇒ ClassifierMeasurementsObject[ Classifier: RandomForest
                                Number of test examples 6 ]
⇒ cm["Accuracy"]
⇐ 1.
⇒ cm["ConfusionMatrixPlot"]
⇐
```

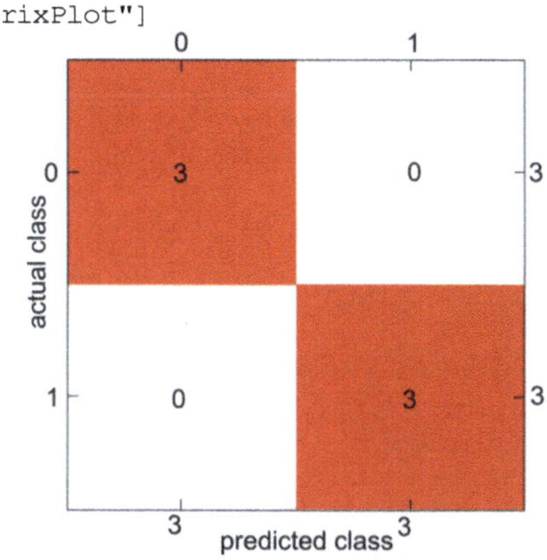

Fig. 2.27 The confusion matrix for the test set.

Python

First, we save the data sets for *Python*,

```
⇒ Xtrain=Map[#[[1]]&,trainingData];
⇒ ytrain=Map[#[[2]]&,trainingData];
⇒ Xtest=Map[#[[1]]&,dataTest];
⇒ ytest=Map[#[[2]]&,dataTest];
⇒ Export["Xtr.mtx",Xtrain];
⇒ Export["Xte.mtx",Xtest];
⇒ Export["ytr.mtx",{ytrain}];
⇒ Export["yte.mtx",{ytest}];
```

This is a temporary saving!

Preparation for reading *. *mtx* file into *Python*,

```
from numpy import array, matrix
from scipy.io import mmread, mmwrite
```

```
Xtrain=mmread('Xtr.mtx')
ytra=mmread('ytr.mtx')
```

Since *mtx* handles matrices, to get a vector, we consider as the "first" element of a matrix

```
ytrain=ytra[0]
```

Similarly for the test set,

```
Xtest=mmread('Xte.mtx')
ytre=mmread('yte.mtx')
```

```
ytest=ytre[0]
```

Employing Random Forest Classifier,

```
from sklearn.ensemble import RandomForestClassifier
rf_fit = RandomForestClassifier(n_estimators=5000,
 criterion="gini",max_depth=5,min_samples_split=2,
 bootstrap=True,max_features='auto',random_state=42,
 min_samples_leaf=1).fit(Xtrain,ytrain)
```

Carry out the prediction on the training set

```
ytr=rf_fit.predict(Xtrain)
```

```
ytr
```

```
⇐ {1,1,1,1,1,1,1,0,0,0,0,0,0,0}
⇒ y=%;
```

The error of the classifier on the training set

```
⇒ Norm[ytrain-y]
⇐ 0
```

Similarly on the test set

```
yte=rf_fit.predict(Xtest)
```

```
yte
```

```
⇐ {1,1,1,0,0,0}
⇒ y=%;
```

The error of the classifier

```
⇒ Norm[ytest-y]
⇐ 0
```

Alternatively we can employ scoring.

```
print("Training set score:
 {:.2f}".format(rf_fit.score(Xtrain, ytrain)))
```

```
Training set score: 1.00
```

```
print("Test set score:
 {:.2f}".format(rf_fit.score(Xtest, ytest)))
```

```
Test set score: 1.00
```

2.4 Support Vector Classification

Basic theory

Support Vector Classification (SVC) is basically a binary classification method used for linear as well nonlinear cases. It is especially useful when the feature vectors to be classified are long vectors representing many features (Christianini et al, (2000).

Linear Classifier

The classifier can be expressed as

$$y = w^T x,$$

where x represents the feature vector to be classified, y is the label standing for a class and w is the vector of weights as parameter of the classifier.

The method can be applied to linearly separable sets. In case of the AND (x_1, x_2) logical function, we have a linearly separable problem. The decision line is $y = \alpha + \beta x_1 + \gamma x_2$, see Fig. 2.28,

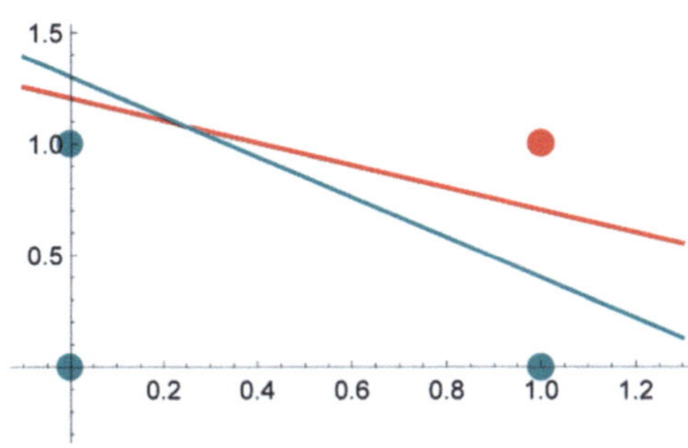

Fig. 2.28 AND problem is linearly separable.

SVC provides *margin maximization,* see Fig. 2.29. From the infinite decision boundaries we select the one, which is most robust, less insensitive on the measurement errors, see Fig. 2.29 (Kecman, 2011).

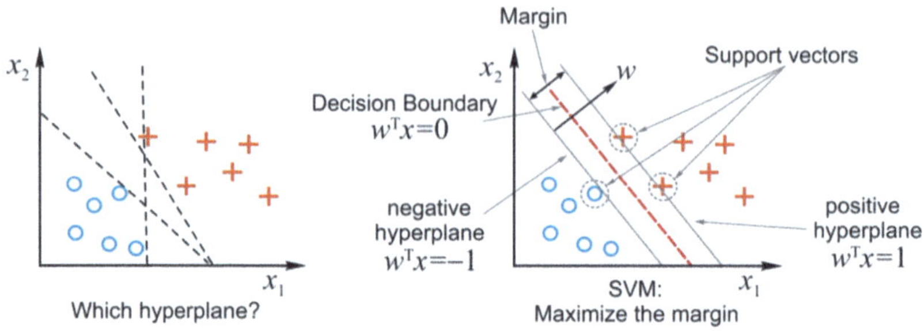

Fig. 2.29 Linear support vector classifier.

Non-Linear Classifier

SVC can be employed for nonlinear problems too. In case of the XOR (x_1, x_2), see Table 2.3.

Table 2.3 XOR logical table

x_1	x_2	y
0	0	0
0	1	1
1	0	1
1	1	0

we have a linearly non-separable problem, see Fig.2.30

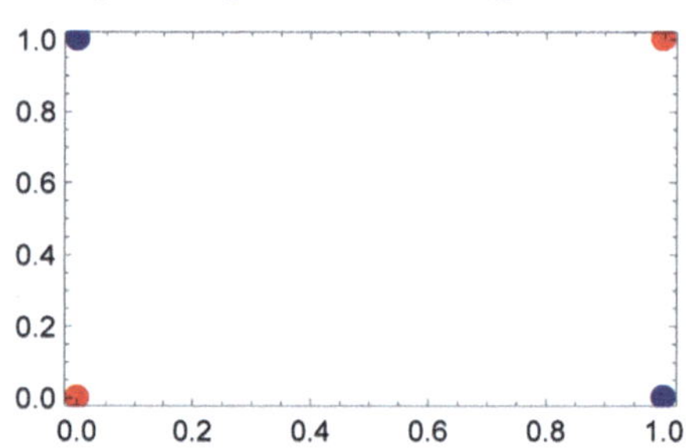

Fig. 2.30 XOR problem is linearly non-separable.

However to get a linear problem, let us introduce a new variable $x_3 = x_1 x_2$, see Table 2.4

Table 2.4 XOR logical table with an additional variable x3

x_1	x_2	x_3	y
0	0	0	0
0	1	0	1
1	0	0	1
1	1	1	0

We now have a 3D problem. Therefore the linear separation is possible with a 2D hyper plane, see Fig.2.31

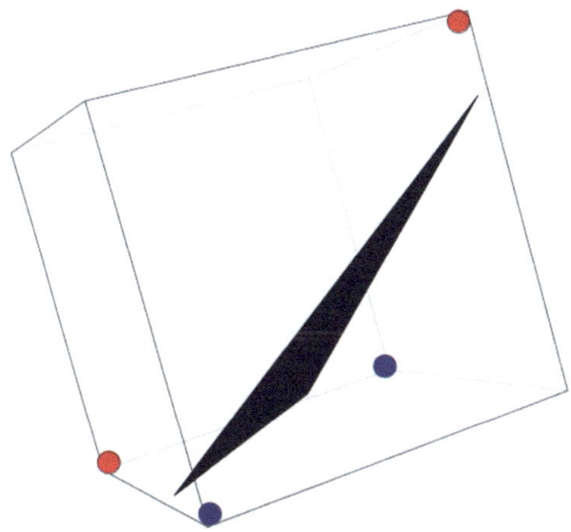

Fig. 2.31 Solution of the XOR problem in 3D.

Many times a nonlinear problem represented in a higher dimensional space can be considered as a linear one. This is the basic idea of the *kernel trick*. The *kernel method* employ a projection into a higher dimension, where the problem is a linearly separable one

$$K(x_1, x_2) = (x_1, x_2, x_1 x_2)$$

There are different types of kernels, for example:

RBF kernel

$$K(x_1, x_2) = \exp\left(-\gamma |x_1 - x_2|^2\right)$$

Polynomial kernel

$$K(x_1, x_2) = \gamma (c + x_1 x_2)^d$$

Sigmoid kernel

$$K(x_1, x_2) = \tanh(c + \gamma x_1 x_2)$$

Characterizing the method

Pros: High accuracy, insensitive to outliers, no assumptions about data, practically it is efficient for all types of problems

Cons: Computationally expensive, requires a lot of memory

Works with: Numeric values, nominal values

2.4.1 Margin maximization

For the 2D vectors to be classified, first we demonstrate the margin maximization in case of linearly separable problem. Let us consider a small data set from *Python* repository *mglearn*. See Fig. 2.32.

```
import mglearn
X,y= mglearn.datasets.make_forge()
X
```

```
⇐ {{9.96347,4.59677},{11.033,-0.168167},{11.5416,5.21116},
  {8.69289,1.54322},{8.10623,4.28696},{8.30989,4.80624},
  {11.9303,4.64866},{9.67285,-0.202832},{8.3481,5.13416},
  {8.67495,4.47573},{9.17748,5.09283},{10.2403,2.45544},
  {8.68937,1.4871},{8.9223,-0.639932},{9.49123,4.33225},
  {9.25694,5.13285},{7.99815,4.85251},{8.18378,1.29564},
  {8.73371,2.49162},{9.32298,5.09841},{10.0639,0.990781},
  {9.50049,-0.264303},{8.34469,1.63824},{9.50169,1.93825},
  {9.15072,5.49832},{11.564,1.33894}}
⇒ u=%;
```

The labels are

```
y
```

```
⇐ {1,0,1,0,0,1,1,0,1,1,1,1,0,0,1,1,1,0,0,1,0,0,0,0,1,0}
⇒ v=%;
```

Mathematica

Let us visualize them, see Fig. 2.33

```
⇒ class1={};class2={};
⇒ MapThread[If[#1==0,AppendTo[class1,#2],
  AppendTo[class2,#2]]&,{v,u}];
⇒ p0=ListPlot[{class1,class2},PlotStyle → {Green,Red},
  Frame → True,Axes → None,PlotMarkers → {Automatic,Large},
  AspectRatio → 1]
```

⇐

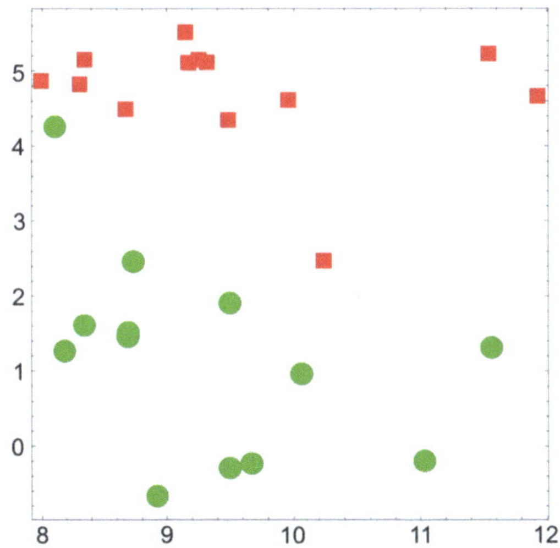

Fig. 2.32 Data points for binary linear classification.

Let us employ Logistic Regression. In order to demonstrate the efficiency of the SVC, first let us employ

```
⇒ trainingData=MapThread[#1 → #2&,{u,v}];
⇒ c1=Classify[trainingData,Method → "LogisticRegression",
    PerformanceGoal → "Quality"];
```

The result can be seen in Fig. 2.34

```
⇒ p1=Show[{DensityPlot[c1[{x,y}],{x,7.5,12},{y,-1,6},
    ColorFunction → "CMYKColors",PlotPoints → 50],p0}]
```

⇐

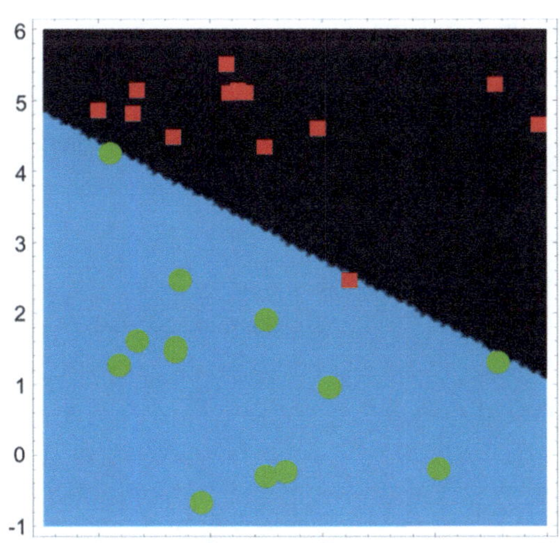

Fig. 2.33 Logistic classification seems optimal but not robust.

This looks like an optimal solution however, it is very sensitive for measurement errors since some data points belonging to the different classes are very close to the decision line! Employing SVC with linear kernel, we can get classification error, however the solution is robust, see Fig. 2.34

```
⇒ c2=Classify[trainingData,Method → {"SupportVectorMachine",
   "KernelType" → "Linear"},PerformanceGoal → "Quality"];
⇒ p2=Show[{DensityPlot[c2[{x,y}],{x,7.5,12},{y,-1,6},
   ColorFunction → "CMYKColors",PlotPoints → 80],p0}]
```

⇐

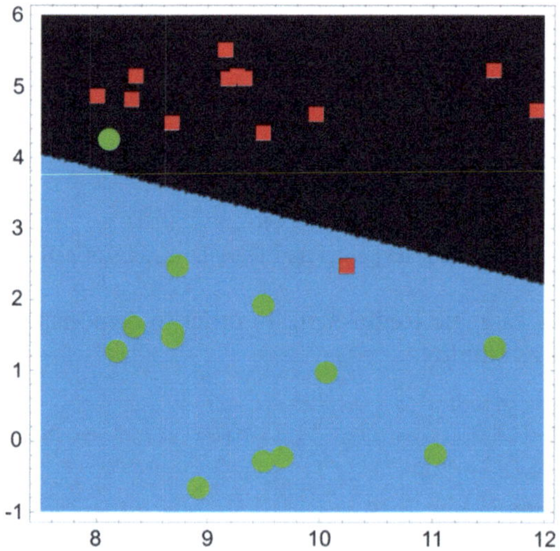

Fig. 2.34 Linear SVC is not perfect but it is much more robust since it maximizes the margin.

Now, we use *Python*.

Python

Training and using prediction, we get the labels (Malik, 2018).

```
from sklearn.svm import LinearSVC
svcreg=LinearSVC(C=1000).fit(X,y)
pu=svcreg.predict(X)
pu
```

```
⇐ {1,0,1,0,0,1,1,0,1,1,1,0,0,0,1,1,1,0,0,1,0,0,0,0,1,0}
⇒ vP=%;
```

Error of Linear SVC of *Python*

```
⇒ Norm[v-vP]
⇐ 1
```

Error of *Mathematica*

```
⇒ Norm[v-vM]
```

⇐ $\sqrt{2}$

Now let us consider a linearly non-separable problem, see Fig. 2.35.

2.4.2 Feature Space Mapping

Creating the two sets, see Fig. 2.35.

```
⇒ data1=RandomVariate[MultinormalDistribution[{0,0},
    IdentityMatrix[2]], 500];
  data2 =Transpose[
    {#radius *Cos[#angle],#radius* Sin[#angle]}&@<|
     "radius" → RandomReal[{5, 8},500],
     "angle" → RandomReal[{0, 2 Pi},500]|>];
  totalset=Join[Map[# → 1&,data1],Map[# → 2&,data2]];
```

Let us visualize them.

```
⇒ p0=ListPlot[{data1,data2},PlotStyle → {Blue,Red},
    PlotMarkers → {Automatic,Small},Frame → True,AspectRatio → 1]
```

⇐

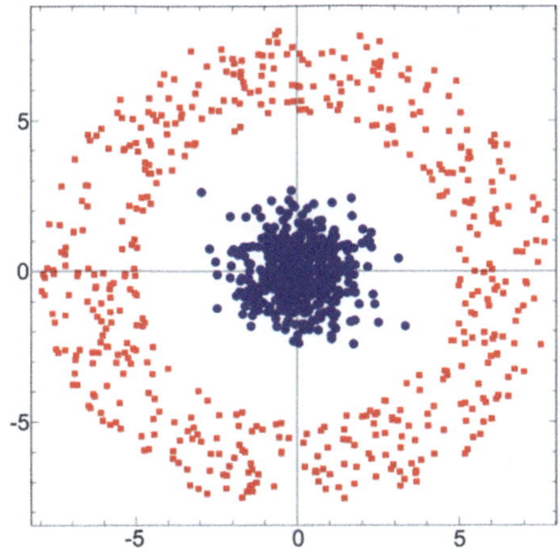

Fig. 2.35 Nonlinear problem.

The total data set is divided into training and test sets.

```
⇒ Length[totalset]
⇐ 1000
⇒ SeedRandom[1234]
⇒ trainingset=RandomSample[totalset,700];
⇐ testset=RandomSample[totalset,300];
```

Mathematica

Let us use RBF kernel.

```
⇒ c2=Classify[trainingset, Method → {"SupportVectorMachine",
   "KernelType" → "RadialBasisFunction",
   "GammaScalingParameter" → 1},PerformanceGoal → "Quality"]
⇐ ClassifierFunction[ Input type: NumericalVector (Length: 2)
                      Classes: 1, 2 ]
```

Computing the accuracy and the confusion matrix on the training set

```
⇒ ctraining=ClassifierMeasurements[c2,trainingset]
⇒ ClassifierMeasurementsObject[ Classifier: SupportVectorMachine
                                Number of test examples: 700 ]
⇒ ctraining["Accuracy"]
⇒ 0.998571
⇒ ctraining["ConfusionMatrixPlot"]
```

⇐

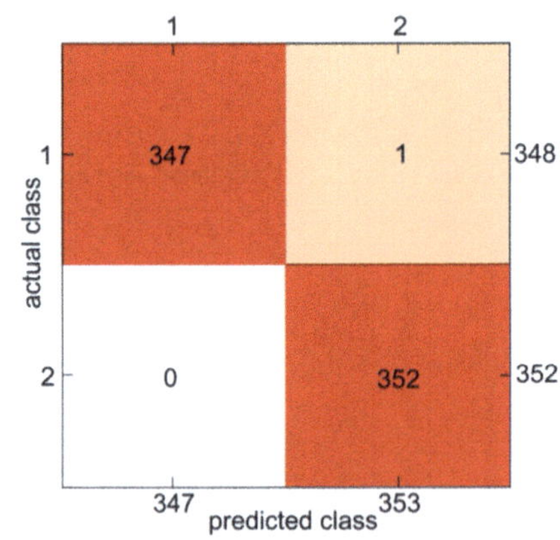

Fig. 2.36 Confusion matrix for the training set.

Similarly for the test set,

```
⇒ ctest=ClassifierMeasurements[c2,testset]
⇒ ClassifierMeasurementsObject[ Classifier: SupportVectorMachine
                                Number of test examples: 300 ]
⇒ ctest["Accuracy"]
⇒ 1
⇒ ctest["ConfusionMatrixPlot"]
```

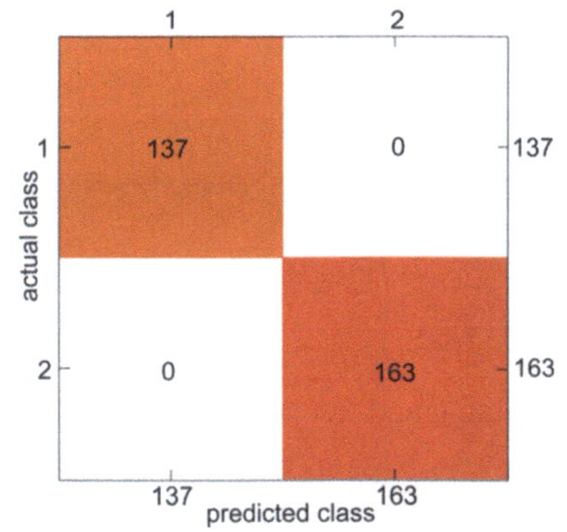

⇐

Fig. 2.37 Confusion matrix for the test set.

Let us visualize the result of the classification, see Fig. 2.38.

```
⇒ p2=Show[{DensityPlot[c2[{x,y}],{x,-8,8},{y,-9,9},
    ColorFunction → "CMYKColors",PlotPoints->50],p0},
    AspectRatio → 1.1]
```

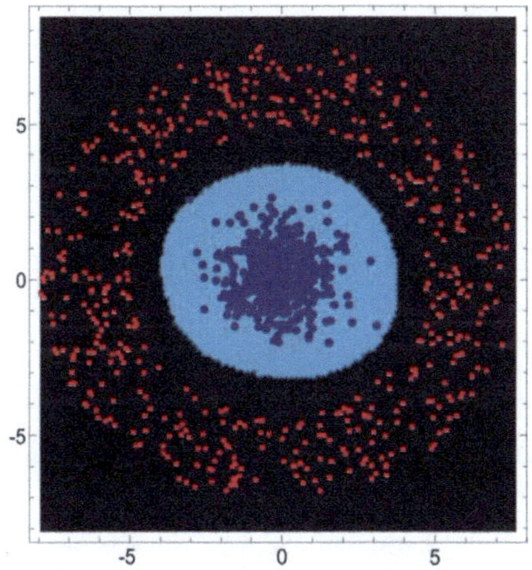
⇐

Fig. 2.38 Result of SVC with RBF kernel.

Now we can solve the same problem using *Python*.

Python

Save the randomly generated data sets for *Python*,

```
⇒ X=Map[#[[1]]&,totalset];
⇒ y=Map[#[[2]]&,totalset];
⇒ Export["G:\\dataX.txt",X];
⇒ Export["G:\\datay.txt",y];
```

Read data from txt files

```
X=[i for i in range(1000)]
f=open('G:\\dataX.txt','r')
for i in range(0,1000):
 X[i]=list(ast.literal_eval(f.readline()))
```

```
y=[i for i in range(1000)]
f=open('G:\\datay.txt','r')
for i in range(0,1000):
 y[i]=int(f.readline())
```

Splitting data set into training and test set,

```
from sklearn.metrics import
 accuracy_score,classification_report
from sklearn.model_selection import train_test_split
x_train,x_test,y_train,y_test =
 train_test_split(X,y,train_size = 0.7,random_state=42)
```

This is a general class of SVC supporting linear, polynomial and RBF kernels, not only the linear one (see earlier LinearSVC) (Ilango, 2017).

```
# RBF Kernel
from sklearn.svm import SVC
svm_rbf_fit =
 SVC(kernel='rbf',C=1.0,gamma=0.1).fit(x_train,y_train)
```

Information about the training

```
print ("\nSVM RBF Kernel Classifier - Train accuracy:",round
 (accuracy_score(y_train,svm_rbf_fit.predict(x_train)),3))
print ("\nSVM RBF Kernel Classifier - Train
 Classification Report\n",classification_report
  (y_train,svm_rbf_fit.predict(x_train)))
```

```
SVM RBF Kernel Classifier - Train accuracy:
1

SVM RBF Kernel Classifier - Train Classification Report

             precision    recall  f1-score   support
          1       1.00      1.00      1.00       350
          2       1.00      1.00      1.00       350
avg / total       1.00      1.00      1.00       700
```

Information about the test set

```
print ("\nSVM RBF Classifier - Test accuracy:",round
 (accuracy_score(y_test,svm_rbf_fit.predict(x_test)),3))
print ("\nSVM RBF Classifier - Test Classification Report\n",
 classification_report(y_test,svm_rbf_fit.predict(x_test)))
```

```
SVM RBF Kernel Classifier - Test accuracy:
1.0

SVM RBF Kernel Classifier - Test Classification Report
```

```
              precision    recall  f1-score   support
           1       1.00      1.00      1.00       150
           2       1.00      1.00      1.00       150
 avg / total       1.00      1.00      1.00       300
```

2.4.3 Learning Chess Board Fields

Let us consider a 2×2 chess board. The training points are generated by uniformly distributed random numbers from the interval [−1,1]×[−1, 1], see Fig. 2.39.

```
⇒ M={{1,-1},{-1,1}};
⇒ p1=
   Rotate[MatrixPlot[M,ColorRules → {1 → Black,-1 → White}],Pi/2]
```

⇐

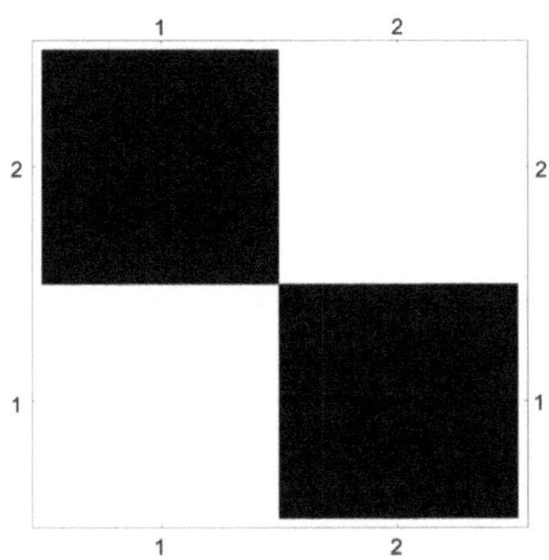

Fig. 2.39 The 2×2 chess board problem.

Creating the training set using 400 random samples, see Fig. 2.40.

```
⇒ xym={};zm={};
⇒ SeedRandom[2538]
⇒ Do[x1=Random[Real,{-0.9995,0.9995}];
   x2=Random[Real,{-0.9995,0.9995}];
   If[x1 x2>0,z=1,z=-1];AppendTo[xym,{x1,x2}];
   AppendTo[zm,z],{k,1,400}];
```

Preparation of the data to display them

```
⇒ data=Transpose[Join[Transpose[xym],{zm}]];
⇒ data1=Map[Drop[#,-1]&,Select[data,#[[3]]>0&]];
  data2=Map[Drop[#,-1]&,Select[data,#[[3]]<0&]];
⇒ p2=ListPlot[{data1,data2},PlotMarkers → {"\[FilledUpTriangle]",
   "\[FilledRectangle]"},PlotStyle → {Hue[.0],Hue[.7]},
   Frame → True,AspectRatio → 1]
```

⇐

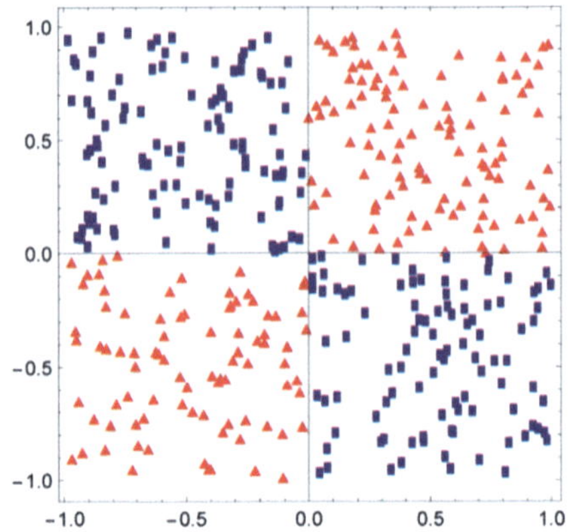

Fig. 2.40 The generated random points of the chess board problem.

Let us employ the same Gaussian (RBF) kernel with gain $\beta = 20$

$$K(u,\ v)\ =\ Exp(-\beta\ (u-v)(u-v))$$

```
⇒ β=20;
```

The training data

```
⇒ trainingData=MapThread[#1 → #2&,{xym,zm}];
```

Mathematica

Then the training classifier,

```
⇒ ChessSVC=Classify[trainingData,Method → {"SupportVectorMachine",
    "KernelType" → "RadialBasisFunction",
    "GammaScalingParameter" → β},PerformanceGoal → "Quality"]
⇐ ClassifierFunction[ Input type: NumericalVector (Length: 2)
                      Classes: -1, 1 ]
```

Testing the classifier on the training set

```
⇒ ctesting=ClassifierMeasurements[ChessSVC,trainingData]
⇐ ClassifierMeasurementsObject[
    Classifier: SupportVectorMachine
    Number of test examples: 400 ]
⇒ ctesting["Accuracy"]
⇐ 1.
⇒ ctesting["ConfusionMatrixPlot"]
```

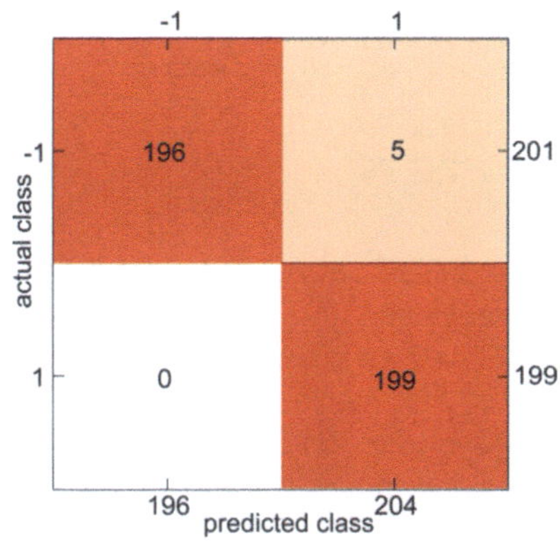

⇐

Fig. 2.41 The confusion matrix on the testing set.

Now let us visualize the solution. The zero contour lines represent the boundary of the clusters, see Fig. 2.42

```
⇒ p4=DensityPlot[-Sign[ChessSVC[{u,v}]],{u,-1,1},{v,-1,1},
   PlotPoints → 100,ColorFunction → GrayLevel,Mesh → False,
   Axes → True,AxesStyle → {{Green,Thick},{Green,Thick}}];
⇐ Show[{p4,p2}]
```

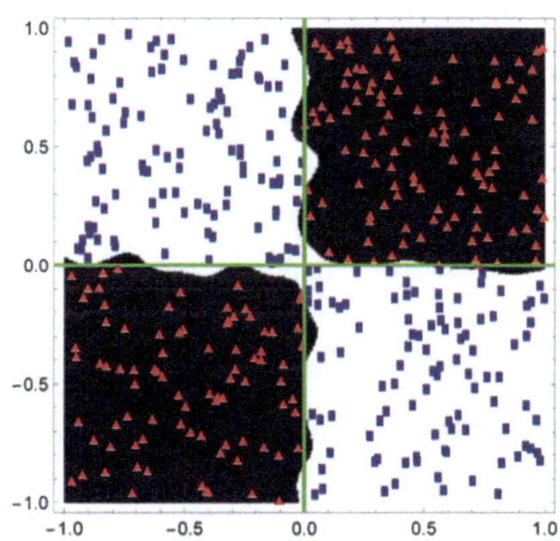

⇐

Fig. 2.42 The solution of the chess board problem.

Saving data for *Python*

```
⇒ dataX=xym;
⇒ dataX=xym;
⇒ Export["dataX.mtx",dataX];
⇒ Export["datay.mtx",{datay}];
```

Python

```
# RBF Kernel
from sklearn.svm import SVC
```

```
from numpy import array, matrix
from scipy.io import mmread, mmwrite
```

```
dataX=mmread('dataX.mtx')
datay=mmread('datay.mtx')
```

```
datay=datay[0]
```

```
svm_rbf_fit = SVC(kernel='rbf',C=100.0,gamma=20.).
 fit(dataX,datay)
```

```
print("Training set score:
 {:.2f}".format(svm_rbf_fit.score(dataX, datay)))
Training set score: 1.00
```

Let us see the error (mislabeling) if there is any

```
prediction=svm_rbf_fit.predict(dataX)
prediction
```

```
⇐ {1,1,-1,1,1,-1,1,-1,1,1,-1,-1,1,1,-1,-1,-1,-1,-1,-1,-1,-1,
   -1,-1,1,-1,1,-1,1,-1,1,-1,1,-1,-1,-1,1,1,1,-1,1,-1,-1,-1,-1,
   -1,-1,1,-1,1,1,1,1,-1,1,1,-1,-1,-1,1,1,-1,-1,1,-1,1,-1,
   1,1,1,1,-1,-1,1,1,-1,-1,-1,1,-1,1,1,-1,1,-1,-1,1,1,1,-1,
   1,-1,-1,1,-1,-1,-1,1,-1,1,-1,1,-1,1,1,-1,-1,-1,1,-1,1,1,
   ....
   -1,-1,1,1,-1,1,1,-1,-1,1,1,1,-1,-1,-1,-1,1,1,-1,1,1,-1,
   -1,-1,1,1,1,1,-1,-1,-1,1,-1,-1,-1,-1,1,-1,-1,-1,-1,-1,-1}
```

```
⇒ yP=%;
⇒ Norm[Flatten[datay]-yP]
⇐ 0
```

2.5 Naive Bayes Classifier

Basic theory

Bayes theorem provides a way of calculating posterior probability $P\ (c\,|\,x)$ from $P(c)$, $P(x)$ and $P\ (x\,|\,c)$ (James et al, 2013). Look at the equation below:

$$P\ (c\,|\,x) = \frac{P(x|c)P(c)}{P(x)}$$

$P\ (c\,|\,x)$ is the posterior probability of class (c, target) given predictor (x, attributes).

$P\ (c)$ is the prior probability of class.

$P(x \mid c)$ is the likelihood which is the probability of predictor given class.

$P(x)$ is the prior probability of predictor.

Explanatory Example

Will John play tennis or not, when the *weather is Sunny*?

2.5.1 Playing Tennis Today?

The computation can be followed considering Tables 2.5/a, b, and c

Tables 2.5 a, b and c

Weather	Play
Sunny	No
Overcast	Yes
Rainy	Yes
Sunny	Yes
Sunny	Yes
Overcast	Yes
Rainy	No
Rainy	No
Sunny	Yes
Rainy	Yes
Sunny	No
Overcast	Yes
Overcast	Yes
Rainy	No

Frequency table		
Weather	No	Yes
Overcast		4
Rainy	3	2
Sunny	2	3
Grand Total	5	9

Likelihood table				
Weather	No	Yes		
Overcast		4	=4/14	0.29
Rainy	3	2	=5/14	0.36
Sunny	2	3	=5/14	0.36
All	5	9		
	=5/14	=9/14		
	0.36	0.64		

Feature → Weather: Sunny, Rainy, Overcast
Classes → No, Yes

Step 1: Convert data set into Frequency table
Step 2: Create Likelihood table
Step 3: Now, use Naive Bayesian equation to calculate the posterior probability for each class. The class with the highest posterior probability is the outcome of prediction, for example,

What is the probability that they will play on a sunny day?
P(Yes|Sunny) = P(Sunny|Yes)×(Yes)/P(Sunny) = ?

P(Sunny|Yes) = 3/9= 0.33 - they played 9 times and 3 of these days were sunny
P(Sunny) = 5/14 = 0.36 - there were 5 sunny days from the total 14 days
P(Yes) = 9/14 = 0.64 they played 9 times during the 14 days
then

P(Yes|Sunny) = 0.33×0.64/0.36 = 0.60

Let us employ *Mathematica* to solve the problem.

Mathematica

```
⇒ dataTraining={"Sunny" → "No","Overcast" → "Yes","Rainy" → "Yes",
   "Sunny" → "Yes","Sunny" → "Yes","Overcast" → "Yes",
   "Rainy" → "No","Rainy" → "No","Sunny" → "Yes","Rainy" → "Yes",
   "Sunny" → "No","Overcast" → "Yes","Overcast" → "Yes",
   "Rainy" → "No"};
⇒ TableForm[dataTraining,TableAlignments → {Left,Center}]
⇐  Sunny → No
   Overcast → Yes
   Rainy → Yes
   Sunny → Yes
   Sunny → Yes
   Overcast → Yes
   Rainy → No
   Rainy → No
   Sunny → Yes
   Rainy → Yes
   Sunny → No
   Overcast → Yes
   Overcast → Yes
   Rainy → No
```

Training the classifier

```
⇒ c=Classify[dataTraining,Method → {"NaiveBayes"},
   PerformanceGoal → "Quality"]
⇐ ClassifierFunction[ Input type: Nominal
                      Classes: No, Yes ]
```

Give our data as input for the classifier,

```
⇒ c[{"Sunny"}]
⇐ {Yes}
```

Out of Yes or No, we can get the probabilities, too

```
⇒ c[{"Sunny"},"TopProbabilities"]
⇐ {{Yes → 0.58296,No → 0.41704}}
```

Testing the classifier

```
⇒ cm=ClassifierMeasurements[c,dataTraining]
⇐ ClassifierMeasurementsObject[ Classifier: NaiveBayes
                                Number of test examples: 14 ]
⇒ cm["Accuracy"]
```

```
⇐ 0.714286
⇒ ctesting["ConfusionMatrixPlot"]
```

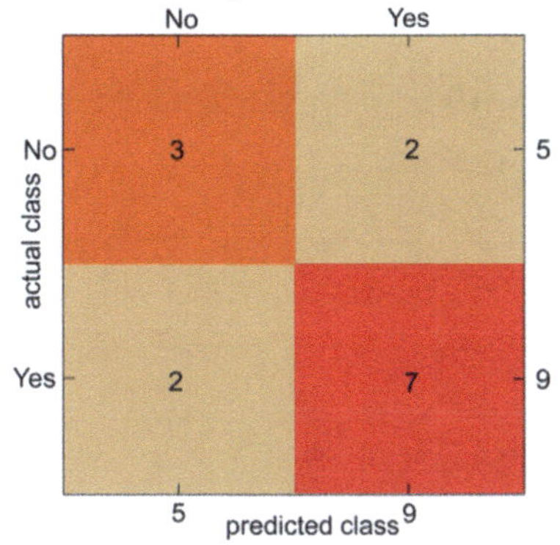

⇐

Fig. 2.43 Confusion matrix.

Error, misclassified data are 4, indeed (1, 3, 10, 11).
Computed labels

```
⇒ Map[c[#[[1]]]&,dataTraining]
⇐ {Yes,Yes,No,Yes,Yes,Yes,No,No,Yes,No,Yes,Yes,Yes,No}
```

Original labels

```
⇒ Map[#[[2]]&,dataTraining]
⇐ {No,Yes,Yes,Yes,Yes,Yes,No,No,Yes,Yes,No,Yes,Yes,No}
```

Python

In order to solve the problem with *Python*, we transform the dataset from nominal values into numerical ones

```
⇒ dataX=Map[#[[1]]&,dataTraining]/.
   {"Sunny" → 0,"Overcast" → 1,"Rainy" → 2}
⇐ {0,1,2,0,0,1,2,2,0,2,0,1,1,2}
⇒ datay=Map[#[[2]]&,dataTraining]/.{"Yes" → 1,"No" → 0}
⇐ {0,1,1,1,1,1,0,0,1,1,0,1,1,0}
```

```
import numpy as np
```

Then the input and output data

```
X=np.array
 ([[0],[1],[2],[0],[0],[1],[2],[2],[0],[2],[0],[1],[1],[2]])
```

```
y=np.array([0,1,1,1,1,1,0,0,1,1,0,1,1,0])
```

Loading classifier

```
from sklearn.naive_bayes import GaussianNB
```

Training classifier (Sunil, 2017)

```
clf=GaussianNB().fit(X,y)
```

In our case: Input: Sunny→0

```
prediction=clf.predict(0)
prediction
```

⇐ 1

Output is: 1→Play.

The probabilities are,

```
prediction=clf.predict_proba(0)
prediction
```

⇐ {{0.290031,0.709969}}

2.5.2 Zebra, Gorilla, Horse and Penguin

Let us consider images of animals, see Fig. 2.44

Fig. 2.44. A collection of images of different animals

We have four classes.

Mathematica

```
⇒ animals=Join[zebra,gorilla,penguin,hourse];
⇐ trainingSet=MapThread[#1 → #2&,
    {animals,{Z,Z,Z,Z,Z,G,G,G,G,G,P,P,P,P,P,H,H,H,H,H}}];
```

Now, we employ GPU,

```
⇒ c=Classify[trainingSet,Method → "NaiveBayes" ,
    PerformanceGoal → "Quality",TargetDevice → "GPU"]
⇐ ClassifierFunction[
    SummaryEmbedGrid[
     SummaryPanel[ Input type: Image
                   Classes:  G, H, P, Z ] , ] ]
⇒ ClassifierInformation[c]
```

The accuracy of the classification is

```
⇒ ctesting = ClassifierMeasurements[c, trainingSet]
⇐ ClassifierMeasurementsObject[
    SummaryEmbedGrid[
     SummaryPanel[ Classifier: Naive Bayes
                   Number of test examples:  20 ], ] ]
⇒ ctesting["Accuracy"]
⇐ 1.
```

The confusion matrix is

```
⇒ ctesting["ConfusionMatrixPlot"]
⇐
```

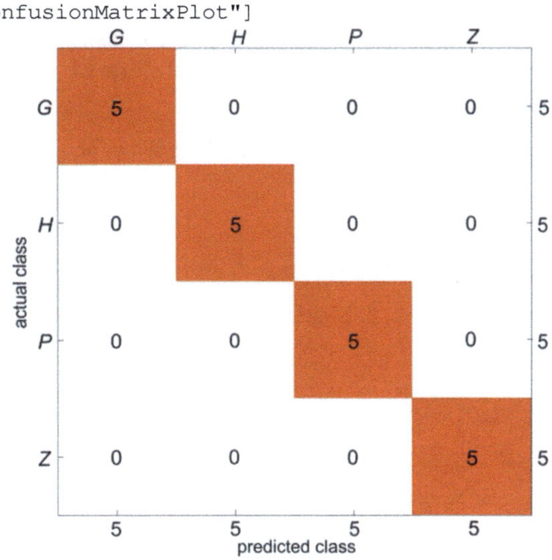

Fig. 2.45 The confusion matrix.

Let us test our classifier with images, see Fig. 2.46

⇒ imgNew01=  ; imgNew01=  ;

imgNew03=  ;

Fig. 2.46 Test images.

```
⇒ c[imgNew01, "Probabilities"]
⇐ <|G → 2.68033*10^-287,H → 0.,P → 1.,Z → 0.|>
⇒ c[imgNew02, "Probabilities"]
⇐ <|G → 4.24751*10^-141,H → 3.00681*10^-116,P → 1.,Z → 5.07884*10^-142|>
⇒ c[imgNew03, "Probabilities"]
⇐ <|G → 9.40641*10^-119,H → 1.,P → 3.53518*10^-95,Z → 8.81794*10^-162|>
```

The pelican is an odd-one-out of classes and similar to the penguin class. The horse with star dress is recognized as horse.

Python

In order to employ *Python*, we need image preprocessing. We shall employ the vector to represent the feature of an image. Therefore we need to reduce the dimensions of the images (Hsu et al, 2017).

```
⇒ animalsReduced=Map[ImageResize[#,{256,256}]&,animals];
⇒ dataX= DimensionReduce[animalsReduced, Method → "TSNE",
   PerformanceGoal → "Quality"];
```

the classes are Z,G,P,H→{0,1,2,3}

```
⇒ datay={{0,0,0,0,0,1,1,1,1,1,2,2,2,2,2,3,3,3,3,3}};
```

Data for *Python*

```
⇒ Export["dataX.mtx",dataX];
⇒ Export["datay.mtx",datay];
```

```
from numpy import array, matrix
from scipy.io import mmread, mmwrite
```

```
dataX=mmread('dataX.mtx')
datay=mmread('datay.mtx')
```

```
datay=datay[0]
```

```
from sklearn.naive_bayes import GaussianNB
clf=GaussianNB().fit(dataX,datay)
```

```
print("Training set score:
  {:.2f}".format(clf.score(dataX, datay)))
```

```
⇐ Training set score: 1 .00
```

Let us see the error (mislabeling) if there is any

```
prediction=clf.predict(dataX)
prediction
```

```
⇐ {0,0,0,0,0,1,1,1,1,1,2,2,2,2,2,3,3,3,3,3}

⇒ yP=%;
⇒ Norm[Flatten[datay]-yP]
⇐ 0
```

Now we check the test images.

```
⇒ {t1,t2,t3} = Map[ImageResize[#,{256,256}]&,
    {imgNew01,imgNew02,imgNew03}];
⇒ testXT = DimensionReduce[{t1,t2,t3},Method → "TSNE",
    PerformanceGoal → "Quality"];
⇒ Export["testXTk.mtx",testXT];
```

```
testXT=mmread('testXTk.mtx')
```

```
prediction=clf.predict(testXT)
prediction
```

```
⇐ {3,3,3}
```

2.6 Comparison of Classification Methods

We consider here four painting styles: modernism, impressionism, abstract and surrealism. There are 15 -15 images of all classes, and 10 - 10 randomly selected images from each class were used as training set, the total set were used to test the created classifiers. Five classifying methods were considered and compared on the bases of the running time of teaching and of the quality of their performance on the testing set.

Even for a human beings, it is not an easy task, but some of the machine learning algorithms can perform quite well.

⇒ Modernism={

};

Fig. 2.47 Collection of modernism type paintings.

Fig. 2.48 Collection of impressionism type paintings.

⇒ Abstract={ , 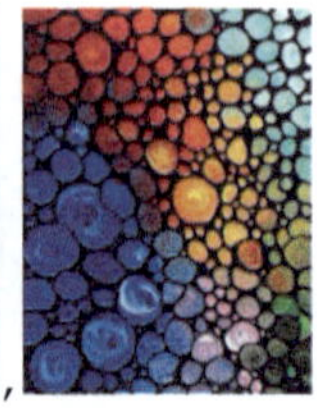, ,
, , ,
, 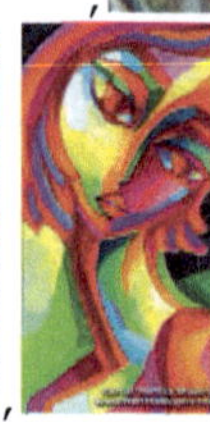, 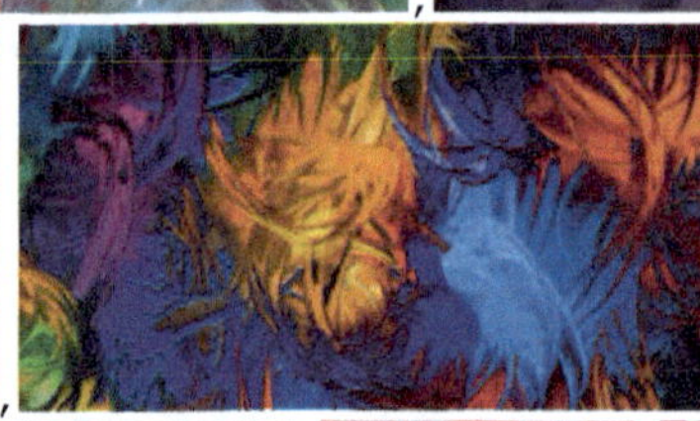,
, , , 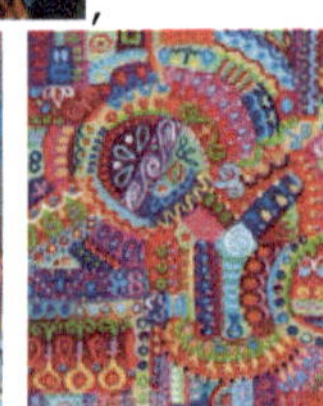,
, };

Fig. 2.49 Collection of abstract type paintings.

```
⇒ Surrealism={
```

```
};
```

Fig. 2.50 Collection of surrealism type paintings.

Preparation of the training set,

```
⇒ s=Table[Table[i,{i,1,4}],{j,1,10}]//Transpose//Join//Flatten
⇐ {1,1,1,1,1,1,1,1,1,1,2,2,2,2,2,2,2,2,2,2,
   3,3,3,3,3,3,3,3,3,3,4,4,4,4,4,4,4,4,4,4}
⇒ v=Map[RandomSample[#,10]&,
   {Modernism,Impressionism,Abstract,Surrealism}]//Flatten;
⇒ data=MapThread[#1 → #2&,{v,s}];
```

Classification process,

```
⇒ AbsoluteTiming[c=Classify[data,Method → "LogisticRegression"]]
```

⇐ {9.71596,ClassifierFunction[Input type: Image Classes: 1, 2, 3, 4]}

Preparing the test set,

```
⇒ s=Table[Table[i,{i,1,4}],{j,1,15}]//Transpose//Join//Flatten
⇐ {1,1,1,1,1,1,1,1,1,1,1,1,1,1,1,2,2,2,2,2,2,2,2,2,2,2,2,2,2,2,
   3,3,3,3,3,3,3,3,3,3,3,3,3,3,3,4,4,4,4,4,4,4,4,4,4,4,4,4,4,4}
⇒ v=Join[{Modernism,Impressionism,Abstract,Surrealism}]//Flatten;
⇒ dataTotal=MapThread[#1 → #2&,{v,s}];
```

Testing process,

```
⇒ cm = ClassifierMeasurements[c, dataTotal]
⇐ ClassifierMeasurementsObject[ Classifier: LogisticRegression Number of test examples: 60 ]
```

The accuracy of the classifying function on the test set

```
⇒ cm["Accuracy"]
⇐ 0.9
```

Confusion matrix,

```
⇒ cm["ConfusionMatrixPlot"]
```

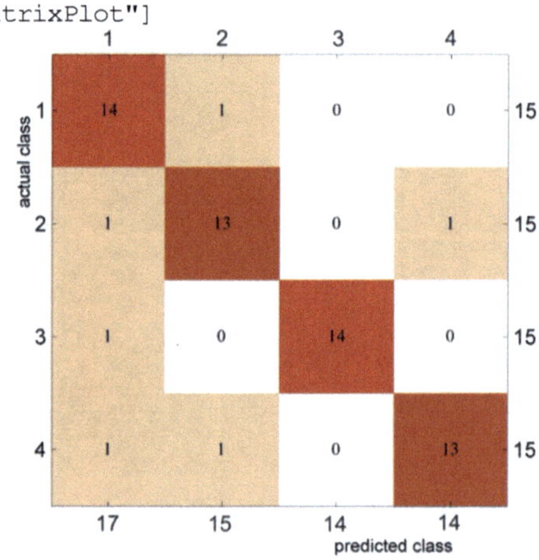

⇐

Fig. 2.51 Confusion matrix of the classification function on the total dataset.

Probability of the membership of a single object randomly selected from the test set,

```
⇒ p=RandomSample[Modernism,1]
```

⇐

Fig. 2.52 Randomly selected painting from the modernism collection.

```
⇒ Map[c[p,{"Probability",#}]&,{1,2,3,4}]//Flatten
⇐ {0.758568,0.0887136,0.100931,0.0517867}
```

The result is the probability value of the membership function of the different collections.

Table 2.6 shows the result of the different methods on the test set,

Table 2.6 Comparison of Classification Methods

Method	Running Time [s]	Accuracy
Nearest Neigbours	9.0	0.82
Logistic Regression	9.9	0.90
Decision Trees	9.0	0.82
Support Vector Classification	9.9	0.87
Naive Bayes	9.0	0.82

We can see that the performances of the Logistic regression and Support Vector Classification are the best however their running times of the training process is somewhat longer.

References

Brownlee J (2016): Logistic Regression for Machine Learning, https://machinelearningmastery.com/logistic-regression-for-machine-learning/

Christianini N, Shawe-Taylor J (2000): Support Vector Machines and other kernel -based, learning methods, Cambridge University Press, Cambridge, New York

Dangeti P (2017): Statistics for Machine Learning, Packt, Birmingham-Mumbai

Fortuner B (2019): Logistic Regression, in Machine Learning Cheatsheet, https://ml-cheatsheet.readthedocs.io/en/latest/logistic_regression.html

Goswami A (2018): Intro to image classification with KNN, https://medium.com/@YearsOfNoLight/intro-to-image-classification-with-knn-987bc112f0c2

Hsu SC, Chen IC, Huang CL (2017): Image Classification Using Naive Bayes Classifier With Pairwise Local Observations, Journal of Information Science and Engineering 32, https://pdfs.semanticscholar.org/0dc3/4e186e8680336e88c3b5e73cde911a8774b8.pdf

Ilango G (2017): Image Classification using Python and Scikit-learn https://gogul09.github.io/software/image-classification-python

James D, Witten D, Hastie T, Tibshirani R (2013): An Introduction to Statistical Learning with Applications in R, Springer, Heidelberg, New York

Kecman V (2001): Learning and Soft Computing, The MIT Press, Cambridge, Massachusetts, London

Malik U (2018): Implementing SVM and Kernel SVM with Python's Scikit-Learn https://stackabuse.com/implementing-svm-and-kernel-svm-with-pythons-scikit-learn/

Müller AC, Guido S (2017): Introduction to Machine Learning with Python, O' Reilly Media Inc. Sevastopol, CA 95472 USA

Navlani A (2018): KNN Classification using Scikit-learn, https://www.datacamp.com/community/tutorials/k-nearest-neighbor-classification-scikit-learn

Prince SJD (2012): Computer Vision: models, learning and inference, Cambridge University

Skerritt B (2018): What is a Decision Tree in Machine Learning? https://hackernoon.com/what-is-a-decision-tree-in-machine-learning-15ce51dc445d

Sunil R (2017): 6 Easy Steps to Learn Naive Bayes Algorithm (with codes in Python and R) https://www.analyticsvidhya.com/blog/2017/09/naive-bayes-explained/

Chapter 3
Clustering

3.1 KMeans Clustering

Basic theory

The goal of unsupervised learning is to discover the hidden patterns or structures of the data in which no target variable exists to perform either classification or regression methods. Unsupervised learning methods are often more challenging, as the outcomes are subjective and there is no simple goal for the analysis, such as predicting the class or continuous variable. These methods are performed as part of exploratory data analysis. In addition, since there is no universally accepted mechanism for performing the validation of the results, assessing the results obtained from unsupervised learning methods can be hard (Han et al. 2012). Nevertheless, unsupervised learning methods are nowadays growing in importance in various fields as evidenced by many researchers currently actively working on the subject.

The *K-Means* clustering algorithm is an iterative process of moving the centers of clusters or centroids to the mean position of their constituent points, and reassigning instances to their closest clusters iteratively until there is no significant change in the number of cluster centers possible or number of iterations reached. Prototype-based clustering means that each cluster is represented by a prototype, which can either be the centroid (average) of similar points with continuous features, or the medoids (the most representative or most frequently occurring point) in the case of categorical features. While *K-Means* is very good at identifying clusters with a spherical shape, one of the drawbacks of this clustering algorithm is that *one has to specify the number of clusters, K, a priori*. An inappropriate choice of *K* can result in poor clustering performance. Later on in this chapter, the *silhouette method for computing silhouette coefficient as indicator for outliers*, which are useful techniques to evaluate the quality of a clustering to help in determining the optimal number of clusters *K* will be discussed (VanderPlas 2016).

J. Awange et al., *Hybrid Imaging and Visualization*,
https://doi.org/10.1007/978-3-030-26153-5_3

The problem with the *K-Means* is that one or more clusters can be empty. Note that this problem does not exist for *K-Medoids* or *Fuzzy C-means*. This method can be employed when there are very many samples, not too many clusters and the geometry of the elements are flat, which basically means that the cluster can be easily separated. When applying *K-Means* to real-world data using Euclidean distance metric, the motivation is to make sure that the features are measured on the same scale and standardized or min-max scaled if necessary (*Standardized* in Python). In short, *K-Means* can be useful in case of many samples and few cluster and flat geometry.

3.1.1 Small Data Set

Simple introductory example employing a small data set from Python repository, see Fig. 3.1.

Python

Starting Python session in *Mathematica,*

```
⇒ session=
   StartExternalSession[<|"System" → "Python",
    "Version" → "3.5.4","Executable" →
     "C:\Users\Ben\AppData\Local\Programs\Python\Python35\
       python.exe"|>]//Quiet
```

```
⇐ ExternalSessionObject[
   SummaryPanel[ System: Python   EvaluationCount: None
                 UUID: 7b76966c2-ebb7-4ee8-bb25-02322d1456a8 ]]
```

Reading the necessary procedures,

```
from numpy import array, matrix
from scipy.io import mmread, mmwrite
import numpy as np
```

Importing K-Means clustering function.

```
from sklearn.datasets import make_blobs
from sklearn.cluster import KMeans
```

Reading data set and clustering in case of supposing three clusters,

```
X, y = make_blobs(random_state=1)
kmeans = KMeans(n_clusters=3).fit(X)
pre= kmeans.predict(X)
pred=np.array([pre])
```

Write data and the resulting labels into files for *Mathematica.*

```
mmwrite('labels.mtx',pred)
mmwrite('dataX',X)
```

The center coordinates of the clusters,

```
cent= kmeans.cluster_centers_
cent
```

⇐ {{-6.58197,-8.17239},{-1.47108,4.33722},{-10.0494,-3.85954}}

Reading Python results into *Mathematica* for visualization see Fig. 3.1.

```
⇒ centers=%;
⇒ labels=Import["labels.mtx"]//Flatten;
⇒ datap=Import["dataX.mtx"];
⇒ p0=ListPlot[datap,
    Frame → True,PlotMarkers → "\[DifferenceDelta]",Axes → None]
```

⇐

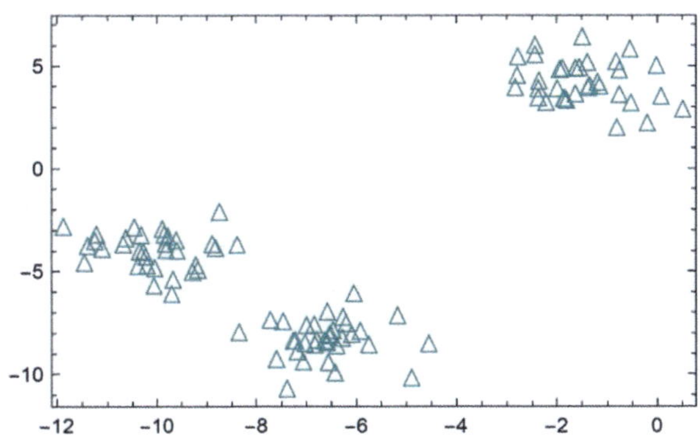

Fig. 3.1 Data points for clustering

Computing the elements belonging to the different clusters,

```
⇒ dataC=MapThread[Flatten[{#1,#2}]&,{datap,labels}];
⇒ dataC1=Map[Most,Select[dataC,#[[3]]==0&]];
⇒ dataC2=Map[Most,Select[dataC,#[[3]]==1&]];
⇒ dataC3=Map[Most,Select[dataC,#[[3]]==2&]];
```

Visualizing different clusters with their centers, see Fig. 3.2,

```
⇒ Show[{p0,ListPlot[{dataC1,dataC2,dataC3},Frame → True,
    PlotMarkers → Automatic,Axes → None],
   ListPlot[centers,PlotMarkers → "♠",
    PlotStyle → {Black,Large}]},AspectRatio → 0.7]
```

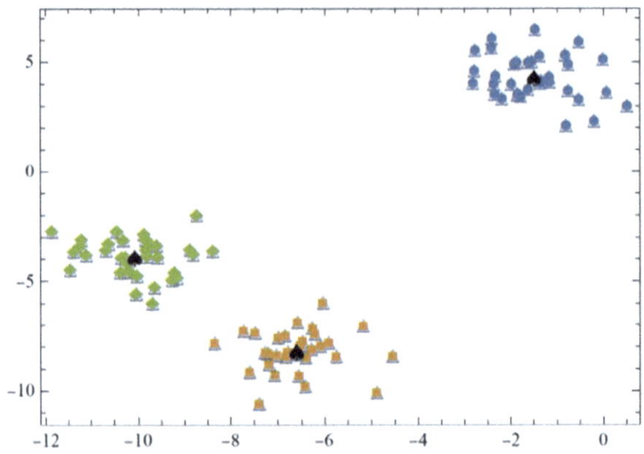

⇐

Fig. 3.2 The results of the clustering with the centers of the clusters employing *Python.*

The clustering performance can be evaluated by computing the *silhouette coefficient*,

$$s = (b - a) / \max(a, b),$$

where a is the mean distance between a sample and all other points in the same clusters, and b is the mean distance between a sample and all the other points of the nearest cluster. *Python* has a built in function.

```
from sklearn import metrics
s=metrics.silhouette_score(X,prediction,'euclidean')
s
```

⇐ 0.769709

A more detailed explanation about silhouette coefficient, see Section 3.5 later in this Chapter.

Now let us employ *Mathematica* for computing the clusters.

Mathematica

Again assuming three clusters,

```
⇒ clusters=FindClusters[datap,3,Method → "KMeans",
   PerformanceGoal → "Quality"];
```

The centers of the clusters are,

```
⇒ centers=Map[Mean[#]&,clusters]
⇐ {{-1.47108,4.33722},{-10.0494,-3.85954},{-6.58197,-8.17239}}
```

Let us visualize the clusters and their centers, see Fig. 3.3.

```
⇒ Show[{p0,ListPlot[clusters,Frame → True,
   PlotMarkers → Automatic,Axes → None],ListPlot[centers,
   PlotMarkers → "♠",PlotStyle → {Black,Large}]},
   AspectRatio → 0.7]
```

⇐

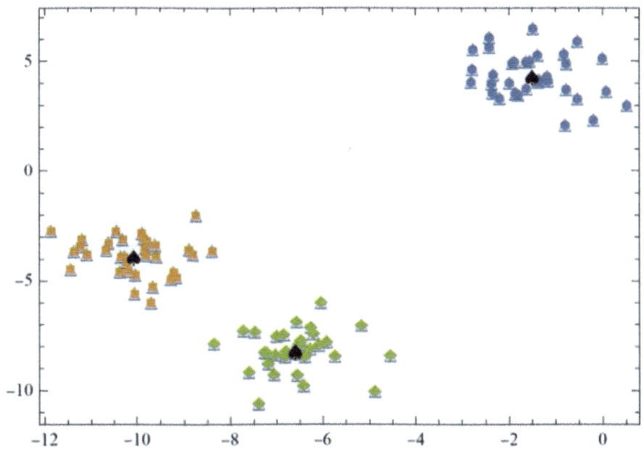

Fig. 3.3 The results of the clustering with the centers of the clusters using *Mathematica.*

A more sophisticated method can be employed to find the boundary of the different clusters. We can define a clustering function, like

```
⇒ clustersF=ClusterClassify[datap,3,Method → "KMeans",
   PerformanceGoal → "Quality"]
⇐ ClassifierFunction[ Input type: NumericalVector (Length: 2)
                      Classes: 1, 2, 3 ]
```

Let us test it in two points (−6, −8) and (−5, −1). The first point with probability 1 belongs to the first cluster,

```
⇒ clustersF[{-6,-8},"Probabilities"]//Normal
```

⇐ $1 \to 1., 2 \to 3.92839 \times 10^{-36}, 3 \to 4.92367 \times 10^{-10}$ [

while the second point belongs partly to the first (0.31) and partly to the second (0.67) cluster. These probabilities can be considered as membership function values.

```
⇒ clustersF[{-5,-1},"Probabilities"]//Normal
⇐ {1 → 0.305066,2 → 0.665971,3 → 0.0289628}[
```

Let us visualize the cluster regions, see Fig. 3.4

```
⇒ pD=DensityPlot[clustersF[{x,y}],{x,-12,2},{y,-12,7},
   PlotPoints → 100];
⇒ Show[{pD,p0,ListPlot[clusters,Frame → True,
   PlotMarkers → Automatic,Axes → None],
  ListPlot[centers,PlotMarkers → "♠",PlotStyle → {Black,Large}]},
   AspectRatio → 0.7]
```

⇐

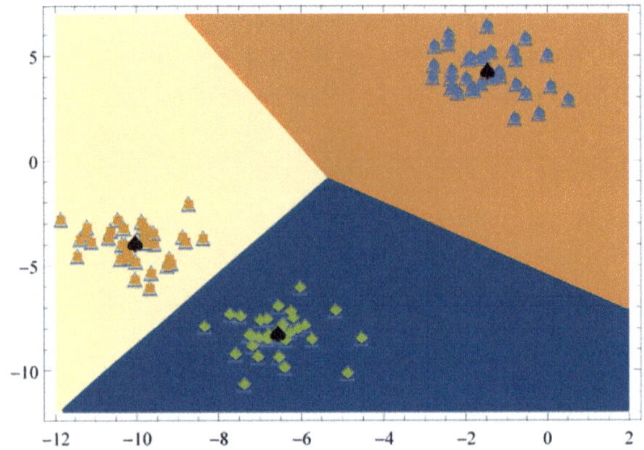

Fig. 3.4 The results of the clustering with the boundary of the cluster.

There is only one element with membership (1/3,1/3,1/3) in the common vertex of the three regions.

In order to compute the silhouette coefficient for the *Mathematica*'s result, we save its labels in a file,

```
⇒ index=Map[#-1&,ClusteringComponents[datap,3,1,Method → "KMeans",
    PerformanceGoal → "Quality"]]
⇐ {0,1,1,1,2,2,2,1,0,0,1,1,2,0,2,2,2,0,1,1,2,1,2,0,1,
    2,2,0,0,2,0,0,2,0,1,2,1,1,1,2,2,1,0,1,1,2,0,0,0,0,
    1,2,2,2,0,2,1,1,0,0,1,2,2,1,1,2,0,2,0,1,1,1,2,0,0,
    1,2,2,0,1,0,1,1,2,0,0,0,0,1,0,2,0,0,1,1,2,2,0,2,0}
```

```
⇒ Export["pred.mtx",{index}]
⇐ pred.mtx
```

and read in into *Python*,

```
index=mmread('pred.mtx')
```

Then we can employ the built-in function of *Python*,

```
ind=index[0]
from sklearn import metrics
s=metrics.silhouette_score(X,ind,'euclidean')
s
```

```
⇐ 0.769709
```

It means that Python and Mathematica provided the same quality for clustering in case of three clusters.

Now, let us try to carry out the clustering with five clusters.

```
⇒ clusters=FindClusters[datap,5,Method → "KMeans",
    PerformanceGoal → "Quality"];
```

The centers of the clusters,

```
⇒ centers=Map[Mean[#]&,clusters]
⇐ {{-1.47108,4.33722},{-9.8649,-4.74746},{-11.0221,-3.37753},
   {-9.3663,-3.32909},{-6.58197,-8.17239}}
```

Fig. 3.5 shows the results,

```
⇒ Show[{p0,ListPlot[clusters,Frame → True,PlotMarkers → Automatic,
   Axes → None],
  ListPlot[centers,PlotMarkers → "♠",PlotStyle → {Black,Large}]},
   AspectRatio → 0.7]
⇐
```

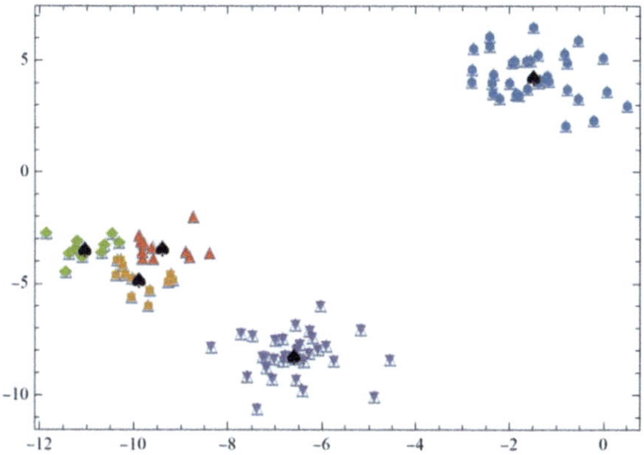

Fig. 3.5 The results of the clustering in case of five clusters.

In order to visualize the results, we carry out the same steps we did in case of three clusters,

```
⇒ clustersF=ClusterClassify[datap,5,Method → "KMeans",
   PerformanceGoal → "Quality"]
⇐ ClassifierFunction[ Input type: NumericalVector (Length: 2)
                      Classes: 1, 2, 3, 4, 5 ]
```

```
⇒ clustersF[{-6,-8},"Probabilities"]//Normal
⇐ {1 → 1.,2 → 3.34515 × 10^-57,3 → 9.54873 × 10^-23,
   4 → 2.77988 × 10^-14,5 → 3.27983 × 10^-13}
```

```
⇒ clustersF[{-5,-1},"Probabilities"]//Normal
⇐ {1 → 0.00221919,2 → 0.00757256,
   3 → 1.97708 × 10^-11,4 → 0.990203,5 → 5.05125 × 10^-6 }
```

```
⇒ pD=DensityPlot[clustersF[{x,y}],{x,-12,2},{y,-12,7},
   PlotPoints → 100];
```

Fig. 3.6 shows the results,

```
⇒ Show[{pD,p0,ListPlot[clusters,Frame → True,
   PlotMarkers → Automatic,Axes → None],
  ListPlot[centers,PlotMarkers → "♠",PlotStyle → {Black,Large}]},
   AspectRatio → 0.7]
```

⇐

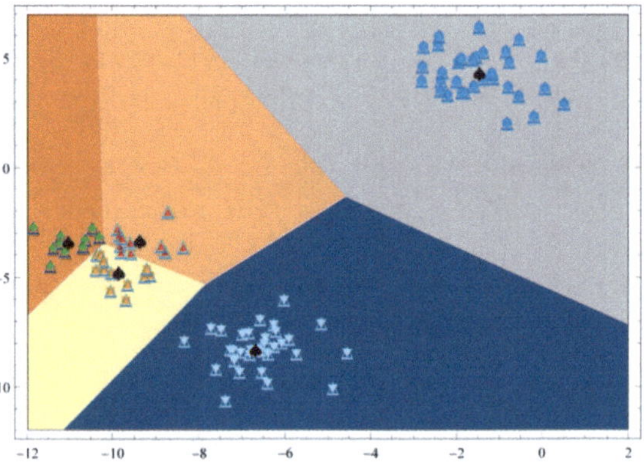

Fig. 3.6 The results of the clustering in case of five clusters with boundaries.

Now, the value of the silhouette coefficient is smaller than in case of three clusters, indicating worse quality of clustering, less separated clusters.

```
⇒ index=Map[#-1&,ClusteringComponents[datap,5,1,
    Method → "KMeans",PerformanceGoal → "Quality"]];
⇒ Export["pred.mtx",{index}]
⇐ pred.mtx
```

```
index=mmread('pred.mtx')
```

```
ind=index[0]
s=metrics.silhouette_score(X,ind,'euclidean')
s
```

⇐ 0.638162

As we see, the quality of the clustering is not so good compared to the case of the three clusters. Consequently silhouette coefficient is an indicator for finding the proper number of clusters (Okabe and Yamada, 2018).

3.1.2 Clustering Images

As an imaging example, let us consider images of three species: zebras, gorillas and penguins, see Fig. 3.7.

⇒ zebra={  };

⇒ gorilla={  };

⇒ penguin={,,,,};

Fig. 3.7. Images of three species of animals to be clustered.

First, let us employ *Mathematica.*

Mathematica

In order to employ numerical data we employ feature extraction to reduce dimensions using *t - SNE* (*t* distributed Stochastic Neighbor Embedding). There is a class of algorithms for visualization called manifold learning algorithms that allow for much more complex mappings, and often provide better visualization. A particularly useful one is the *t - SNE* algorithm. This computes a new representation of the training data, but does not allow for the transformation of new data. This means this algorithm cannot be applied to a test set, rather, it can only transform the data it is trained for, see Chapter 1. Manifold learning can be useful for exploratory data analysis, but is rarely used if the final goal is supervised learning.

Now, the idea behind *t - SNE* is to find a two-dimensional representation of the data that preserves the distances between points as best as possible.

```
⇒ animals=Join[zebra,gorilla,penquin];
```

Since the images have different sizes, we resize them to size of 396×512,

```
⇒ animalsReduced=Map[ImageResize[#,{392,512}]&,animals];
```

Let us represent them with two dimensional vectors, see Fig. 3.8. Since the dimension reduction algorithm is time consuming, GPU is used instead of CPU.

```
⇒ reduced=
    DimensionReduce[animals,2,Method → "TSNE",TargetDevice → "GPU"];
```

Then visualizing the results,

```
⇒ pani=ListPlot[MapThread[Labeled[#1,#2]&,{reduced,
    animalsReduced}],Frame → True,PlotRange → All]
```

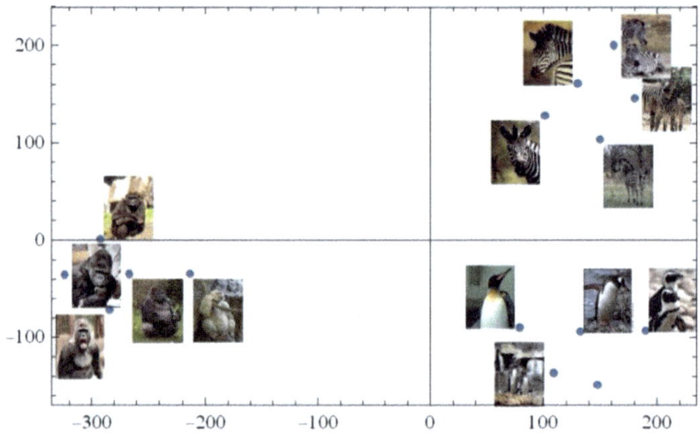

⇐

Fig. 3.8 The map of images after employing 2D dimension reduction.

As it can be seen that the different species are clustered in different groups quite transparently. The centers of the groups can easily be computed,

```
⇒ centersM=Map[Mean[#]&,Partition[reduced,5]]
⇐ {{144.597,147.594},{-276.146,-35.0111},{131.549,-112.583}}
```

So here we did not employed any clustering, since the clusters are so transparent.

Let us save the reduced coordinates of the images for *Python*.

```
⇒ X=reduced
⇐ {{180.106,145.609},{130.228,160.8},{149.597,103.534},
   {101.513,128.057},{161.541,199.97},{-324.505,-35.4624},
   {-266.282,-34.5745},{-292.198,0.989372},{-284.038,-71.5398},
   {-213.707,-34.4679},{79.1827,-89.9688},{189.736,-93.374},
   {132.477,-94.0187},{109.039,-136.691},{147.311,-148.862}}
```

```
⇒ Export["dataX.mtx",X]
⇐ dataX.mtx
```

and read the file into *Python*.

```
X=mmread('dataX.mtx')
```

In Python we employ three clusters,

```
import numpy as np
```

The same result are obtained as those of *Mathematica*,

```
kmeans = KMeans(n_clusters=3).fit(X)
prediction= kmeans.predict(X)
prediction
```

```
⇐ {0,0,0,0,0,1,1,1,1,1,2,2,2,2,2}
```

```
cent= kmeans.cluster_centers_
cent
```

⇐ {{144.597,147.594},{-276.146,-35.0111},{131.549,-112.583}}

```
⇒ centersP=%;
```

The quality of the clustering can be indicated again by the silhouette coefficient,

```
from sklearn import metrics
s=metrics.silhouette_score(X,prediction,'euclidean')
s
```

⇐ 0.779621

The result of the *Python* can be seen in Fig. 3.9.

```
⇒ Show[{pani,ListPlot[centersM,PlotMarkers → "♠",
   PlotStyle → {Blue,Large}],ListPlot[centersP,
   PlotMarkers → "▼",PlotStyle → {Red,Large}]},AspectRatio → 0.7]
```

⇐

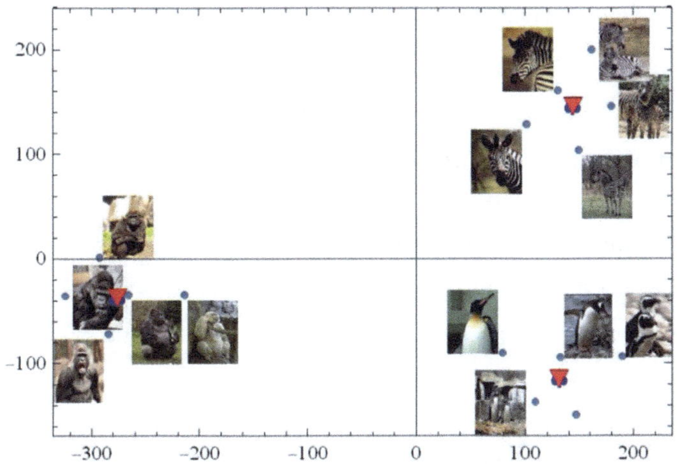

Fig. 3.9 The results of the *Python* using *KMeans Clustering* in case $K = 3$.

3.2 Hierarchical Clustering

Basic theory

Hierarchical clustering is an alternative approach identifying groups in a dataset. It does not require specification of the number of clusters to be generated as in case of K-Means Clustering method. Hierarchical or agglomerate clustering groups data over a variety of scales by creating a *cluster tree, dendrogram*. This dendrogram, see Fig. 3.10, represents all possible clusters from one cluster (Step

4 - {a,b,c,d,e}) down to *n* clusters (n is the number of elements to be clustered), (Step 0 -{a},{b},{c},{d},{e}).

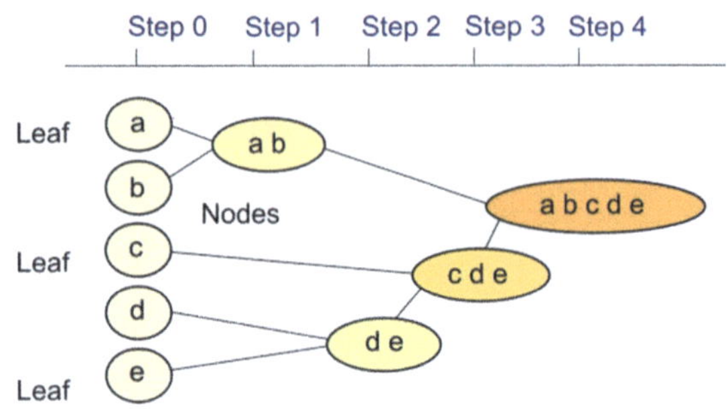

⇐

Fig. 3.10 The dendrogram representing all of the clusters of a data set.

The user can decide the *level of linkage* of the clusters, which will determine the number of clusters (James et al. 2013).

First let us employ the same data set what we used in the previous section.

3.2.1 Dendrogram for Small Data Set

Load data set and the clustering function from Python depository.

```
from sklearn.datasets import make_blobs
from scipy.cluster.hierarchy import linkage
import numpy as np
from sklearn.cluster import AgglomerativeClustering
```

The data set

```
X, y = make_blobs(random_state=1)
```

```
X
```

```
⇐ {{-0.794152,2.10495},{-9.15155,-4.81286},{-11.4418,-4.45781},
  {-9.76762,-3.19134},{-4.53656,-8.40186},{-6.26302,-8.10666},
  {-6.38481,-8.47303},{-9.20491,-4.57688},{-2.76018,5.55121},
  {-1.17104,4.33092},{-10.0364,-5.56912},{-9.87589,-2.82386},
  ...
  {-8.87629,-3.54445},{-6.02606,-5.96625},{-7.04747,-9.27525},
  {-1.37397,5.29163},{-6.25393,-7.10879},{0.0852519,3.64528}}
```

```
⇒ datap=%;
```

In order to illustrate the dendrogram we can employ *Mathematica* function,

```
⇒ Needs["HierarchicalClustering`"]
```

⇒ DendrogramPlot[datap,TruncateDendrogram → 18,HighlightLevel → 3,
Orientation → Left]

⇐

Fig. 3.11 The dendrogram of the make_blobs data set shading the elements in case of three clusters.

Fig. 3.11 shows clearly that the linkage for one cluster is very high, for three clusters (shaded) is definitely lower.

Let us employ first *Python*, after the analysis of the dendrogram, and decide three clusters.

Python

Employing Python code, we get the labels of the three clusters,

```
X, y = make_blobs(random_state=1)
ac = AgglomerativeClustering(n_clusters=3,
 affinity='euclidean',linkage='complete')
labels = ac.fit_predict(X)
prediction=labels
prediction
```

```
⇐ {1,2,2,2,0,0,0,2,1,1,2,2,0,1,0,0,0,1,2,2,0,2,0,1,2,0,
  0,1,1,0,1,1,0,1,2,0,2,2,2,0,0,2,1,2,2,0,1,1,1,1,2,
  0,0,0,1,0,2,2,1,1,2,0,0,2,2,0,1,0,1,2,2,2,0,1,1,2,
  0,0,1,2,1,2,2,0,1,1,1,1,2,1,0,1,1,2,2,0,0,1,0,1}
```

```
⇒ labels=%;
```

Let us visualize the clusters, see Fig. 3.12

```
⇒ p0=ListPlot[datap,Frame → True,PlotMarkers → "Δ",Axes → None]
⇒ dataC=MapThread[Flatten[{#1,#2}]&,{datap,labels}];
⇒ dataC1=Map[{#[[1]],#[[2]]}&,Select[dataC,#[[3]]==0&]];
⇒ dataC2=Map[{#[[1]],#[[2]]}&,Select[dataC,#[[3]]==1&]];
⇒ dataC3=Map[{#[[1]],#[[2]]}&,Select[dataC,#[[3]]==2&]];
⇒ Show[{p0,ListPlot[{dataC1,dataC2,dataC3},Frame → True,
  PlotMarkers → Automatic,Axes → None]},AspectRatio → 0.7]
```

⇐

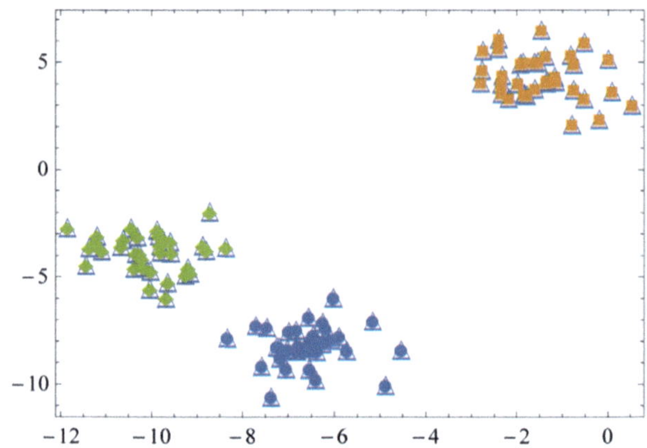

Fig. 3.12 The elements to be clustered.

The quality of the clustering,

```
from sklearn import metrics
s=metrics.silhouette_score(X,prediction,'euclidean')
s
```

⇐ 0.769709

Now let us suppose two clusters, see Fig. 3.13.

```
⇒ DendrogramPlot[datap,TruncateDendrogram → 18,HighlightLevel → 2,
   Orientation → Left]
```

⇐

Fig. 3.13 The dendrogram assuming two clusters.

Using Python prediction for the labels,

```
X, y = make_blobs(random_state=1)
ac = AgglomerativeClustering(n_clusters=2,
 affinity='euclidean',linkage='complete')
labels = ac.fit_predict(X)
prediction=labels
prediction
```

```
⇐ {1,0,0,0,0,0,0,0,1,1,0,0,0,1,0,0,0,1,0,0,0,0,0,1,0,
   0,0,1,1,0,1,1,0,1,0,0,0,0,0,0,0,0,1,0,0,0,1,1,1,1,
   0,0,0,0,1,0,0,0,1,1,0,0,0,0,0,0,1,0,1,0,0,0,0,1,1,
   0,0,0,1,0,1,0,0,0,1,1,1,1,0,1,0,1,1,0,0,0,0,1,0,1}
⇒ labels=%;
```

Visualizing the result, see Fig. 3.14.

```
⇒ dataC=MapThread[Flatten[{#1,#2}]&,{datap,labels}];
⇒ dataC1=Map[{#[[1]],#[[2]]}&,Select[dataC,#[[3]]==0&]];
⇒ dataC2=Map[{#[[1]],#[[2]]}&,Select[dataC,#[[3]]==1&]];
⇒ Show[{p0,ListPlot[{dataC1,dataC2},Frame → True,
   PlotMarkers → Automatic,Axes → None]},AspectRatio → 0.7]
```

⇐ 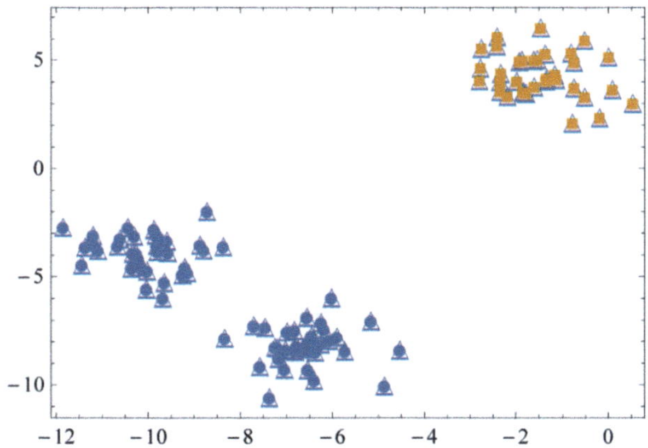

Fig. 3.14 The results of the Hierarchical Clustering using Python in case of $k = 2$.

The quality of the clustering is nearly the same, a bit worse,

```python
from sklearn import metrics
s=metrics.silhouette_score(X,prediction,'euclidean')
s
```

```
⇐ 0.765825
```

However employing four clusters, one of the clusters contains only one element, see Fig. 3.15.

```
⇒ DendrogramPlot[datap,TruncateDendrogram → 18,HighlightLevel → 4,
   Orientation → Left]
```

⇐

Fig. 3.15 The dendrogram assuming four clusters.

```python
X, y = make_blobs(random_state=1)
ac = AgglomerativeClustering(n_clusters=4,
 affinity='euclidean',linkage='complete')
labels = ac.fit_predict(X)
prediction=labels
prediction
```

⇐ {0,2,2,2,1,3,3,2,0,0,2,2,1,0,1,3,3,0,2,2,3,2,3,0,2,
1,3,0,0,1,0,0,1,0,2,3,2,2,2,3,3,2,0,2,2,3,0,0,0,0,
2,1,3,3,0,3,2,2,0,0,2,3,3,2,2,3,0,3,0,2,2,2,3,0,0,
2,3,3,0,2,0,2,2,3,0,0,0,0,2,0,3,0,0,2,2,3,1,0,3,0}

Fig. 3.16 shows the results, which indicate that the linearly separable problem became a non-linearly separable one.

```
⇒ labels=%;
⇒ dataC=MapThread[Flatten[{#1,#2}]&,{datap,labels}];
⇒ dataC1=Map[{#[[1]],#[[2]]}&,Select[dataC,#[[3]]==0&]];
⇒ dataC2=Map[{#[[1]],#[[2]]}&,Select[dataC,#[[3]]==1&]];
⇒ dataC3=Map[{#[[1]],#[[2]]}&,Select[dataC,#[[3]]==2&]];
⇒ dataC4=Map[{#[[1]],#[[2]]}&,Select[dataC,#[[3]]==3&]];
⇒ Show[{p0,ListPlot[{dataC1,dataC2,dataC3,dataC4},Frame → True,
  PlotMarkers → Automatic,Axes → None]},AspectRatio → 0.7]
```

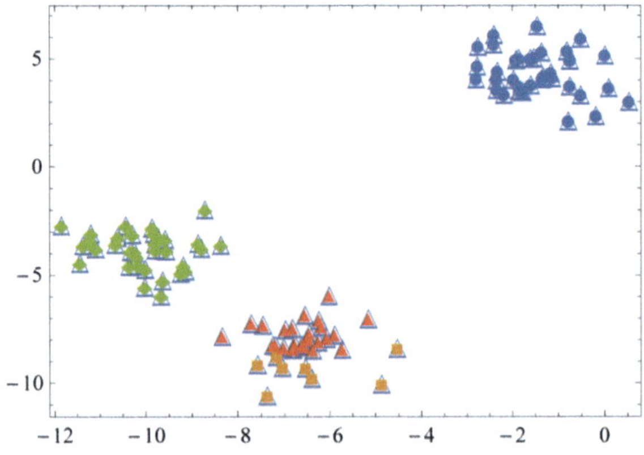

⇐

Fig. 3.16 The results of the Hierarchical Clustering using Python in case of $k = 4$.

The quality of the clustering now is seemingly worse,

```
from sklearn import metrics
s=metrics.silhouette_score(X,prediction,'euclidean')
s
```

⇐ 0.651076

3.2.2 Image Segmentation

Our problem is to compute the sizes of the different areas of the cross section of a test-piece of a rock, which indicate the concentration of the different mineral components. After using preparation employing acid, basically there are three clusters: blue, yellow and brown). However the intensities of the pixels are different, see Fig. 3.17.

⇒ img= 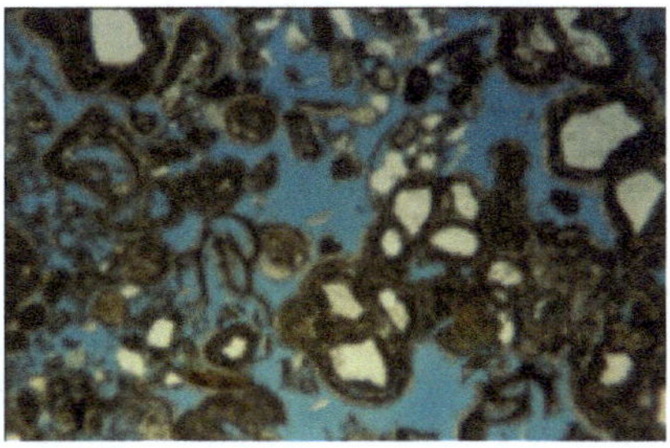 ;

Fig. 3.17 The digital image of the prepared cross section of the test-piece of the rock.

Now, we shall employ *Hierarchical Clustering* to find the three distinct clusters, and the number of the elements which are proportional to the areas of the different mineral components. It goes without saying that this is a 2D approach, and provides only a rough estimation for the real 3D situation.

First, let us display the RGB vectors of the different pixels, see Fig. 3.18.

```
⇒ pv=ImageData[img];
⇒ X=Flatten[pv,1];ListPointPlot3D[X,PlotStyle → {PointSize[0.001]},
  AxesLabel → {Red,Green,Blue},BoxRatios → {1,1,1}]
```

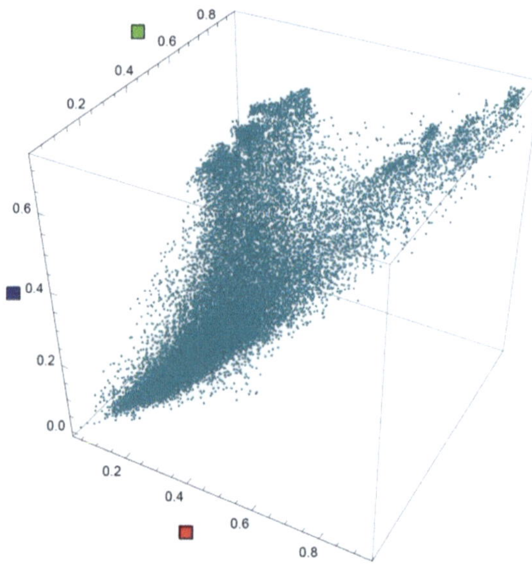

⇐

Fig. 3.18 The pixel vectors of the image in the RGB color space.

Mathematica

In *Mathematica* we can solve the problem using a built-in function.

```
⇒ c0=FindClusters[X,3];
```

Fig. 3.19 shows the three distinct sets of pixels in the RGB space.

```
⇒ ListPointPlot3D[c0,PlotStyle → {PointSize[0.001]},
    AxesLabel → {Red,Green,Blue},BoxRatios → {1,1,1}]
```

⇐

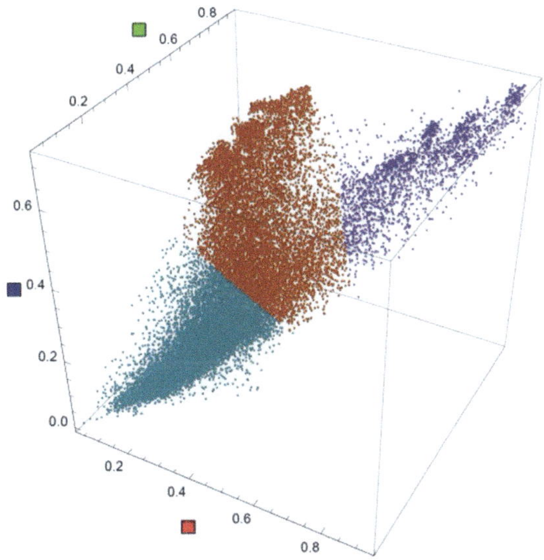

Fig. 3.19 The clustered pixel vectors in the RGB color space.

The number of pixels belonging to the different clusters can easily be computed.

```
⇒ n1=Length[c0[[1]]]
⇐ 14 002
```

```
⇒ n2=Length[c0[[2]]]
⇐ 10 057
```

```
⇒ n3=Length[c0[[3]]]
⇐ 2341
```

The total number of pixels,

```
⇒ n=n1+n2+n3
⇐ 26 400
```

Double check

```
⇒ n==200 × 132
⇐ True
```

Then the ratio of the areas can be computed, too

```
⇒ {n1,n2,n3}/n//N
⇐ {0.530379,0.380947,0.0886742}
```

In order to visualize the segmented image, the pixels should be labeled

```
⇒ Xn=MapThread[(#1 → #2)&,{X,Map[ToString[#]&,Range[n]]}];
```

Now, we carry out the clustering again together with the corresponding labels

```
⇒ c1=FindClusters[Xn,3];
```

The three clusters are

```
⇒ cl1=Map[ToExpression[#]&,cl[[1]]];
⇒ cl2=Map[ToExpression[#]&,cl[[2]]];
⇒ cl3=Map[ToExpression[#]&,cl[[3]]];
```

We use the following colors for visualization of the different clusters: cluster 1 → Red, cluster 2 → Green, and cluster 3 → Blue

Then we color the pixels accordingly

```
⇒ pc=Table[0,{n}];
⇒ Do[If[MemberQ[cl1,i],pc[[i]]={1,0.,0.}];
 If[MemberQ[cl2,i],pc[[i]]={0.,1,0}];
 If[MemberQ[cl3,i],pc[[i]]={0,0,1}],{i,1,n}]
```

In matrix form

```
⇒ pcdata=Partition[pc,200];
⇒ Dimensions[pcdata]
⇐ {132,200,3}
```

Then the segmented image is, see Fig.3.20.

```
⇒ pgnew=Image[pcdata]
```

⇐ 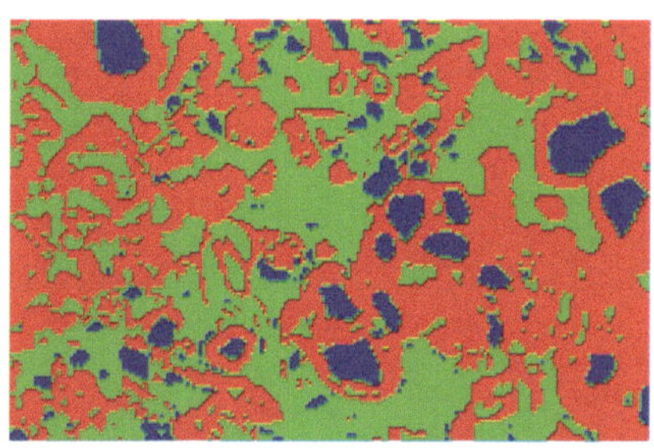

Fig. 3.20 The segmented image.

The original and the segmented images can be seen on Fig.3.21.

```
⇒ {img,pgnew}
```

⇐ { 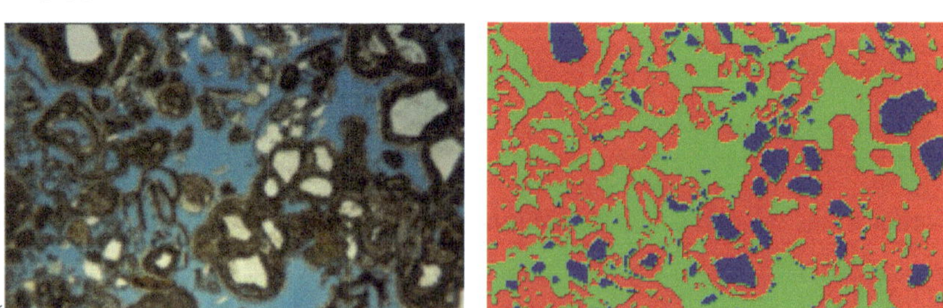 }

Fig. 3.21 The original and the segmented images.

Python

To solve the problem in *Python*, we export the set of the RGB vectors of the pixels in a file

```
⇒ Export["dataX.mtx",X]
⇐ dataX.mtx
```

Loading the necessary procedures

```
from numpy import array, matrix
from scipy.io import mmread, mmwrite
import numpy as np
```

Loading the clustering function (Pestunova et al. 2015),

```
from scipy.cluster.hierarchy import linkage
import numpy as np
from sklearn.cluster import AgglomerativeClustering
```

Reading the data set

```
X=mmread('dataX.mtx')
```

Predicting the labels of the three clusters

```
ac = AgglomerativeClustering(n_clusters=3,
  affinity='euclidean',linkage='complete')
labels = ac.fit_predict(X)
prediction=labels
prediction
```

```
⇐ {0,0,0,0,0,0,0,0,0,0,0,0,0,0,0,0,0,0,0,0,0,0,0,0,0,0,0,1,
  2,2,2,2,2,2,2,2,1,1,0,0,0,0,0,0,0,0,0,0,0,0,0,0,0,0,0,0,
  1,1,0,1,1,1,1,1,1,1,0,0,0,0,0,0,0,0,0,0,0,0,0,0,0,1,2,1,
  1, ... 26231 ...,
  0,0,0,0,0,0,0,0,0,1,1,1,1,0,1,1,1,1,1,1,1,0,0,0,0,0,0,0,
  0,0,1,1,1,1,1,1,1,1,1,1,1,1,1,1,1,1,1,1,1,1,1,0,0,0,0,
  0,0,0,0,0,0,0,0,0,1,1,0,0,0,0,0,0,0,0,0,0,0,0,0,0,0,0,0}
  large output  show less  show more  show all  set size limit...
```

```
⇒ total=%;
```

counting the elements in the different sets,

```
⇒ c1=Select[total,#==0&]; Length[c1]
⇐ 17 531
⇒ c2=Select[total,#==1&]; Length[c2]
⇐ 7517
⇒ c3=Select[total,#==2&]; Length[c3]
⇐ 1352
```

The result is different from that of *Mathematica*.

```
⇒ Length[total]
⇐ 26 400
```

Creating segmented image,

```
⇒ cluster1={};cluster2={};cluster3={};
⇒ Do[Which[total[[i]]==0,AppendTo[cluster1,i],total[[i]]==1,
    AppendTo[cluster2,i],total[[i]]==2,AppendTo[cluster3,i]],
    {i,1,26400}];
⇒ pc=Table[0,{n}];n
⇐ 26 400
⇒ Do[If[MemberQ[cluster1,i],pc[[i]]={1,0.,0.}];
    If[MemberQ[cluster2,i],pc[[i]]={0.,1,0}];
    If[MemberQ[cluster3,i],pc[[i]]={0,0,1}],{i,1,n}]
```

In matrix form

```
⇒ pcdata=Partition[pc,200];
⇒ Dimensions[pcdata]
⇐ {132,200,3}
```

Then the segmented image is

```
⇒ pgnew=Image[pcdata]
⇐
```

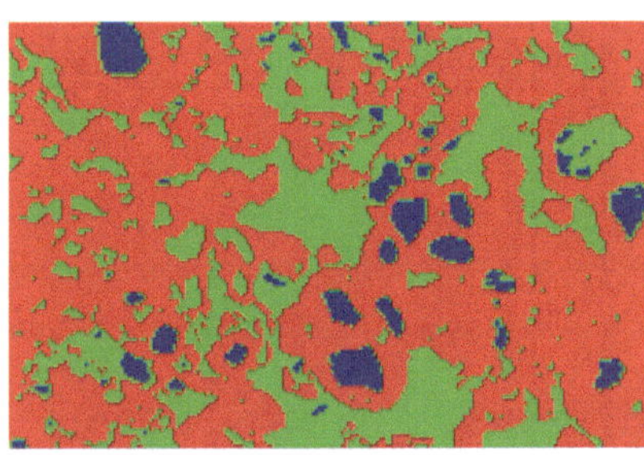

Fig. 3.22 The segmented image via *Python*.

The original and the segmented images.

```
⇒ {img,pgnew}
⇐ {
```

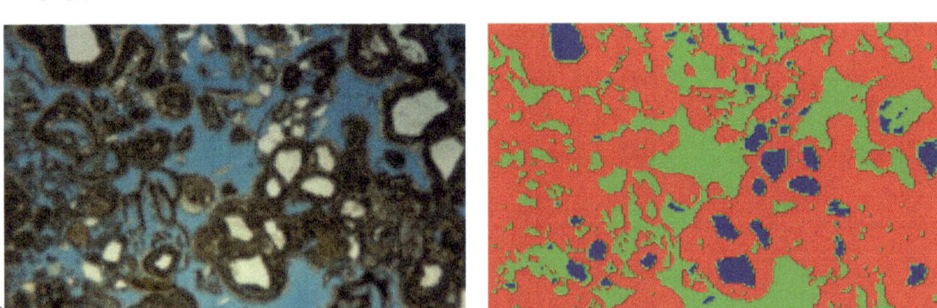

}

Fig. 3.23 The original and the segmented image via *Python*.

The results of Python seems more "consistent", less diverse, compare Figs. 3.23 and 3.24

⇐

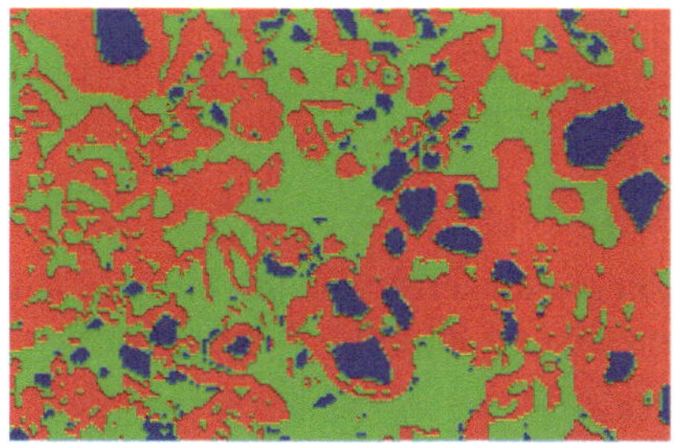

Fig. 3.24 The segmented image via *Mathematica.*

In general, Hierarchical Clustering can be successfully employed for image segmentation, (Galbiati, 2009).

3.3 Density-Based Spatial Clustering of Applications with Noise

Basic theory

The *Density-Based Spatial Clustering of Applications with Noise* (DBSCAN) is a clustering algorithm that groups together points, see Fig. 3.25, that are closely packed (core points, red A) or reachable from the core points (yellow points, B) together marking as outliers points (blue N) that lie alone in low-density regions.

It can be useful in case of not too many clusters, not-flat geometry but arbitrary shaped clusters and in case of outliers.

There are two parameters, the radius of the red circles and *the minimal number of the red* (*A*) *points*.

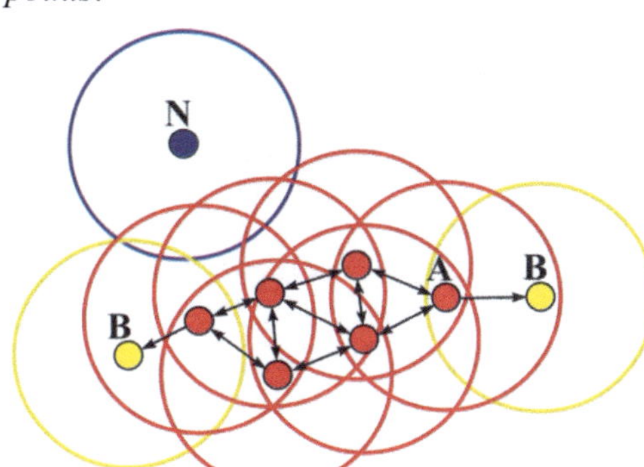

Fig. 3.25 Topology of the DBSCAN method.

These parameters are the weak point of the method, since the result is very sensitive on these parameter values (Müller and Guido, 2017).

3.3.1 Data Set Moons

Simple introductory example demonstrates the *positive features* of the method. Let us consider the moon data set from the Python repository,

```
from sklearn.datasets import make_moons
X, y =make_moons(n_samples=200,noise=0.05,random_state=0)
```

The 2D data points are

```
X
```

```
⇐ {{0.816805,0.521645},{1.6186,-0.379829},{-0.0212695,0.273728},
  {-1.02181,-0.07544},{1.76655,-0.170699},{1.88203,-0.0423845},
  {0.974816,0.209994},{0.887988,-0.489367},{0.898652,0.366378},
  {1.11639,-0.534604},{-0.3638,0.827902},{0.247024,-0.238567},
  ....
  {0.424479,0.932688},{0.808614,0.535999},{0.940009,0.271114},
  {-0.0160918,0.373696},{-0.536334,0.860268},{1.88282,0.244356},
  {0.175752,-0.007231},{0.124236,1.0079},{1.62153,-0.223285}}
```

```
⇒ trainX=%;
```

In order to visualize them we use their labels,

```
y
```

```
⇐ {0,1,1,0,1,1,0,1,0,1,0,1,1,1,0,0,0,1,0,0,1,1,0,1,0,1,1,1,1,
  0,0,0,1,1,0,1,1,0,0,1,1,0,0,1,1,0,0,0,1,1,0,1,1,0,1,0,0,1,
  0,0,1,0,1,0,1,0,0,1,0,0,1,0,1,1,1,0,1,0,0,1,1,0,1,1,1,0,0,
  0,1,1,0,0,1,0,1,1,1,1,0,1,1,1,0,0,0,1,0,0,1,0,0,0,0,0,0,1,
  0,1,1,0,0,0,1,0,1,0,0,1,1,1,0,0,0,1,1,1,1,0,1,0,1,1,0,0,0,
  0,1,1,0,1,1,1,0,0,1,0,1,1,0,0,1,1,0,1,1,1,0,1,1,1,0,0,0,0,
  1,1,1,0,0,0,1,0,1,1,1,0,0,1,0,0,0,0,0,0,1,0,1,1,0,1}
```

```
⇒ clusters=%;
⇒ total=MapThread[{#1,#2}&,{trainX,clusters}];
⇒ clust1=Select[total,#[[2]]==0&];
⇒ clust2=Select[total,#[[2]]==1&];
⇒ pclust1=Map[#[[1]]&,clust1];
⇒ pclust2=Map[#[[1]]&,clust2];
```

Then Fig. 3.26 shows the two data sets,

⇒ `p0=ListPlot[{pclust1,pclust2},PlotStyle → {Pink,Green}]`

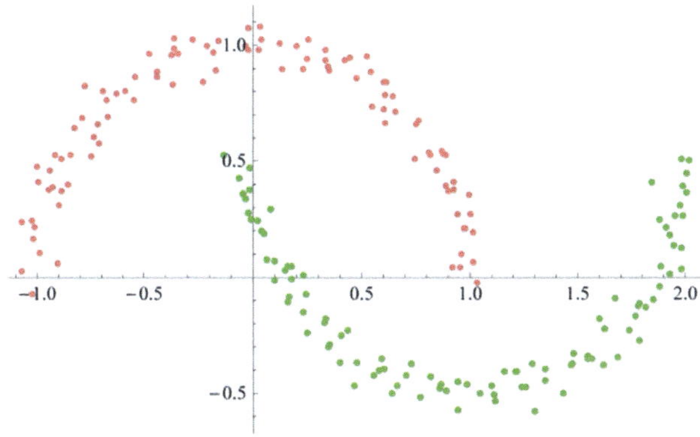

⇐

Fig. 3.26 The moon data sets.

First let us try to separate the two sets via *K-Means Clustering*. In this case the number of clusters should be given.

Mathematica

Using *Mathematica* we get,

⇒ `c=FindClusters[trainX,2,Method → "KMeans"];`

Fig. 3.27 shows that this method provides a linear separation only,

⇒ `ListPlot[{c[[1]],c[[2]]}]`

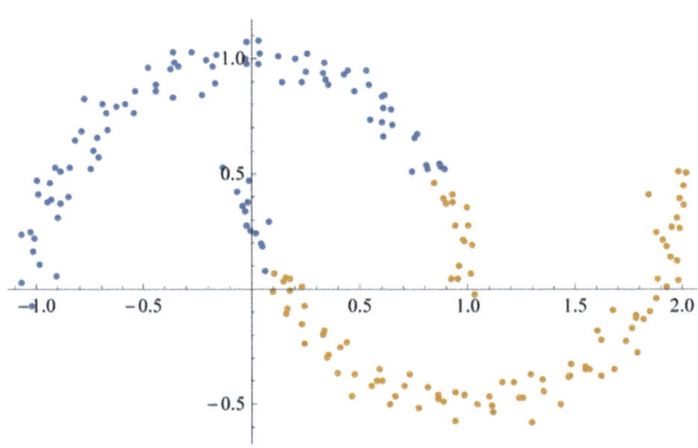

⇐

Fig. 3.27 The moon data sets clustered by K-Mean Clustering.

In case of *Hierarchical Clustering* we do not need to specify the number of clusters preliminary,

⇒ `Needs["HierarchicalClustering`"]`

⇒ `DendrogramPlot[trainX,TruncateDendrogram → 10,HighlightLevel → 2]`

⇐ 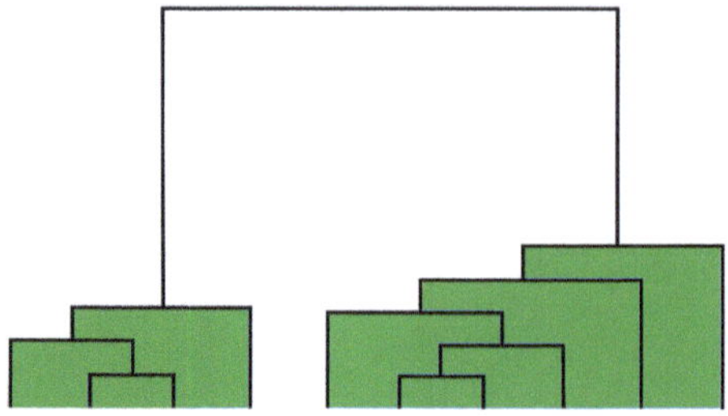

Fig. 3.28 The dendrogram of the clustered moon data set.

```
⇒ XN=MapThread[(#1 → ToString[#2])&,{trainX,Range[Length[trainX]]}];
```

Employing two clusters, we get again basically linear separation, see Fig. 3.29.

```
⇒ c=FindClusters[XN,2]
⇐ {{1,3,4,11,15,17,19,20,23,25,30,31,32,34,35,39,40,42,43,46,48,51,
   54,56,57,59,60,62,64,66,67,69,70,72,74,76,77,79,82,85,86,88,91,
   94,99,100,103,105,106,107,108,110,111,112,113,114,117,120,121,
   122,123,124,125,126,132,133,138,142,143,144,145,149,150,154,156,
   159,160,163,167,171,172,173,174,178,179,180,182,186,187,189,190,
   191,192,193,195,196,199},
   {2,5,6,7,8,9,10,12,13,14,16,18,21,22,24,26,27,28,29,33,36,37,38,
   41,44,45,47,49,50,52,53,55,58,61,63,65,68,71,73,75,78,80,81,83,
   84,87,89,90,92,93,95,96,97,98,101,102,104,109,115,116,118,119,
   127,128,129,130,131,134,135,136,137,139,140,141,146,147,148,151,
   152,153,155,157,158,161,162,164,165,166,168,169,170,175,176,177,
   181,183,184,185,188,194,197,198,200}}
```

Let us visualize it, see Fig. 3.29,

```
⇒ c1=Map[ToExpression[#]&,c[[1]]]; c2=Map[ToExpression[#]&,c[[2]]];
⇒ C1=Map[trainX[[#]]&,c1];
⇒ C2=Map[trainX[[#]]&,c2];
⇒ ListPlot[{C1,C2}]
```

⇐ 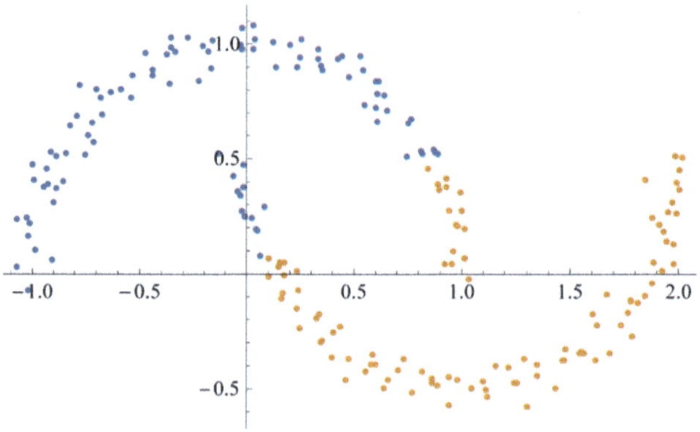

Fig. 3.29 Results of the Hierarchical Clustering.

For DBSCAN< again one does not need to know the number of clusters,

```
⇒ c=FindClusters[trainX,Method → "DBSCAN"];
⇒ Length[c]
⇐ 2
```

The result is nearly perfect, see Fig. 3.30.

```
⇒ ListPlot[{c[[1]],c[[2]]}]
```

⇐

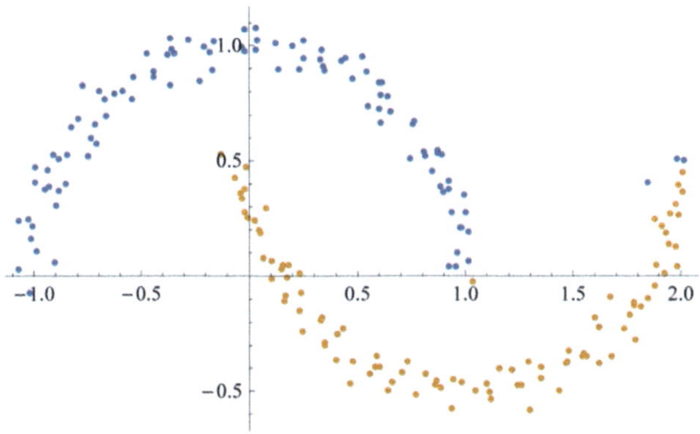

Fig. 3.30 Results of the DBSCAN Clustering.

Some "misclustered" points can be considered as noises, namely outliers.

Let us visualize the regions and the boundaries of the clusters. The labels of the data elements are,

```
⇒ index=Table[0,{i,1,200}];
⇒ Do[If[MemberQ[c[[2]],trainX[[i]]],index[[i]]=1],{i,1,200}];
⇒ index
⇐ {0,1,1,0,1,1,0,1,0,1,0,1,1,1,0,0,0,1,0,0,1,1,0,1,0,1,1,1,1,
  0,0,0,1,1,0,1,1,0,0,1,1,0,0,1,1,0,0,0,1,1,0,1,1,0,1,0,0,1,
  0,0,1,0,1,0,1,0,0,1,0,0,1,0,1,1,1,0,1,0,0,1,1,0,1,1,1,0,0,
  0,1,1,0,0,1,0,1,1,1,1,0,1,1,1,0,0,0,1,0,0,1,0,0,0,0,0,0,1,
  0,1,1,0,0,0,1,0,1,0,0,1,1,1,0,0,0,1,1,1,1,0,1,0,1,1,0,0,0,
  0,1,1,0,1,1,1,0,0,1,0,1,1,0,0,1,1,0,1,1,1,0,1,1,1,0,0,0,0,
  1,1,1,0,0,0,1,0,1,1,1,0,0,1,0,0,0,0,0,0,1,0,1,1,0,1}
```

Now we define a clustering function,

```
⇒ f=ClusterClassify[trainX,Method → "DBSCAN"]
⇐ ClassifierFunction[ Input type: NumericalVector (Length: 2)
                      Classes: 1, 2 ]
```

Then Fig. 3.31 shows the result

```
⇒ Show[{DensityPlot[f[{u,v}],{u,-1.1,2.1},{v,-0.7,1.2},
  PlotPoints → 50],p0}]
```

⇐

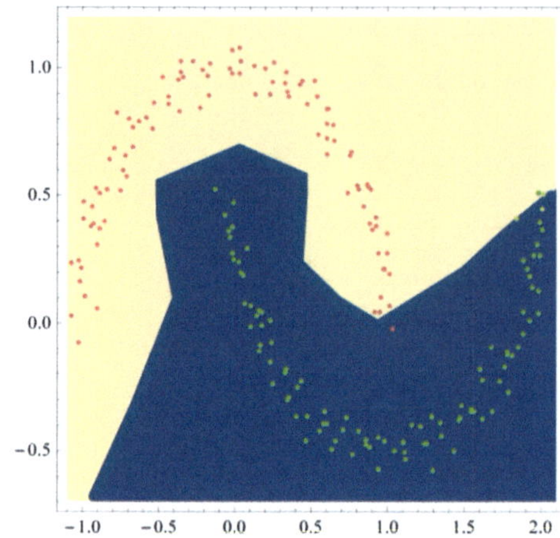

Fig. 3.31 The region of the clusters in case of the DBSCAN Clustering.

Unfortunately the margin between the two clusters for the first coordinates $x \geq 1$ is practically zero, which makes the method very sensitive on the noises or measuring errors.

We shall see in the next section that the *Spectral Clustering* can maximize this margin, resulting in a robust clustering.

Python

Now, we employ Python with parameters $\epsilon = 0.2$ and min_samples = 5. Let us load the clustering function,

```
from sklearn.cluster import DBSCAN
db = DBSCAN(eps=0.2, min_samples=5,metric='euclidean')
```

Training the method,

```
y_db=db.fit_predict(X)
```

The resulting labels are,

```
y_db
```

```
⇐ {0,1,1,0,1,1,0,1,0,1,0,1,1,1,0,0,0,1,0,0,1,1,0,1,0,1,1,1,1,
   0,0,0,1,1,0,1,1,0,0,1,1,0,0,1,1,0,0,0,1,1,0,1,1,0,1,0,0,1,
   0,0,1,0,1,0,1,0,0,1,0,0,1,0,1,1,1,0,1,0,0,1,1,0,1,1,1,0,0,
   0,1,1,0,0,1,0,1,1,1,1,0,1,1,1,0,0,0,1,0,0,1,0,0,0,0,0,0,1,
   0,1,1,0,0,0,1,0,1,0,0,1,1,1,0,0,0,1,1,1,1,0,1,0,1,1,0,0,0,
   0,1,1,0,1,1,1,0,0,1,0,1,1,0,0,1,1,0,1,1,1,0,1,1,1,0,0,0,0,
   1,1,1,0,0,0,1,0,1,1,1,0,0,1,0,0,0,0,0,0,1,0,1,1,0,1}
⇒ yP=%;
```

We can compare the errors of the two codes.

Python error

```
⇒ Norm[yP-y]
⇐ 0
```

Mathematica error

```
⇒ Norm[index-y]
⇐ 2
```

As expected, *Mathematica* provides four erroneous, misclustered points.

3.3.2 Segmentation of MRI of Brain

Medical image processing is the most challenging and emerging field nowadays. Magnetic Resonance Images (MRI) acts as the source for the development of classification system. The extraction, identification and segmentation of infected region from Magnetic Resonance (MR) brain image is of significant concern but a dreary and time-consuming task performed by radiologists or clinical experts, and the final classification accuracy depends on their experience only. To overcome these limitations, it is necessary to use computer-aided techniques. Improving the efficiency of classification accuracy and reducing the recognition complexity play important role in the medical imaging.

Let us consider a brain MRI, see Fig. 3.32.

⇒ img= 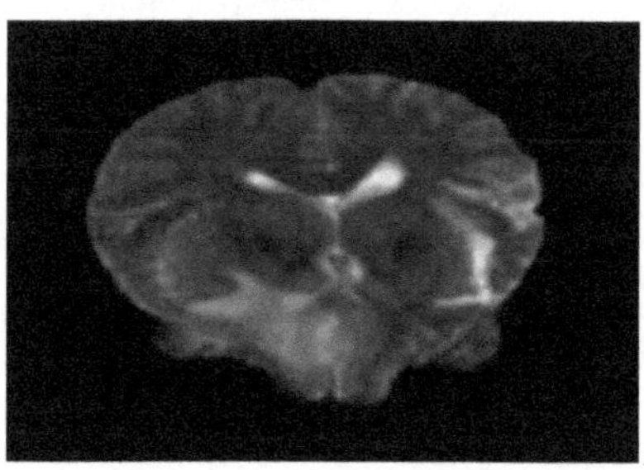 ;

Fig. 3.32 An MRI of a brain.

Let us employ different parameter values of the DBSCAN method in order to study their effect on the clustering (Qixiang et al. 2003).

Too low neighborhood radius (ϵ) and low numbers of neighbors results into too many clusters, see Fig. 3.33.

```
⇒ ClusteringComponents[img01,Method → {"DBSCAN",
   "NeighborsNumber" → 3,"NeighborhoodRadius" → 0.01}]//Colorize
```

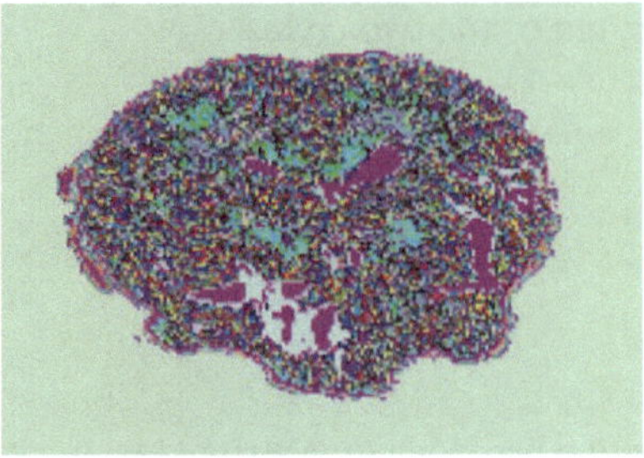

⇐

Fig. 3.33 Result of DSBCAN method with $\epsilon = 0.01$ and NeighborsNumber = 3.

Increasing the minimal number of neighbors results even in more fragmented clusters. The optimal value for minimal neighbors is about the data dimension plus one, or higher. In our case let it be 2+2= 4 up to $\epsilon = 0.1$, see Fig. 3.34.

```
⇒ ClusteringComponents[img01,Method → {"DBSCAN",
   "NeighborsNumber" → 4,"NeighborhoodRadius" → 0.1}]//Colorize
```

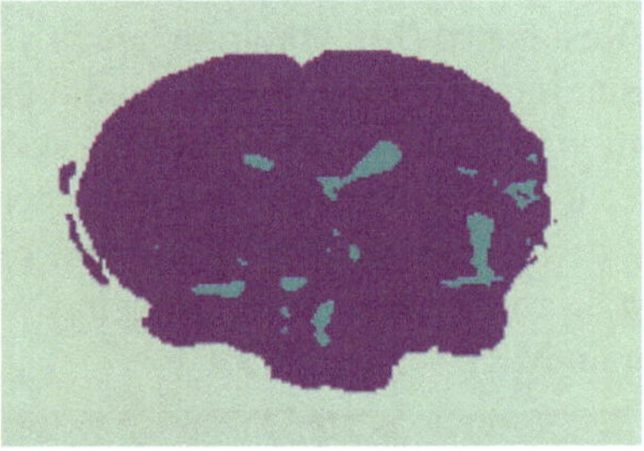

⇐

Fig. 3.34 Result of DSBCAN method with $\epsilon = 0.1$ and NeighborsNumber = 4.

Now we have got only a few clusters. However, decreasing $\epsilon = 0.053$, we can get a realistic segmentation, see Fig. 3.35.

```
⇒ ClusteringComponents[img01,Method → {"DBSCAN",
   "NeighborsNumber" → 4,"NeighborhoodRadius" → 0.053}]//Colorize
```

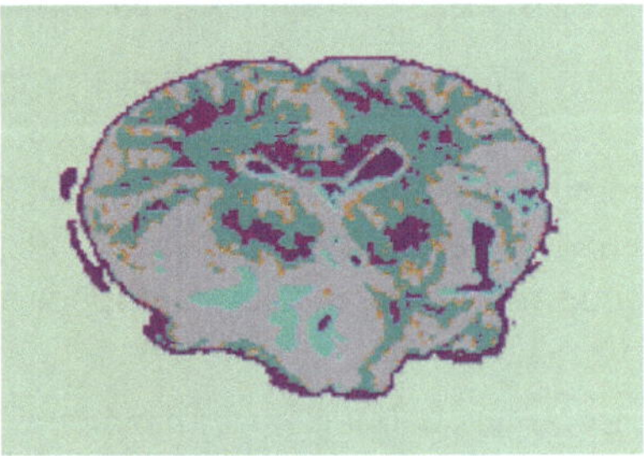

⇐

Fig. 3.35 Result of DSBCAN method with $\epsilon = 0.053$ and minimal neighbors = 4.

Undoubtedly, the proper adjustment of the parameters of the DBSCAN method is its weakest point.

3.4 Spectral Clustering

Basic theory

It makes use of the spectrum (eigenvalues) of the similarity (affinity) matrix of the data to perform dimensionality reduction before clustering in fewer dimensions. To perform a spectral clustering we need three main steps (Han et al. 2017):

1. Create a similarity graph between our N objects to cluster.
2. Compute the first k eigenvectors of its affinity matrix to define a feature vector for each object.
3. Run k-means on these features to separate objects into k classes.

Fig. 3.36 shows the general framework for spectral clustering approaches.

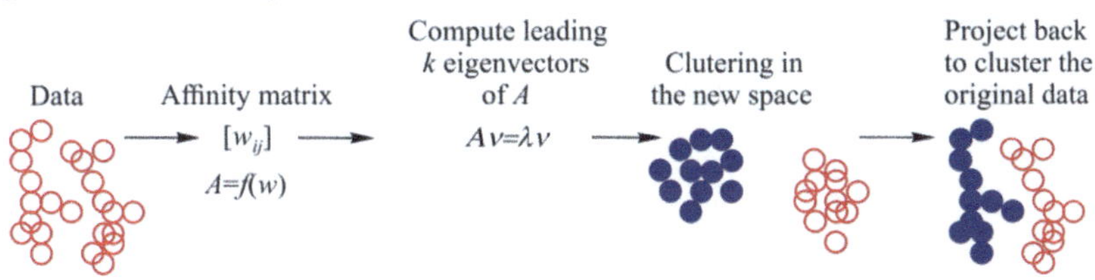

Fig. 3.36 The framework of spectral clustering approaches.

Spectral clustering is effective in high-dimensional applications such as image processing. Theoretically, it works well when certain conditions apply. Scalability, however, is a challenge. Computing eigenvectors on a large matrix is costly. Spectral clustering can be combined with other clustering methods, such as biclustering (Aoullay, 2018).

3.4.1 Nonlinear Data Set Moons

Now we are going to solve the moon clustering problem discussed in the previous section.

Loading the necessary procedures,

```
from numpy import array, matrix
from scipy.io import mmread, mmwrite
import numpy as np
```

Loading data set. Now we consider more data.

```
from sklearn.datasets import make_moons
X, y = make_moons(n_samples=1500,noise=0.05,random_state=0)
```

Providing the data for *Mathematica*,

```
mmwrite('pubi.mtx',X)
```

```
mmwrite('puba.mtx',[y])
```

Reading data for *Mathematica*

```
⇒ trainX=Import["pubi.mtx"];
⇒ clusters=Import["puba.mtx"];
```

Preparation data for visualization,

```
⇒ total=MapThread[{#1,#2}&,{trainX,First[clusters]}];
⇒ clust1=Select[total,#[[2]]==0&];
⇒ clust2=Select[total,#[[2]]==1&];
⇒ pclust1=Map[#[[1]]&,clust1];
⇒ pclust2=Map[#[[1]]&,clust2];
```

The two clusters can be seen in Fig. 3.37.

```
⇒ p0=ListPlot[{pclust1,pclust2},PlotStyle → {Pink,Green}]
```

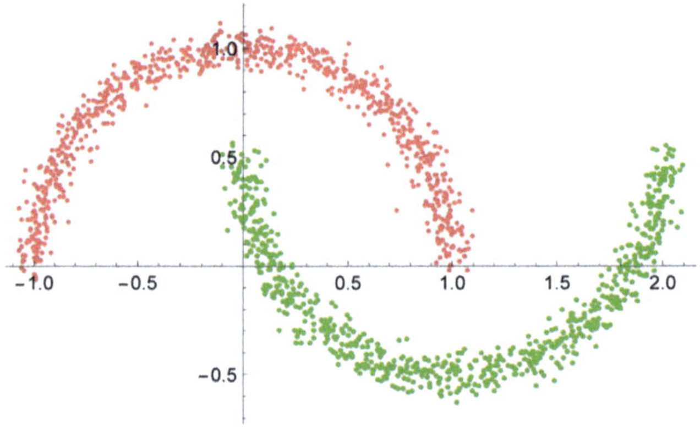

⇐

Fig. 3.37 The moon data set.

Let us try to separate the two sets via *Spectral Clustering*. In this case the number of clusters should be given.

Mathematica

In case of this method we do not need to give the number of clusters. To construct similarity graph, a parameter ε-neighborhood can be prespecified. Each vertex is connected to vertices falling inside a ball of radius ε, which is a real value that has to be tuned in order to catch the local structure of data.

```
⇒ c=FindClusters[trainX,
    Method → {"Spectral","NeighborhoodRadius" → 0.047}];
⇒ n=Length[c]
⇐ 3
⇒ ListPlot[Table[c[[i]],{i,1,n}]]
```

⇐

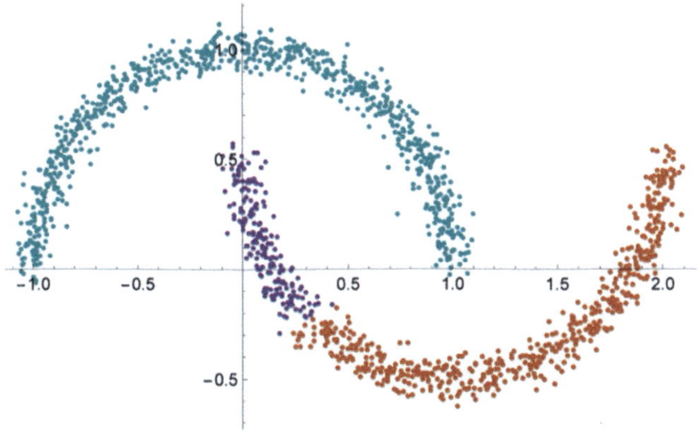

Fig. 3.38 The clustered moon data set in case $\epsilon = 0.047$.

However, only this parameter cannot control the process fully, see Fig. 3.38, since *K-Means Clustering* is involved. Therefore specifying the number of the clusters can be more efficient, see Fig. 3.39.

```
⇒ c=FindClusters[trainX,2,Method → "Spectral"];
⇒ n=Length[c]
⇐ 2
⇒ ListPlot[Table[c[[i]],{i,1,n}]]
```

⇐

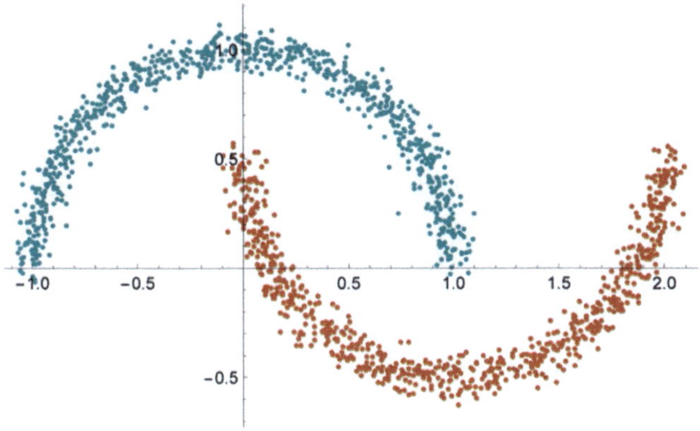

Fig. 3.39 The clustered moon data set in case of two clusters, $n = 2$.

In order to visualize the regions and the boundary of the clusters, we compute the labels of the clustered data elements.

```
⇒ index=Table[0,{i,1,1500}];
⇒ Do[If[MemberQ[c[[2]],trainX[[i]]],index[[i]]=1],{i,1,1500}];
⇒ cuki=MapThread[Flatten[{#1,#2}]&,{trainX,index}];
```

Fig. 3.40 shows the regions of the two clusters. It can be seen that the margin is maximized. Consequently, this result is more robust than that provided by DBSCAN method in the previous section.

```
⇒ Show[{p0,ListDensityPlot[cuki,Mesh → None,
   InterpolationOrder → 0,ColorFunction → "SouthwestColors"],p0}]
```

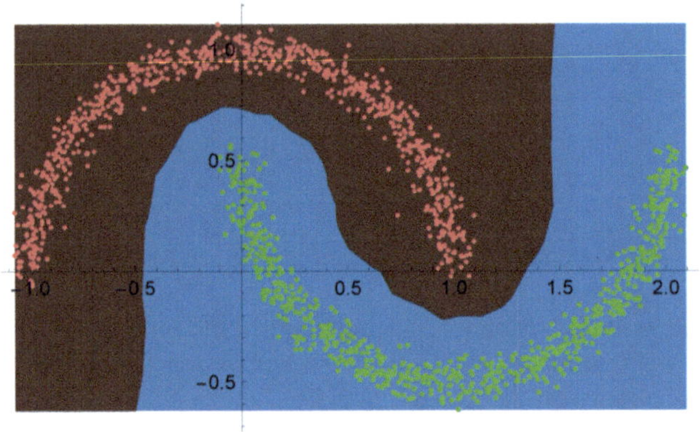

⇐

Fig. 3.40 The clustered regions of the moon data set in case of two clusters, $n = 2$.

Python

Now, let us load the corresponding *Python* clustering procedure,

```
from sklearn import cluster
spectral = cluster.SpectralClustering(n_clusters=2,
 eigen_solver='arpack',affinity="nearest_neighbors")
```

Computing eigenvalues on a large matrix frequently requires scaling (Luxburg, 2007). So scaling and then carry out prediction, we get

```
from sklearn.preprocessing import StandardScaler
X = StandardScaler().fit_transform(X)
spectral.fit(X)
y_pred = spectral.labels_.astype(np.int)
```

The norm of the misclustered elements,

```
import numpy
```

```
from numpy import linalg as LA
```

```
> LA.norm(y_pred-y)
```

```
⇐ 0.
```

3.4.2 Image Coloring

As an application of the Spectral Clustering, let us consider a coloring problem. Fig. 3.41 shows a grayscale image, which we would like to color.

```
⇒ img=
```

```
;
```

Fig. 3.41 A grayscale image.

This can be done using *Mathematica*'s built-in function, see Fig. 3.42.

```
⇒ Colorize[img]
⇐
```

Fig. 3.42 Colorized grayscale image via *Mathematica* built-in function.

However, we would like to solve the problem using *Spectral Clustering* method. Let us consider the data matrix of the image

```
⇒ imgData=ImageData[img];
⇒ imgData[[1,2]]
⇐ 0.619608
⇒ {m,n}=Dimensions[imgData]
⇐ {107,150}
```

We collect the pixel values

```
⇒ index={};value={};
⇒ Do[AppendTo[index,{i,j}];
    AppendTo[value,imgData[[i,j]]],{i,1,m},{j,1,n}]
⇒ mn=Dimensions[value]
⇐ {16050}
```

Then let us cluster them, creating a clustering function

```
⇒ c=ClusterClassify[value,Method → "Spectral"]
⇐ ClassifierFunction[ Input type: Numerical
                       Number of classes: 7 ]
```

We assign the labels to the data elements

```
⇒ cucu=Map[c[#]&,value];
```

Employing the following colors corresponding to the seven clusters,

```
⇒ colorVector={{0.1`,0.5`,0.3`},{1,0.9`,0.32`},{0.75`,0.65`,0.1`},
   {0.45`,0.53`,0.4`},{0.86`,0.33`,0.41`},{0.14`,0.84`,0.63`},
   {0.42`,0.36`,0.75`}};
⇒ szin=Map[RGBColor[#]&,colorVector]
```

⇐ 

The different colors can be assigned to the data elements according to their clusters,

```
⇒ colorData={};
⇒ Do[AppendTo[colorData,colorVector[[cucu[[i]]]]],{i,1,First[mn]}];
```

Arranging the colored pixels in matrix form of 107×150

```
⇒ cacaData=Partition[colorData,150];
```

We get the colorized form of the grayscale image, see Fig. 3.4.8

```
⇒ Image[cacaData]
```

⇐

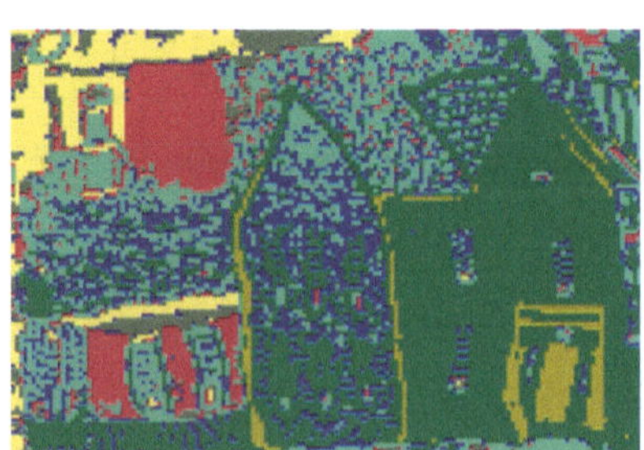

Fig. 3.43 Colorized grayscale image via Spectral Clustering.

3.5 Comparison of Clustering Methods

3.5.1 Measurement of Quality of Cluster Analysis

Silhouette analysis can be used to study the separation distance between the resulting clusters. The silhouette plot displays a measure of how close each point in one cluster is to points in the neighboring clusters and thus provides a way to assess parameters like number of clusters visually. This measure has a range of [−1, 1].

The clustering performance can be evaluated by computing,

$$s_i = \frac{b - a_i}{\max(a_i, b)},$$

where a_i is the mean distance between the *i*-th sample and all other points in the same clusters, and b is the mean distance between a sample and all the other points of the nearest cluster. The average value of the silhouette coefficient is called silhouette score. *Python* has a built in function

```
from sklearn import metrics
s=metrics.silhouette_score(X,prediction,'euclidean')
s
```

Silhouette coefficients (as these values are referred to) near 1 indicate that the sample is far away from the neighboring clusters. A value of 0 indicates that the sample is on or very close to the decision boundary between two neighboring clusters and negative values indicate that those samples might have been assigned to the wrong cluster.

3.5.2 Optimal Number of Clusters

In this example the silhouette analysis is used to choose an optimal value for the number of clusters. The silhouette plots (Fig. 3.44-3.48) show that the $n_{clusters}$ values of 3, 5 and 6 are a bad pick for the given data due to the presence of clusters with below average silhouette scores and also due to wide fluctuations in the size of the silhouette plots. Silhouette analysis is more ambivalent in deciding between 2 and 4.

Also from the thickness of the silhouette plot, the cluster size can be visualized. The silhouette plot for cluster 0 when $n_{clusters}$ is equal to 2, is bigger in size owing to the grouping of the 3 sub clusters into one big cluster. However when the $n_{clusters}$ is equal to 4, all the plots are more or less of similar thickness and hence are of similar sizes as can be also verified from the labeled scatter plot on the right.

```
from __future__ import print_function

from sklearn.datasets import make_blobs
from sklearn.cluster import KMeans
from sklearn.metrics
 import silhouette_samples,silhouette_score

import matplotlib.pyplot as plt
import matplotlib.cm as cm
import numpy as np

# Generating the sample data from make_blobs
# This particular setting has one distinct cluster and
  3 clusters placed close
# together.
X, y = make_blobs(n_samples=500,
                  n_features=2,
                  centers=4,
                  cluster_std=1,
                  center_box=(-10.0, 10.0),
                  shuffle=True,
                  random_state=1)  # For reproducibility

range_n_clusters = [2, 3, 4, 5, 6]

for n_clusters in range_n_clusters:
    # Create a subplot with 1 row and 2 columns
    fig, (ax1, ax2) = plt.subplots(1, 2)
    fig.set_size_inches(18, 7)

  # The 1st subplot is the silhouette plot
    # The silhouette coefficient can range
      from -1, 1 but in this example all
    # lie within [-0.1, 1]
    ax1.set_xlim([-0.1, 1])
    # The (n_clusters+1)*10 is for inserting blank space
      between silhouette
    # plots of individual clusters, to demarcate them clearly.
    ax1.set_ylim([0, len(X) + (n_clusters + 1) * 10])

  # Initialize the clusterer with n_clusters value and
    a random generator
    # seed of 10 for reproducibility.
    clusterer = KMeans(n_clusters=n_clusters, random_state=10)
    cluster_labels = clusterer.fit_predict(X)

    # The silhouette_score gives the average value for
      all the samples.
    # This gives a perspective into the density and
      separation of the formed
    # clusters
    silhouette_avg = silhouette_score(X, cluster_labels)
    print("For n_clusters =", n_clusters,
          "The average silhouette_score is :", silhouette_avg)
```

```
    # Compute the silhouette scores for each sample
    sample_silhouette_values =
      silhouette_samples(X, cluster_labels)

    y_lower = 10
    for i in range(n_clusters):
        # Aggregate the silhouette scores for samples
          belonging to
        # cluster i, and sort them
        ith_cluster_silhouette_values = \
            sample_silhouette_values[cluster_labels == i]

        ith_cluster_silhouette_values.sort()

        size_cluster_i=ith_cluster_silhouette_values.shape[0]
        y_upper = y_lower + size_cluster_i

        color = cm.nipy_spectral(float(i) / n_clusters)
        ax1.fill_betweenx(np.arange(y_lower, y_upper),
                 0, ith_cluster_silhouette_values,
                 facecolor=color, edgecolor=color, alpha=0.7)


        # Label the silhouette plots with their cluster
          numbers at the middle
        ax1.text(-0.05, y_lower + 0.5 × size_cluster_i,str(i))

        # Compute the new y_lower for next plot
        y_lower = y_upper + 10  # 10 for the 0 samples

    ax1.set_title(
        "The silhouette plot for the various clusters.")
    ax1.set_xlabel("The silhouette coefficient values")
    ax1.set_ylabel("Cluster label")

    # The vertical line for average silhouette score of
      all the values
    ax1.axvline(x=silhouette_avg,color="red",linestyle="--")

    ax1.set_yticks([])  # Clear the yaxis labels / ticks
    ax1.set_xticks([-0.1, 0, 0.2, 0.4, 0.6, 0.8, 1])

    # 2nd Plot showing the actual clusters formed
    colors = cm.nipy_spectral(cluster_labels.astype
             (float) / n_clusters)
    ax2.scatter(X[:,0],X[:,1],marker='.',s=30,lw=0,alpha=0.7,
                c=colors, edgecolor='k')

    # Labeling the clusters
    centers = clusterer.cluster_centers_
    # Draw white circles at cluster centers
    ax2.scatter(centers[:, 0], centers[:, 1], marker='o',
                c="white", alpha=1, s=200, edgecolor='k')


    for i, c in enumerate(centers):
        ax2.scatter(c[0], c[1], marker='$%d$' % i, alpha=1,
                    s=50, edgecolor='k')
```

```
    ax2.set_title("The visualization of the clustered data.")
    ax2.set_xlabel("Feature space for the 1st feature")
    ax2.set_ylabel("Feature space for the 2nd feature")

    plt.suptitle(("Silhouette analysis for KMeans
                   clustering on sample data "
                  "with n_clusters = %d" % n_clusters),
                 fontsize=14, fontweight='bold')

plt.show()
```

```
"For n_clusters ="
2
The average silhouette_score is :
0.7049787496083261

"For n_clusters ="
3
The average silhouette_score is :
0.5882004012129721

"For n_clusters ="
4
The average silhouette_score is :
0.6505186632729437

"For n_clusters ="
5
The average silhouette_score is :
0.56376469026194

"For n_clusters ="
6
The average silhouette_score is :
0.4504666294372765
```

⇒ `Import["M:\\CLUSTERING\\silu_01.png"]`

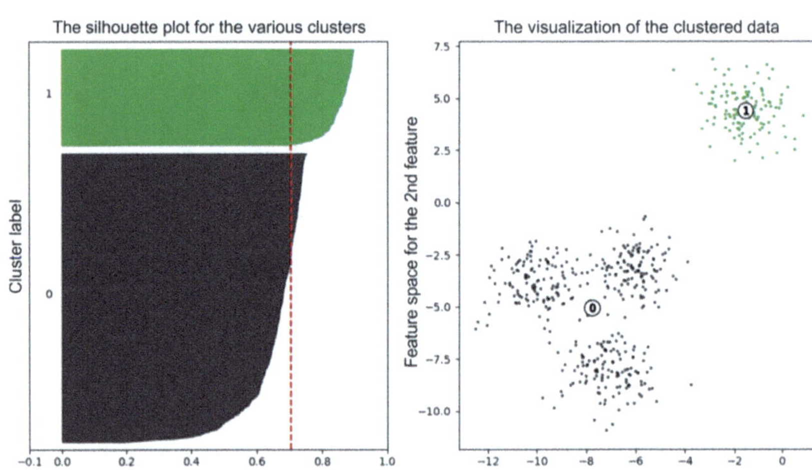

⇐

Fig. 3.44 Silhouette analysis for KMeans clustering on sample data with n_clusters=2.

⇒ `Import["M:\\CLUSTERING\\silu_02.png"]`

⇐

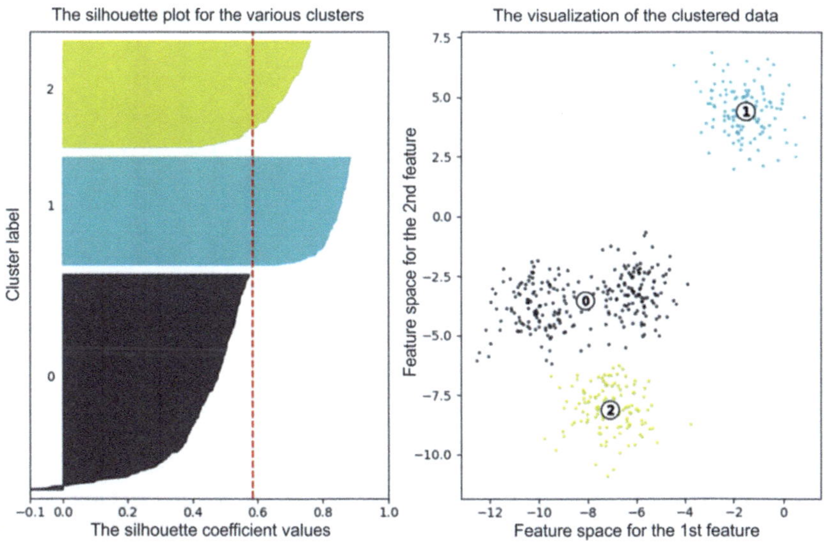

Fig. 3.45 Silhouette analysis for KMeans clustering on sample data with n_clusters=3.

⇒ Import["M:\\CLUSTERING\\silu_03.png"]

⇐

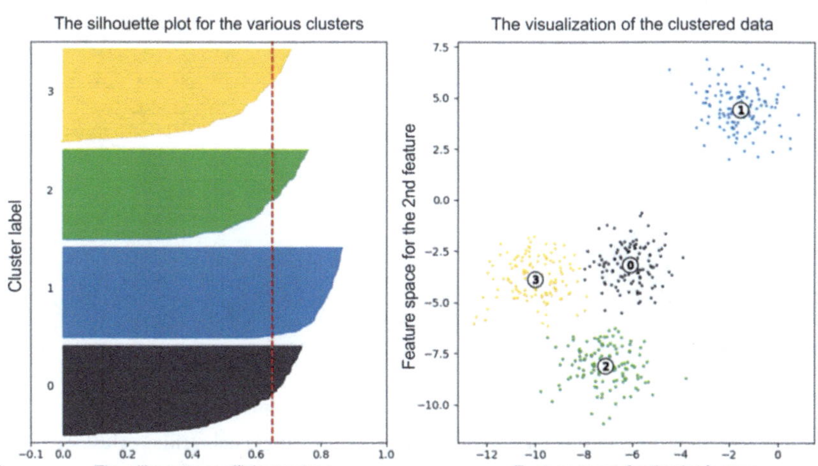

Fig. 3.46 Silhouette analysis for KMeans clustering on sample data with n_clusters=4.

⇒ Import["M:\\CLUSTERING\\silu_04.png"]

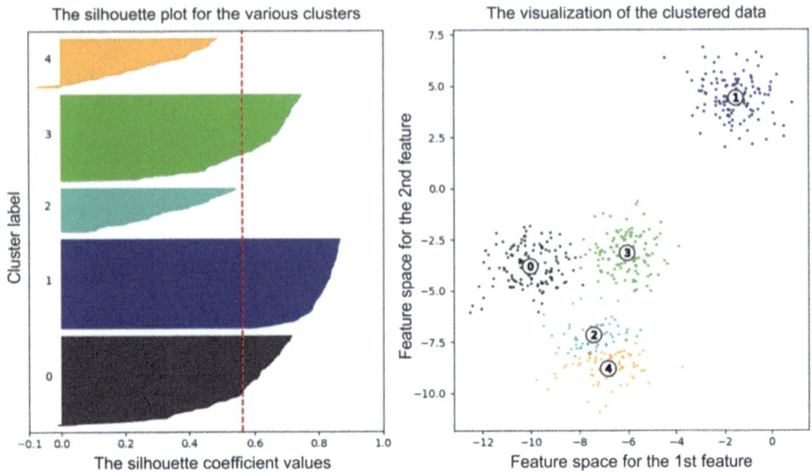

⇐

Fig. 3.47 Silhouette analysis for KMeans clustering on sample data with n_clusters=5.

```
⇒ Import["M:\\CLUSTERING\\silu_05.png"]
```

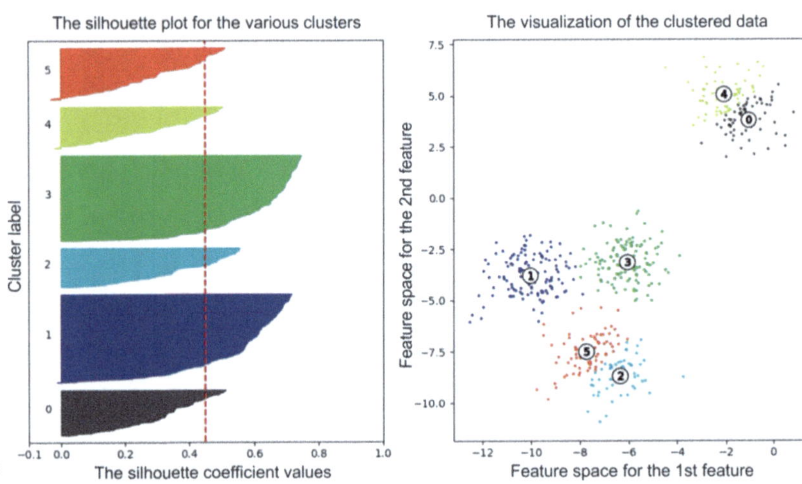

⇐

Fig. 3.48 Silhouette analysis for KMeans clustering on sample data with n_clusters=6.

3.5.3 Segmentation of Parrot Image

Let us consider the following image of a parrot,

```
⇒ img=
```

```
;
```

Fig. 3.49 Parrot image to be segmented.

The segmentation is carried out with pixel clustering employing different methods. The size of the image data

```
⇒ datap=ImageData[img];
⇒ Dimensions[datap]
⇐ {526,800,3}
```

Clustering RGB pixel vectors

```
⇒ X = Flatten[datap, 1];Dimensions[X]
⇐ {420800,3}
```

First we use KMeans method. Let us consider 3 clusters

```
⇒ XC=FindClusters[X,3,Method → "KMeans"];
```

Visualizing the clustered RGB pixel vectors in the RGB color space

```
⇒ ListPointPlot3D[XC, PlotStyle → PointSize[0.001],
  AxesLabel → {Red, Green, Blue}, BoxRatios → {1, 1, 1}]
```

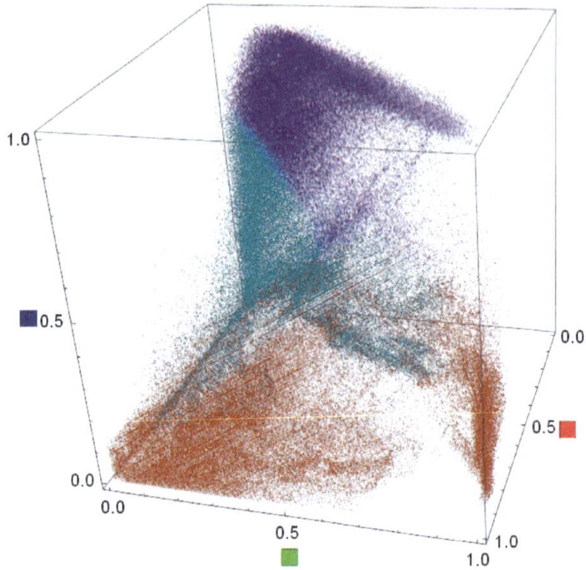

⇐

Fig. 3.50 Representation of the 3 clusters in the RGB color space.

Employing clustering for segmentation

```
⇒ AbsoluteTiming[c=ClusteringComponents[
    img,3,Method → "KMeans",PerformanceGoal → "Quality"];]
⇐ {0.835659,Null}
```

Let us colour the segmented image

```
⇒  Colorize[c, ColorFunction  →  "RoseColors",
     ColorRules  →  {0  →  Black}]
```

⇐

Fig. 3.51 The segmented image using 3 clusters.

The quality of the clustering can be determined by computing the silhouette score of the clustering. Here, we employ the built-in *Python* function.

Labels of the clusters

```
⇒ index=Map[#-1&,c]//Flatten;
```

```
⇒ index//Dimensions
⇐ {420800}
```

Number of clusters

```
⇒ nc=Max[index]+1
⇐ 3
```

Saving data for *Python*

```
⇒ Export["predi.mtx",{index}]
⇐ predi.mtx
```

```
⇒ Export["predX.mtx",X]
⇐ predX.mtx
```

Reading data into *Python*,

```
from numpy import array, matrix
from scipy.io import mmread, mmwrite
import numpy as np
```

```
from sklearn.cluster import KMeans
from sklearn.metrics import silhouette_samples,
    silhouette_score
```

```
index=mmread('predi.mtx')
```

```
X=mmread('predX.mtx')
```

Applying the built-in function of *Python*,

```
ind=index[0]
from sklearn import metrics
```

```
s=silhouette_score(X,ind,'euclidean',sample_size=20000)
s
```

```
⇐ 0.457202
```

Further KMeans clustering with different fixed number of clusters was carried out. In addition, different clustering methods as DBSCAN and Spectral clustering have been applied. In case of these latest methods, the numbers of clusters are determined automatically by the method itself. The results are summarized in Tables 3.1 and 3.2.

Table 3.1 Comparison of the Classification Methods

Method	Running Time [s]	Silhouette score	Number of clusters
KMeans	0.77	0.43	2
KMeans	0.84	0.46	3
KMeans	1.25	0.44	4
KMeans	1.55	0.46	5
DBSCAN	1.63	0.14	2 (automatic)
Spectral	2.83	0.35	2 (automatic)

Table 3.2 Segmented images

Method	Segmented image	Method	Segmented image
KMeans cluster=2		KMeans cluster=3	
KMeans cluster=4		KMeans cluster=5	
DBSCAN cluster=4	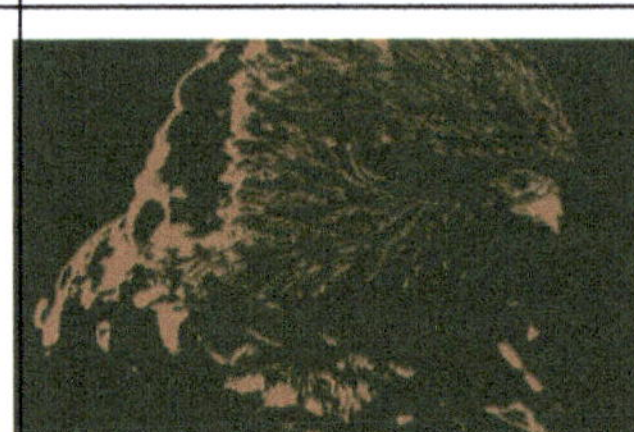	Spectral cluster=5	

Considering the quality of the appearance, the running time of the clustering process and the silhouette score, the KMeans method with 2 clusters seems to be the best choice.

References

Aoullay A (2018): Spectral Clustering for beginners https://towardsdatascience.com/spectral-clustering-for-beginners-d08b7d25b4d8

Galbiati J, Allende H, Becerra C (2009): Dynamic Image Segmentation Method Using Hierarchical Clustering, CIARP 2009: Progress in Pattern Recognition, Image Analysis, Computer Vision, and Applications pp 177-184 https://link.springer.com/content/pdf/10.1007%2F978-3-642-10268-4_21.pdf

Han J, Kamber M, Pei J. (2012): Data Mining, Concepts and Techniques, Third Edition, Elsevier, https://jakevdp.github.io/PythonDataScienceHandbook/05.11-k-means.html

James G, Witten D, Hastie T, Tibshirani R (2013): An Introduction to Statistical Learning, Springer, Berlin, Heidelberg

Luxburg v.U (2007): A Tutorial on Spectral Clustering, Statistics and Computing 17(4), 1- 32 http://www.kyb.mpg.de/fileadmin/user_upload/files/publications/attachments/ Luxburg07_tutorial_4488%5b0%5d.pdf

Müller A C, Guido S (2017): Introduction to Machine Learning with Python, O'RELLY, O'ReillyMedia,Inc., Sevastopol, USA

Okabe M, Yamada S (2018): Clustering Using Boosted Constrained k-Means Algorithm? Front.Robot. AI, 08 March 2018, https://www.frontiersin.org/articles/10.3389/frobt.2018.00018/full

Pestunova I A, Rylova S A, Berikov V B (2015): Hierarchical Clustering Algorithms for Segmentation of Multispectral Images, Optoelectronics, Instrumentation and Data Processing, 51(4), 1-10 https://www.researchgate.net/publication/283762749_Hierarchical_clustering_algorithms_for_ segmentation_of_multispectral_images

Qixiang Y, Wen G, Wei Z (2003): Color image segmentation using density-based clustering, International Conference on 3:II- 401-4 vol.2, August 2003 https : // www.researchgate.net/publication/4028066 _Color _image _segmentation _using _density - based_clustering

VanderPlas J (2016): In Depth: k-Means Clustering, In Python Data Science Handbook, O'Reilly Media, https://jakevdp.github.io/PythonDataScienceHandbook/05.11-k-means.html

Chapter 4
Regression

4.1 KNearest Neighbors Regression

Basic theory

As pointed out in the previous chapter KNearest Neighbors can be employed for clustering as well as for regression. Basically the algorithm has the following steps:

1. Consider the actual data point in the feature space,
2. Compute its distances from every data point,
3. Select the average of the KNearest neighbors, which forms the next predicted value.

In the last step one may use weighted average, where the weights are the inverse distances. This means that the points that are closer have stronger impact on the result of the prediction (Kanevski et al. 2009).

Starting Python session in *Mathematica*,

```
⇒ session=
   StartExternalSession[<|"System" → "Python",
    "Version" → "3.5.4","Executable" →
     "C:\Users\Ben\AppData\Local\Programs\Python\Python35\
       python.exe"|>]//Quiet
```

```
⇐ ExternalSessionObject[
   SummaryPanel[ ⊞ System: Python   EvaluationCount: None
                   UUID: 7b76966c2-ebb7-4ee8-bb25-02322d1456a8 ]]
```


J. Awange et al., *Hybrid Imaging and Visualization*,
https://doi.org/10.1007/978-3-030-26153-5_4

4.1.1 Analysing KNeighbors Regressor

We shall illustrate the effect of the number of neighbours on the prediction. First, we illustrate that this method produces a non-smooth predictor. The training and test data for predicting a function $y=f(X)$.

Let us download a small data set from the *Python* repository and distribute it into training and test sets,

```
import mglearn
X, y = mglearn.datasets.make_wave(n_samples=40)
from sklearn.model_selection import train_test_split
X_train, X_test, y_train,
 y_test= train_test_split(X, y,random_state=0)
```

We assign these values to *Mathematica* variables ensuring their further usage in *Mathematica*, too

```
X_train
```

```
⇐ {{0.085407},{1.85038},{-2.41397},{1.39196},{-0.35908},{0.645269},
  {2.79379},{-1.17455},{-1.1723},{0.60669},{-1.72597},{-2.06403},
  {0.14854},{-1.90905},{2.69331},{2.19706},{-2.60969},{2.70429},
  {-1.80196},{1.99466},{-1.9769},{-0.26358},{-2.6515},{-0.801829},
  {-2.16304},{-1.25263},{1.24844},{1.1054},{0.591951},{-0.752759}}
⇒ Xtrain=%;
```

```
X_test
```

```
⇐ {{-1.24713},{0.671117},{1.71106},{-2.06389},{-2.87649},
  {-1.89957},{0.554487},{2.81946},{-0.40833},{-2.7213}}
⇒ Xtest=%;
```

```
y_train
```

```
⇐ {0.697986,1.87665,-1.41502,0.779321,0.0939886,0.0352788,
  0.868933,0.0844854,0.0945257,1.00032,-1.5137,-2.47196,
  -0.527347,-1.67303,1.53708,1.49417,-0.47411,0.331226,
  -1.13455,0.754188,-2.08582,-0.986181,-1.52731,0.0975635,
  -1.12469,-0.340907,0.229562,0.254389,0.0349788,-0.448221}
⇒ ytrain=%;
```

```
y_test
```

```
⇐ {0.372991,0.217782,0.966954,-1.38774,-1.0598,
  -0.90497,0.436558,0.778964,-0.541146,-0.956521}
⇒ ytest=%;
```

Mathematica

Let us visualize these two sets. The training set

```
⇒ training=Map[Flatten[#,1]&,Transpose[Join[{Xtrain,ytrain}]]];
```

The testing set

```
⇒ test=Map[Flatten[#,1]&,Transpose[Join[{Xtest,ytest}]]];
```

Fig. 4.1 shows the data set for training (green disk) and for test (red squares),

```
⇒ p0=ListPlot[{training,test},PlotStyle → {Green,Red},Frame → True,
    Axes → None,PlotMarkers → {Automatic,Large},AspectRatio → 1]
```

⇐

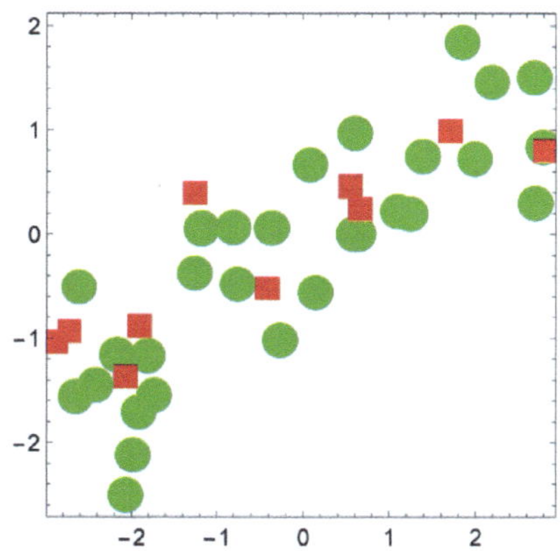

Fig. 4.1 Data points of the training (green) and the test (red) sets in case of KNearest Neighbors Regression.

Employing $k = 3$ neighbors, the prediction function f is,

```
⇒ datatraining=Flatten[Xtrain] → Flatten[ytrain];
⇒ f=Predict[datatraining,Method → {"NearestNeighbors",
    "NeighborsNumber" → 3},PerformanceGoal → "Quality"]
⇐ PredictorFunction[ Input type: Numerical
                     Method: NearestNeighbors ]
```

The predictor function can be seen in Fig 4.2, which shows clearly that the predictor function provides local, stepwise approximation.

```
⇒ Show[{p0,Plot[f[x],{x,-3,3},PlotStyle → {Thin,Blue}]}]
```

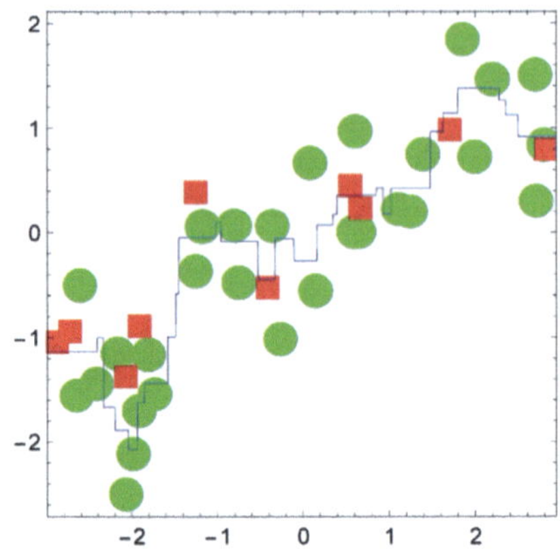

⇐

Fig. 4.2 The predictor function with KNearest Neighbors Regression using $k = 3$ neighbors.

The prediction error on the test set can be visualized, see Fig. 4.3.

```
⇒ ftest=PredictorMeasurements[f,Flatten[Xtest] → ytest];
⇒ ftest["ComparisonPlot"]
```

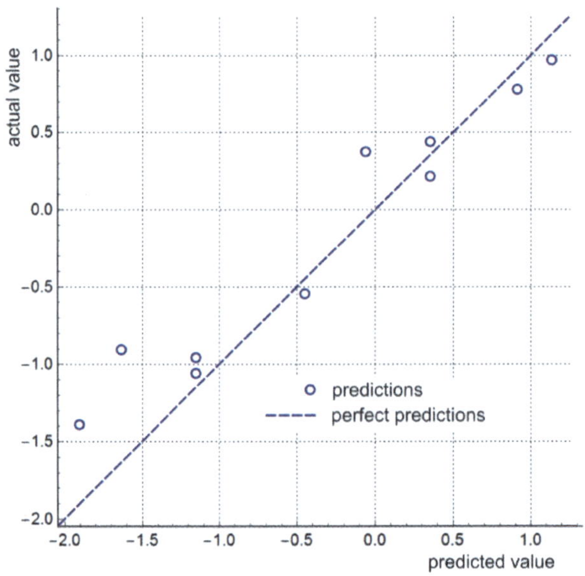

⇐

Fig. 4.3 Prediction error on the test set using $k = 3$ neighbors.

It is usual to measure the effectivity of the predictor function $f(X)$, computing the score of the predictor,

$$\text{Score}(f) = 1 - \frac{\sum_{i=1}^{n}\left(y_i - f(X_i)\right)^2}{\sum_{i=1}^{n}\left(y_i - \text{Mean}(y_i)\right)^2}.$$

The closer the score to 1.00 the better the prediction. This can be computed for the training as well as for the test set.

In our case for the test set,

```
⇒ u=Total[Map[#²&,ytest-Map[f[#]&,Flatten[Xtest]]]]
⇐ 1.08677
```

and

```
⇒ v=Total[Map[#²&,ytest-Mean[ytest]]]
⇐ 6.56328
```

Therefore the score for the test set

```
⇒ 1-u/v
⇐ 0.834417
```

similarly for the training set,

```
⇒ u=Total[Map[#²&,ytrain-Map[f[#]&,Flatten[Xtrain]]]]
⇐ 6.32668
```

and

```
⇒ v=Total[Map[#²&,ytrain-Mean[ytrain]]]
⇐ 35.0381
```

Then the score for the training set

```
⇒ 1-u/v
⇐ 0.819434
```

The two scores are close to each other, so the prediction is proper.

Python

Loading the procedure, and using prediction with $k = 3$ on the training set (Singh 2018),

```
from sklearn.neighbors import KNeighborsRegressor
reg=KNeighborsRegressor(n_neighbors=3).fit(X_train,y_train)
prediction=reg.predict(X_test)
prediction
```

```
⇐ {-0.0539654,0.35686,1.13672,-1.89416,-1.13881,
   -1.63113,0.35686,0.912414,-0.446804,-1.13881}
```

The results are the same as those of the *Mathematica*

```
⇒ Map[f[#] &, Flatten[Xtest]]
⇐ {-0.0539654,0.35686,1.13672,-1.89416,-1.13881,
   -1.63113,0.35686,0.912414,-0.446804,-1.13881}
```

Now let us compute the scores (Chen and Shah, 2018).

```
print("Training set score: {:.2f}".
 format(reg.score(X_train, y_train)))
print("Test set score: {:.2f}".
 format(reg.score(X_test, y_test)))
```

```
Training set score: 0.82
Test set score: 0.83
```

Now, we shall study the effect of the number of neighbors on the prediction quality. Here we compute the test and training scores for neighbors $k = 1$ and $k = 9$.

Mathematica

First using $k = 1$, we get a simple stepwise approximation, where every data point belongs to the predictor function, see Fig. 4.4

```
⇒ f=Predict[datatraining,Method → {"NearestNeighbors",
   "NeighborsNumber" → 1},PerformanceGoal → "Quality"]
⇐ PredictorFunction[ Input type: Numerical
                     Method: NearestNeighbors ]
```

```
⇒ Show[{p0,Plot[f[x],{x,-3,3},PlotStyle → {Thin,Blue}]}]
⇐
```

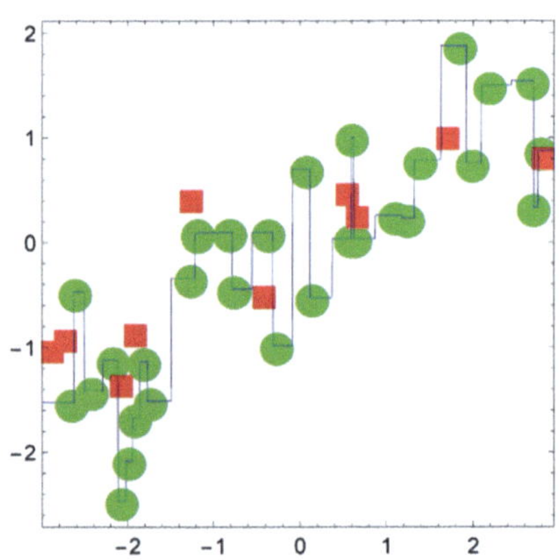

Fig. 4.4 The prediction with KNearest Neighbors Regression with $k = 1$ neighbor.

In this case there is no error on the training set, so its score is 1.

The score on the test set

```
⇒ u=Total[Map[#²&,ytest-Map[f[#]&,Flatten[Xtest]]]]
⇐ 4.25309
```

and

```
⇒ v=Total[Map[#²&,ytest-Mean[ytest]]]
⇐ 6.56328
```

Then

```
⇒ 1-u/v
⇐ 0.351987
```

Now for $k = 9$,

```
⇒ f=Predict[datatraining,Method → {"NearestNeighbors",
    "NeighborsNumber" → 9},PerformanceGoal → "Quality"]
⇐ PredictorFunction[ Input type: Numerical
                     Method: NearestNeighbors ]
```

The result can be seen in Fig. 4.5, which indicates that the increase of k value can somewhat "smoothen" the predictor function.

```
⇒ Show[{p0,Plot[f[x],{x,-3,3},PlotStyle → {Thin,Blue}]}]
```

⇐

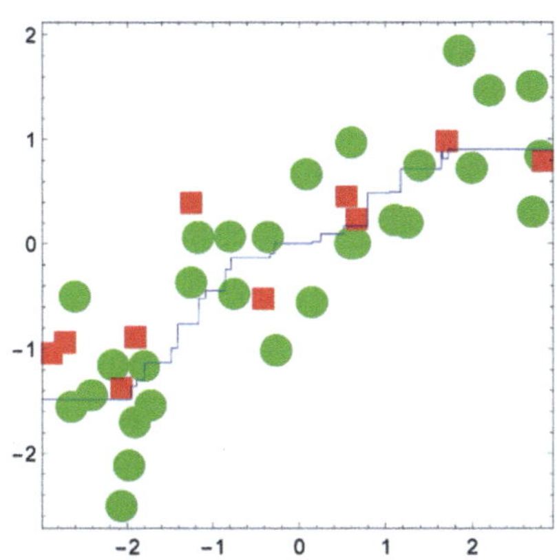

Fig. 4.5 The prediction with KNearest Neighbors using $k = 9$ neighbor.

Score on the test set

```
⇒ u=Total[Map[#²&,ytest-Map[f[#]&,Flatten[Xtest]]]]
⇐ 2.27008
```

and

```
⇒ v=Total[Map[#²&,ytest-Mean[ytest]]]
⇐ 6.56328
```

Then

```
⇒ 1-u/v
⇐ 0.654124
```

similarly for the training set,

```
⇒ u=Total[Map[#^2&,ytrain-Map[f[#]&,Flatten[Xtrain]]]]
⇐ 9.51641
```

and

```
⇒ v=Total[Map[#^2&,ytrain-Mean[ytrain]]]
⇐ 35.0381
```

Then the score for the training set

```
⇒ 1-u/v
⇐ 0.728399
```

So we have a slight *over fitting* again, however, the values of the scores are lower than in the case of $k = 3$.

Python

Employing *Python*, we get the same results. For $k = 1$,

```
reg=KNeighborsRegressor(n_neighbors=1).fit(X_train,y_train)
```

```
print("Training set score: {:.2f}".
 format(reg.score(X_train, y_train)))
print("Test set score: {:.2f}".
 format(reg.score(X_test, y_test)))
```

```
Training set score: 1.00
Test set score: 0.35
```

For $k = 9$,

```
reg=KNeighborsRegressor(n_neighbors=9).fit(X_train,y_train)
```

```
print("Training set score: {:.2f}".
 format(reg.score(X_train, y_train)))
print("Test set score: {:.2f}".
 format(reg.score(X_test, y_test)))
```

```
Training set score: 0.73
Test set score: 0.65
```

4.1.2 Surface Reconstruction

Let us consider a simple surface reconstruction problem.

We consider a general second order surface, see Fig.4.6.

```
⇒ t[x_,y_]:=ax+by+cx^2+dxy+ey^2
```

Employing specific parameter values, we create synthetic data set,

```
⇒ p1=Plot3D[t[x,y]/.{a → 1,b->11,c → 3,d → -4,
    e → -1},{x,-10,10},{y,-10,10},Mesh → 8,ColorFunction → Hue,
    MeshShading → {{Yellow,Orange},{Pink,Red}},BoxRatios → {1,1,1}]
```

⇐

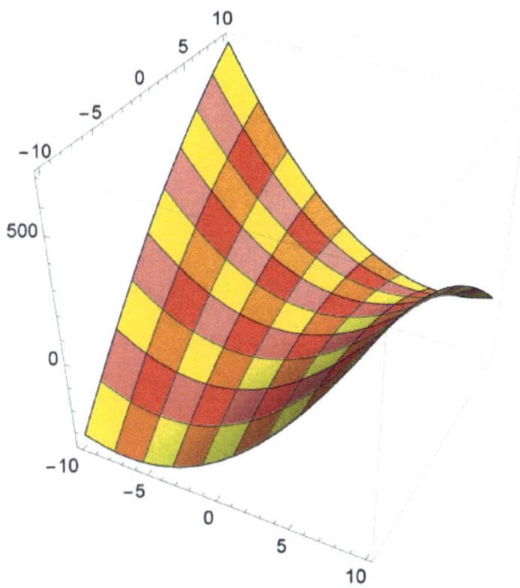

Fig. 4.6 Original surface.

Let us employ measured function data values on a [−10,10]×[−10, 10] region of [x, y] raster with normal random error $N(0,30)$. The error histogram can be seen in Fig. 4.7.

```
⇒ Lerror=RandomVariate[NormalDistribution[0,30],{21,21}];
⇒ Histogram[Flatten[Lerror,1]]
```

⇐

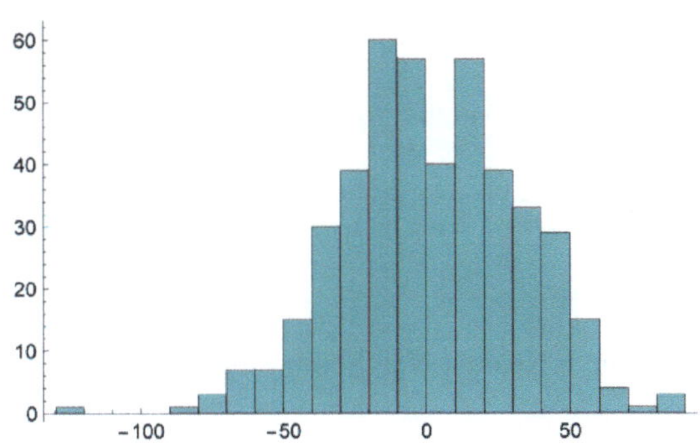

Fig. 4.7 Histogram of the random error.

Then the "measured" data are

```
⇒ dataPoints=Flatten[Table[{i,j,Lerror[[i+11,j+11]]+t[i ,j]/.
    { → 1,b → 11,c → 3,d → -4,e → -1}},{i,-10,10},{j,-10,10}],1];
```

Fig. 4.8 shows the generated noisy data points with the original surface,

```
⇒ p2=ListPointPlot3D[dataPoints,BoxRatios → {1, 1, 1},
    PlotStyle → Blue];
⇒ Show[{p1,p2}]
```

⇐

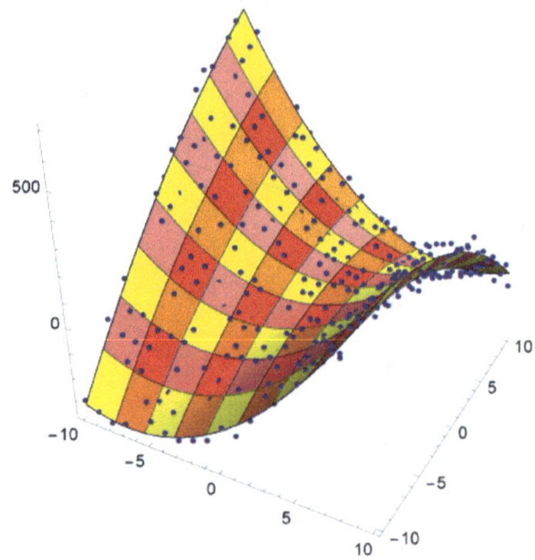

Fig. 4.8 Data values.

Now, let us reconstruct the original surface on bases of these noisy data points. In order to employ linear model we transform the variables,

```
⇒ transform[r_,s_]={r,s,r s,r^2,s^2}
⇐ {r,s,rs,r^2,s^2}
```

Instead of using a nonlinear model with two variables, therefore, we now have a linear model with five variables. Using *Mathematica*, the training data set is

```
⇒ ySet=Table[transform[i,j],{i,-10,10},{j,-10,10}];
⇐ trainingdata=
    MapThread[#1 → #2[[3]]&,{Flatten[ySet,1],dataPoints}];

⇒ trainingdata[[1]]
⇐ {-10,-10,100,100,100} → -303.836
```

Let us now employ the KNearest Neighbors Regression with $k = 3$,

```
⇒ c=Predict[trainingdata,
    Method → {"NearestNeighbors","NeighborsNumber" → 3}]
⇐ PredictorFunction[ Input type: Mixed (Number:5)
                     Method: NearestNeighbors ]
```

This function can be applied to the original data set, for example

```
⇒ c[transform[1,1]]
```

⇐ 20.1957

Fig. 4.9 shows the approximated surface,

```
⇒ p3=Plot3D[c[transform[x,y]],{x,-10,10},{y,-10,10},
   BoxRatios → {1, 1, 1}]
```

⇐

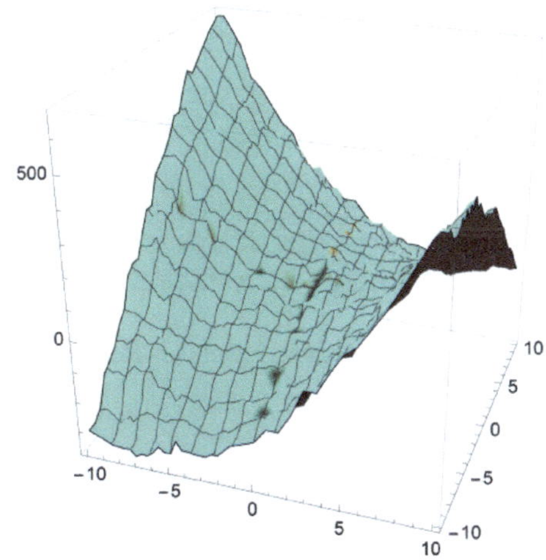

Fig. 4.9 KNearest Neighbors approximation.

We can compare the approximation and the original surface, see Fig. 4.10.

```
⇒ Show[{p1,p3}]
```

⇐

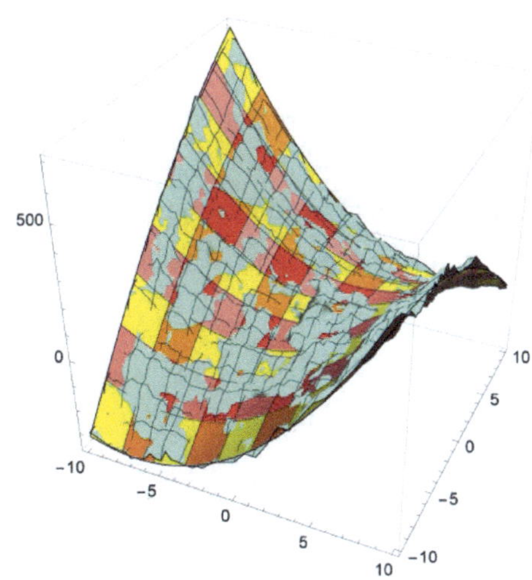

Fig. 4.10 KNearest Neighbors Regression approximation and the original surface.

The histogram of the error distribution of the approximation can be seen in Fig. 4.11.

```
⇒ error=Table[(t[x,y]/.{a → 1,b → 11,c → 3,d → -4,e → -1})-
    c[transform[x,y]],{x,-10,10},{y,-10,10}];
⇒ Histogram[Flatten[Error,1]]
```

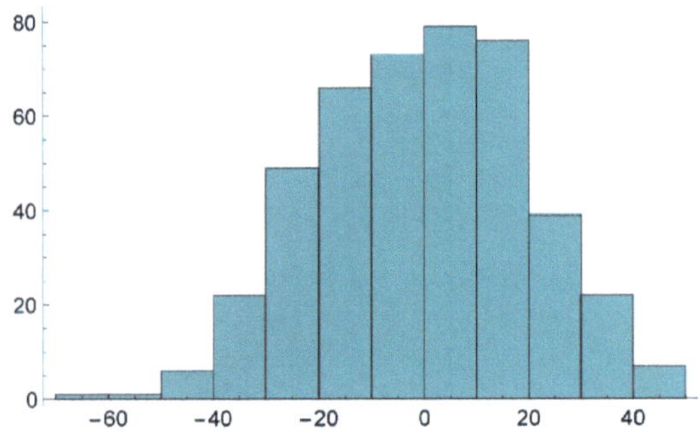

⇐

Fig. 4.11 Approximation error of the KNearest Neighbors Regression method in case of $k = 3$.

Remark

The surface can be smoothed by convolution.

4.2 Linear Regression Models

Basic theory

Let us assume that we have observations $\{y_1, y_2, \ldots, y_n\}$ from a random variable Y which we want to predict, based on the observed values $\{\{x_{11}, x_{21}, \cdots, x_{n,1}\}, \{x_{12}, x_{22}, \cdots, x_{n,2}\}, \cdots, \{x_{1p}, x_{2p}, \cdots, x_{n,p}\}\}$, from p independent explanatory random variables $\{X_1, X_2, \ldots, X_p\}$. Our initial assumption is that the X_i are independent. The linear model in matrix form is represented as:

$$y = \beta X + \epsilon .$$

Then one needs to minimize the following objective,

$$\frac{1}{n}\sum_{i=1}^{n}\epsilon_i^2 = \frac{1}{n}\sum_{i=1}^{n}(y_i - \beta X_i)^2$$

to find the coefficients β. This technique is the *Ordinary Least Squares* (OLS) and can be derived from the *maximum likelihood method* assuming that ϵ_i has a normal distribution with zero mean. Using OLS, difficulties begin to emerge when the assumption of independence no longer applies. Employing *Machine Learning* vocabulary, over learning may occur and it can be cured via *regularization*. We attempt to deal with the problems by introducing a penalty component to the OLS objective function. The idea is to penalize the regression for using too many correlated explanatory variables, as follows (Zelesny, 2016):

$$\frac{1}{n}\sum_{i=1}^{n}(y_i - \beta X_i)^2 + \lambda P_\alpha$$

with

$$P_\alpha = \frac{1}{n}\sum_{j=1}^{p}\frac{1}{2}(1-\alpha)\beta_j^{\,2} + \alpha|\beta_j|$$

In case of $\lambda = 0$ we have *Ordinary Least Square*, otherwise, $\lambda > 0$, and

if $\alpha = 0$ then Ridge Regression (L2 -norm regularization)
if $\alpha = 1$ then Lasso Regression (L1 - norm regularization)
if $\alpha \in (0,1)$ then Elastic Net Regression (L2 and L1 norm mixed regularization)

Practically and generally, it means that we try to restrict the gradient of the fitted function in order to avoid overlearning.

4.2.1 Small Data Set

Let us employ the earlier small data set, but now with 60 elements, splitting into training and test set,

```
import mglearn
from sklearn.model_selection import train_test_split
X, y = mglearn.datasets.make_wave(n_samples=60)
X_train, X_test, y_train,
 y_test= train_test_split(X, y,random_state=0)
```

Let us save these sets for *Mathematica*, too

```
X_train
```

```
⇐ {{2.45592},{-1.89087},{2.19706},{-1.90905},{-2.60969},{2.81751},
   {-2.7213},{-2.79367},{0.5874},{-0.40833},{-2.46904},{0.0854066},
   {-1.89957},{-2.06403},{-1.97686},{-1.17455},{1.6508},{0.671117},
   {2.63699},{0.60669},{-1.72597},{1.71106},{-2.41397},{0.148539},
   {0.975134},{0.280262},{-1.8241},{1.1054},{2.70429},{1.99466},
   {-1.12973},{-0.26358},{-2.6515},{-0.801829},{-1.17232},
   {-2.16304},{-1.25263},{1.24844},{-0.359085},{2.53125},
   {0.591951},{-0.752759},{2.36896},{0.120408},{-1.44732}}
```

```
⇒ Xtrain=%;
```

```
X_test
```

```
⇐ {{-1.80196},{1.85038},{-1.04802},{0.554487},{2.81946},
   {1.39196},{2.79379},{-2.72864},{-2.26777},{-1.24713},
   {-2.06389},{-2.87649},{0.645269},{-0.0289385},{2.69331}}
⇒ Xtest=%;
```

```
> y_train
```

```
⇐ {1.19813,-1.58832,1.2032,-1.27708,-0.83685,0.731414,-0.08137,
   -0.447131,0.652134,-1.25636,-1.72409,0.979232,-0.945758,
   -2.37365,-1.07676,-0.326911,0.250925,0.962506,1.01581,0.292629,
   -1.30838,0.799001,-0.913907,0.261344,0.400123,0.239382,
   -1.54665,0.812638,0.500159,1.07384,-0.025655,-0.385754,
   -0.701173,0.13369,-0.196128,-0.752409,-0.746469,0.449716,
   -0.934165,0.826142,1.17396,-1.18073,1.28948,0.77614,-0.751506}
⇒ ytrain=%;
```

```
> y_test
```

```
⇐ {-1.11948,0.381098,-0.491317,0.658232,1.39516,
   0.137729,0.950818,-1.03732,-1.71132,-0.178514,
   -1.32036,-0.486472,-0.721426,-0.323096,0.709459}
⇒ ytest=%;
```

Then let us visualize these two sets, see Fig. 4.12.

```
⇒ training=Map[Flatten[#,1]&,Transpose[Join[{Xtrain,ytrain}]]];
⇒ test=Map[Flatten[#,1]&,Transpose[Join[{Xtest,ytest}]]];
⇐ p0=ListPlot[{training,test},PlotStyle → {Green,Red},Frame → True,
   Axes → None,PlotMarkers → {Automatic,Large},AspectRatio → 1]
```

⇐

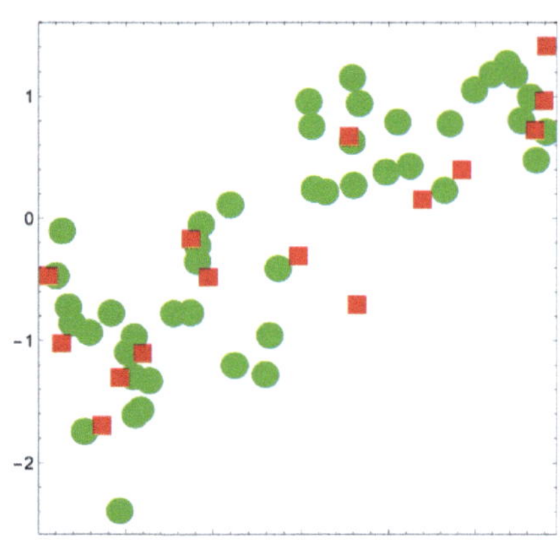

Fig. 4.12 Data points of the training (green) and the test (red) sets.

Mathematica

To fit a regression line we use *Mathematica*'s simple `LinearModelFit` function

```
⇒ lm=LinearModelFit[training,x,x];
⇒ model=lm//Normal
⇐ -0.0171112+0.441537 x
```

Both forms can be employed for computing the approximating values of the model parameters,

```
⇒ lm[0]
⇐ -0.0171112
```

or

```
⇒ model/.x → 0
⇐ -0.0171112
```

Alternatively, one may also employ *Mathematica*'s general prediction model

```
⇒ datatraining=Flatten[Xtrain] → Flatten[ytrain];
⇒ plm=Predict[datatraining,Method → "LinearRegression"]
⇐ PredictorFunction[ Input type: Numerical
                     Method: LinearRegression ]
```

The parameters of the linear model are:

the intercept,

```
⇒ plm[0]
⇐ -0.0193265
```

and the coefficient,

```
⇒ plm[1]-plm[0]
⇐ 0.431938
```

More simple way, as a pure function,

```
⇒ PredictorInformation[plm,"Function"]
⇐ -0.0193265+0.431938 #1&
```

Python

Employing *Python*, we load the model and carry out the prediction (VanderPlas, 2016),

```
from sklearn.linear_model import LinearRegression
lr=LinearRegression().fit(X_train,y_train))
```

The parameters are

```
print(lr.coef_)
```

⇐ [0.44153666]

```
print(lr.intercept_)
```

⇐ -0.01711124414733381

Predicted output from input for example at $x = 0.1$,

```
po=lr.predict(0.1)
po
```

⇐ {0.0270424}

The output can be assigned to a *Mathematica* variable,

```
⇒ s=%;
```

and verified

```
⇒ lm[0.1]==First[s]
⇐ True
```

Let us visualize the linear model with the data points, see in Fig. 4.13

```
⇒ Show[{p0,Plot[model,{x,-3,3}]}]
```

⇐

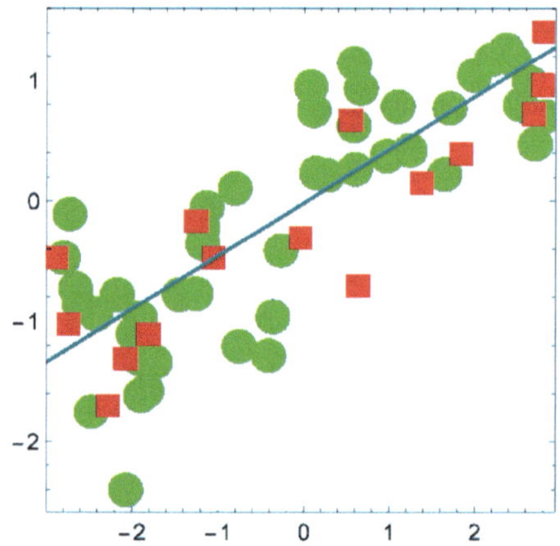

Fig. 4.13 Fitted linear model using ordinary least square.

Comparing the efficiency (score) on the training and test set,

```
print("Training set score: {:.2f}".
 format(reg.score(X_train, y_train)))
print("Test set score: {:.2f}".
 format(reg.score(X_test, y_test))))
```

```
Training set score: 0.66
Test set score: 0.69
```

Compare the data with the predicted values and look at the standard deviation, see Fig. 4.14.

```
⇒ Show[{p0,Plot[{
     plm[x],
     plm[x]+StandardDeviation[plm[x,"Distribution"]],
     plm[x]-StandardDeviation[plm[x,"Distribution"]]},
    {x,-3,3},
    PlotStyle → {Blue,Gray,Gray},
    Filling → {2 → {3}},
    Exclusions → False,
    PerformanceGoal → "Speed",
     PlotLegends → {"Prediction","Confidence Interval"}]}]
```

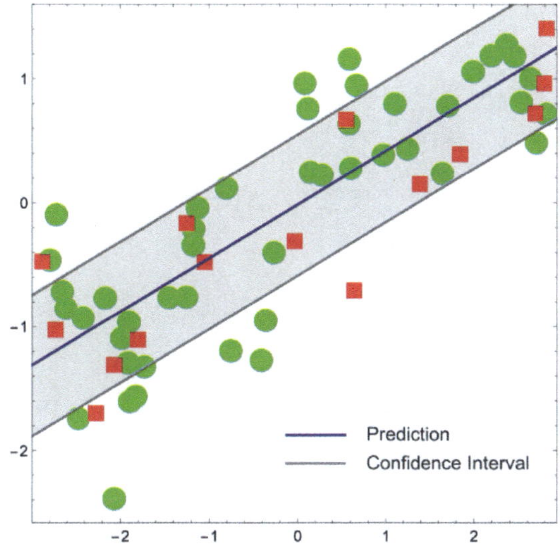

⇐

Fig. 4.14 Fitted linear model with the standard error as confidence interval.

Statistical analysis of the model parameters, see, Table 4.1

```
⇒ lm[{"ParameterConfidenceIntervalTable"},ConfidenceLevel->.99][[1]]
```

Table 4.1 Statistics of the linear model

	Estimate	Standard Error	Confidence Interval
1	−0.0171112	0.0854872	{−0.247508, 0.213285}
x	0.4415370	0.0484137	{ 0.311057, 0.572016}

4.2.2 Generalization of the Ordinary Least Square (OLS)

When the dimension of the feature vector is high, then OLS may perform badly. Now, we load the Boston houses data from the *Python* repository (Sargent and Stachurski 2017),

```
import mglearn
import numpy as np
from sklearn.model_selection import train_test_split
X, y = mglearn.datasets.load_extended_boston()
X_train, X_test, y_train, y_test=
  train_test_split(X, y,random_state=0)
```

```
X_train
```

```
⇐ {{0.00207947,0.22,0.197947,
   OutputSizeLimit`Skeleton[99] ,0.4526,0.213116},
  OutputSizeLimit`Skeleton[377] ,{ OutputSizeLimit`Skeleton[1]}}
 large output | show less | show more | show all | set size limit...
```

Save these data for *Mathematica*,

```
⇒ Xtrain=%;
⇒ Dimensions[Xtrain]
⇐ {379,104}
```

The training set

```
np.savetxt('M:\\daTaXtrain.txt',X_train,fmt='%.5e')
```

```
⇒ XtrainT=Import["M:\\daTaXtrain.txt","Table"];
⇒ Dimensions[XtrainT]
⇐ {379,104}[
```

```
np.savetxt('M:\\daTaytrain.txt',y_train,fmt='%.5e')
```

```
⇒ ytrain=Import["M:\\daTaytrain.txt","Table"];
```

The test set

```
np.savetxt('M:\\daTaXtest.txt',X_test,fmt='%.5e')
```

```
⇒ Xtest=Import["M:\\daTaXtest.txt","Table"];
```

```
np.savetxt('M:\\daTaytest.txt',y_test,fmt='%.5e')
```

```
ytest=Import["M:\\daTaytest.txt","Table"];
```

In this case the number of the features (the dimension of the input vector)

```
⇒ Length[XtrainT[[1]]]
⇐ 104
```

and we have

```
⇒ Length[XtrainT]
⇐ 379
```

elements in the training and

```
⇒ Length[Xtest]
⇐ 127
```

elements in the test set.

Python

Let us try to employ OLS regression

```
from sklearn.linear_model import LinearRegression
lr=LinearRegression().fit(X_train,y_train)
```

Now the coefficient vector contains 104 elements

```
lr.coef_
```

```
⇐ {-402.752,-50.071,-133.317,-12.0021,-12.7107,28.3053,54.492,
   -51.7339,25.2603,36.4991,-10.1039,-19.6289,-21.3677,14.6474,
   2895.05,1510.27,117.995,-26.5658,31.2488,-31.4464,45.2536,
   ⋮
   143.234,-15.6741,-14.9732,-28.613,-31.252,24.5648,-17.8048,
   4.03508,1.71068,34.4735,11.2186,1.14302,3.73717,31.3846}
⇒ coeffOLS=%;
```

```
lr.intercept_
```

```
⇐ 31.6452
⇐ interceptOLS=%;
```

```
print("Training set score: {:.2f}".
 format(lr.score(X_train, y_train)))
print("Test set score: {:.2f}".
 format(lr.score(X_test, y_test)))
```

```
Training set score: 0.95
Test set score: 0.61
```

Mathematica

Let us solve the problem employing *Mathematica*, too.

```
⇒ training=Map[Flatten[#,1]&,Transpose[Join[{XtrainT,ytrain}]]];
⇒ test=Map[Flatten[#,1]&,Transpose[Join[{Xtest,ytest}]]];
```

```
⇒ Dimensions[XtrainT]
⇐ {379,104}
```

Dummy vector for the input variables of the model,

```
⇒ xx=Table[Subscript[x, i],{i,1,104}];
⇒ lm=LinearModelFit[training,xx,xx];
```

```
MessageTemplate[ LinearModelFit , rank ,

The rank of the design matrix 104 is less than the number of terms 105 in the
model. The model and results based upon it may contain significant numerical error.

2 , 64 , 22 , 17 155778988823209065 , Local ]
```

The design matrix is rank deficient, therefore the problem is ill-conditioned.

```
⇒ model=lm//Normal
```

$$\Leftarrow \{31.6407-402.622x_1-50.0681x_2-133.298x_3+18.1894x_4-12.7088x_5+ 28.3039x_6+54.4875x_7-51.726x_8+25.2658x_9+36.4825x_{10}-10.1001x_{11}- 19.6246x_{12}-21.3651x_{13}+14.6492x_{14}+2894.85x_{15}+1509.1x_{16}+118.007x_{17}- \ldots 28.6087x_{94}-31.2507x_{95}+24.5645x_{96}-17.7924x_{97}+4.03026x_{98}+1.71074x_{99}+ 34.4687x_{100}+11.2162x_{101}+1.1435x_{102}+3.73481x_{103}+31.3843x_{104}\}$$

This problem can be cured by employing regularization. There are L2 (ridge regression), L1 (Lasso regression) regularization and the combination of these two (Elastic net regression).

4.2.3 Ridge Regression

In order to introduce L2 regularization, we employ ridge regression with *Python,* which employs regularization parameter *alpha*, which has default value $alpha = 1$. This *alpha* corresponds to λ used in the definition in Section 4.2, see Basic Theory.

Python

```
from sklearn.linear_model import Ridge
ridge=Ridge().fit(X_train,y_train)
```

The model coefficients,

```
ridge.coef_
```

```
⇐ {-1.45195,-1.55626,-1.4585,-0.128253,-0.0852752,8.3226,0.254138,
   -4.94126,3.90318,-1.05389,-1.58274,1.02803,-4.0136,0.436959,
   0.00361746,-0.874004,0.745133,-1.48861,-1.67522,-1.44622,
   ⋮
   1.82299,1.97807,1.82834,-7.13688,1.10266,1.42156,-1.31292,
   -6.76976,1.82608,-2.35757,0.0345818,1.19002,-6.29852,10.3651}
⇒ coeffRidge=%;
```

```python
ridge.intercept_
```

```
⇐ 21.4172
⇒ interceptRidge=%;
```

Comparing the score of the training and test set

```python
print("Training set score: {:.2f}".
 format(ridge.score(X_train, y_train)))
print("Test set score: {:.2f}".
 format(ridge.score(X_test, y_test)))
```

```
Training set score: 0.89
Test set score: 0.75
```

There is a considerable improvement, since the training score and test score are closer to each other, than earlier, see Section 4.2.2.

Let us increase the regularization parameter. Employing ridge regression for *alpha* = 10.

```python
from sklearn.linear_model import Ridge
ridge10=Ridge(alpha=10).fit(X_train,y_train)
```

```python
ridge10.coef_
```

```
⇐ {-0.81369,0.647624,-0.809069,0.311198,-0.685926,4.38593,-0.147446,
   -2.44272,0.846353,-1.14693,-2.33239,1.0695,-3.98476,-0.596589,
   0.00293204,-0.522603,0.140364,-0.647905,-0.76348,-0.749429,
   ⋮
   0.627694,0.235255,1.33926,-1.88592,0.392763,-0.442962,-0.451262,
   -2.05805,-1.49556,-1.76754,-2.03784,1.08037,-3.94939,0.142122}
⇒ coeffRidge10=%;
```

```python
ridge10.intercept_
```

```
⇐ 25.0084
⇒ interceptRidge10=%;
```

```python
print("Training set score: {:.2f}".
 format(ridge10.score(X_train, y_train)))
print("Test set score: {:.2f}".
 format(ridge10.score(X_test, y_test)))
```

```
Training set score: 0.79
```

```
Test set score: 0.64
```

Let us display the coefficients of the different regressors, OLS, Ridge Regression with *alpha* =1 and *alpha* = 10, see Fig. 4.15,

```
⇒ p1=ListPlot[{coeffOLS,coeffRidge,coeffRidge10},
   PlotStyle → {Green,Red,Blue},Frame → True,Axes → None,
   PlotMarkers → {Automatic},AspectRatio → 0.7,
   PlotRange → {-30,30}]
```

⇐

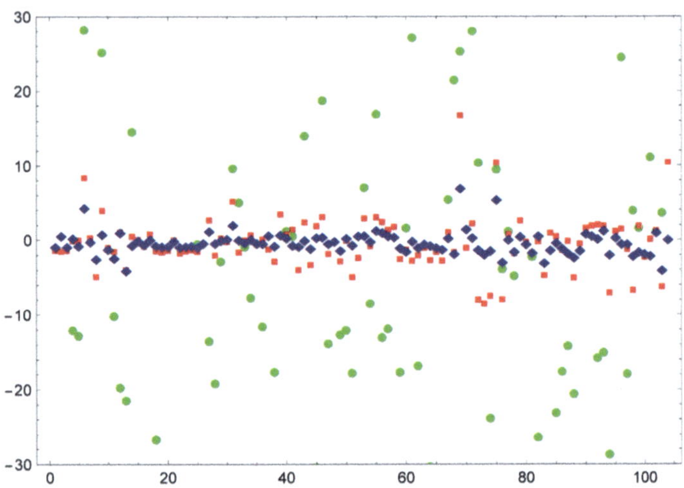

Fig. 4.15 The coefficients of the different regressors, OLS (green), Ridge with alpha = 1 (red) and Ridge with alpha = 10 (blues).

It can be clearly seen, that Ridge Regression keeps the value of coefficients in the narrowest interval.

In *Mathematica* for ridge regression one can use L2 regularization directly, however the magnitudes of regularization parameters in *Python* and *Mathematica* are not consistence. To get similar result, one should use much higher value in *Mathematica* than in *Python*.

Mathematica

The training set

```
⇒ ytrain=Flatten[ytrain];
⇒ datatraining=MapThread[#1 → #2&,{XtrainT,ytrain}];
⇒ pr=Predict[datatraining,
   Method → {"LinearRegression","L2Regularization" → 900}]
⇐ PredictorFunction[
   Input type: NumericalVector (Length: 104)
   Method: LinearRegression ]
```

The score of the training and test sets can be computed step by step.

```
⇒ u=Total[Map[#²&,ytrain-Map[pr[#]&,XtrainT]]]
⇐ 7775.53
```

and

```
⇒ v=Total[Map[#²&,ytrain-Mean[ytrain]]]
⇐ 32331.8
```

Then the score for the training set

```
⇒ 1-u/v
⇐ 0.759509
```

for test set,

```
⇒ ytest=Flatten[ytest];
⇒ u=Total[Map[#²&,ytest-Map[pr[#]&,Xtest]]]
⇐ 3972.83
```

and

```
⇒ v=Total[Map[#²&,ytest-Mean[ytest]]]
⇐ 10375.8
```

Then the score for the test set

```
⇒ 1-u/v
⇐ 0.617105
```

These scores are close to that of the *Python* (*alpha* =10), namely (0.79 and 0.64).

The model parameters,

```
⇒ prI=PredictorInformation[pr,"Function"];
⇒ coeffT=Sort[Map[Reverse[#]&,Partition[Most[
    Cases[prI,_?NumberQ,Infinity]],2]],#1[[1]]<#2[[1]]&];
⇒ coeff=Drop[coeffT,{1,1}]
⇐ {{1,-0.81873},{2,0.386539},{3,-0.412599},{4,0.2587},{5,-0.466501},
    {6,4.71972},{7,-0.2042},{8,-0.674577},{9,0.12333},{10,-0.263636},
    {11,-1.32976},{12,0.672956},{13,-2.15085},{14,-0.652742},
     ⋮
    {96,-0.317175},{97,-0.0253841},{98,-1.32674},{99,-1.09009},
    {100,-0.7421},{101,-2.02978},{102,0.568134},{103,-2.19772}}
```

The intersection values,

```
⇒ intersect=Total[coeffT[[1]]]
⇐ 25.3369
```

It means *Mathematica* with L2 regularization = 900 gives similar result as *Python* with *alpha* =10. However some coefficients are fairly different, see Fig. 4.16.

```
⇒ p1P=ListPlot[{coeffRidge10},PlotStyle → {Blue},Frame → True,
   Axes → None,PlotMarkers → {Automatic},AspectRatio → 0.7,
   PlotRange → {-10,10}];
```

```
⇒ Show[{p1P,ListPlot[Drop[coeff,{1,1}],PlotStyle → Brown,
   Frame → True,Axes → None,PlotMarkers → "▼"
   AspectRatio → 0.7,PlotRange → {-10,10}]}]
```

⇐

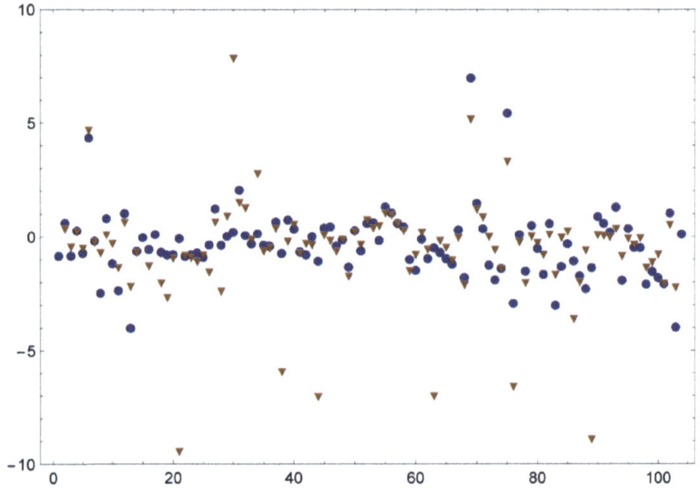

Fig. 4.16 The coefficients of the *Mathematica*'s Ridge = 900 Brown (▼) and *Python* Ridge = 10 Blue (●).

4.2.4 *Lasso Regression*

We have seen that *Lasso Regressions* employs L1 regularization. Now let us employ it using *Python* code,

Python

```
from sklearn.linear_model import Lasso
```

```
lasso=Lasso().fit(X_train,y_train)
```

```
print("Training set score: {:.2f}".
 format(lasso.score(X_train, y_train)))
print("Test set score: {:.2f}".
 format(lasso.score(X_test, y_test)))
```

```
Training set score: 0.29
Test set score: 0.21
```

The result is quite bad. Let us decrease from *alpha* = 1 down to *alpha* = 0.01 and simultaneously set *max_iter* = 10000.

```
lasso=Lasso(alpha=0.01,max_iter=100000).fit(X_train,y_train))
```

```
print("Training set score: {:.2f}".
 format(lasso.score(X_train, y_train)))
print("Test set score: {:.2f}".
 format(lasso.score(X_test, y_test)))
```

```
Training set score: 0.90
Test set score: 0.77
```

The coefficients of the linear model are

```
lasso.coef_
```

```
⇐ {0.,0.,0.,0.,0.,0.,0.,-1.30684,10.9466,0.,0.,0.,-0.315327,0.,0.,
   0.,0.,0.,0.,0.,0.,-8.91227,0.,0.,0.,0.,2.09534,0.,0.,0.,0.,0.,
   0.,0.,0.,0.,0.,0.,0.,0.,0.,0.,0.,-4.0825,0.,6.66788,0.,0.,0.,
   0.,-4.40105,-2.1055,3.7759,0.,4.3838,0.,0.,0.18662,0.,-1.17594,
   -4.29006,0.,0.,-2.21607,0.,-1.8821,0.,0.,29.7405,-2.08168,0.,
   -12.0013,-11.1479,-11.6601,13.1169,-11.1537,0.,0.,3.47563,0.,
   0.,0.,-8.5663,0.,0.,0.,0.,-7.38806,0.,0.,1.00758,0.,0.,-7.5907,
   1.61169,0.,0.,-17.4313,0.,0.,0.,0.28683,-8.15767,17.4965}
```

```
⇒ c=%;
```

The number of the non-zero coefficients is,

```
⇒ Select[c,#!=0.&]//Length
⇐ 33
```

Let us compare the magnitudes of the coefficients of the Ridge regression model with *alpha* = 10 and Lasso regression model above, see Fig. 4.17

```
⇒ Show[{p1P,ListPlot[Drop[c,{1,1}],PlotStyle → Brown,Frame → True,
   Axes → None,PlotMarkers → "\[FilledDownTriangle]",
   AspectRatio → 0.7,PlotRange → {-20,20}]}]
```

⇐ 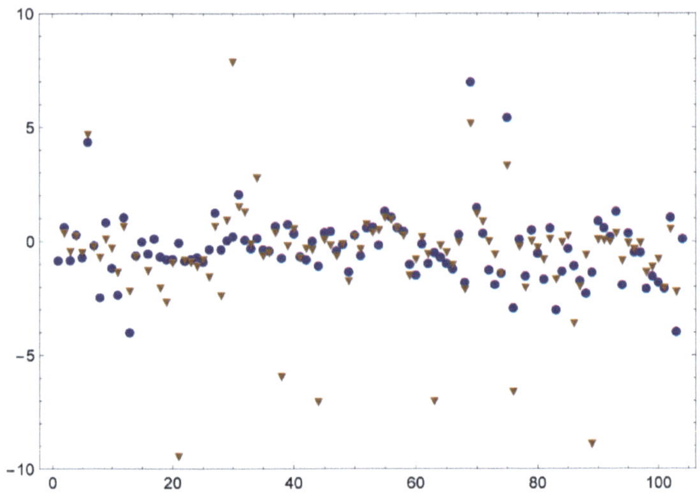

Fig. 4.17 The coefficients of the Python Lasso Regression, Brown (▼) and and Ridge regression with *alpha* = 10 Blue (●).

One can see that Lasso Regression (L1 regularization) eliminates most of the coefficients. From 104 coefficients only 33 which are non-zero, see Fig.4.17. However the absolute value of these non-zero coefficients are higher than in case of Ridge regression with alpha = 10 regression.

Mathematica

Now let us try *Mathematica* with direct *L1* regularization corresponding to Lasso regression,

```
⇒ pL=Predict[datatraining, Method → {"LinearRegression",
  "L1Regularization" → 10},PerformanceGoal → "Quality",
  TimeGoal → 60]
⇐ PredictorFunction[
  Input type: NumericalVector (Length: 104)
  Method: LinearRegression ]
```

Score for training set,

```
⇒ u=Total[Map[#^2&,ytrain-Map[pL[#]&,XtrainT]]]
⇐ 5059.41
```

and

```
⇒ v=Total[Map[#^2&,ytrain-Mean[ytrain]]]
⇐ 32331.8
```

Then the score for the train set

```
⇒ 1-u/v
⇐ 0.843516
```

for test set,

```
⇒ ytest=Flatten[ytest];
⇒ u=Total[Map[#²&,ytest-Map[pL[#]&,Xtest]]]
⇐ 2569.9
```

and

```
⇒ v=Total[Map[#²&,ytest-Mean[ytest]]]
⇐ 10375.8
```

Then the score for the test set

```
⇒ 1-u/v
⇐ 0.710339
```

These values are somewhat worse than those of *Python* (0.9 vs. 0.84 and 0.77 vs. 0.71).

Let us see the coefficients.

```
⇒ f=Information[pL,"Function"]
⇐ 23.5728+3.18472#6-0.304832#8+0.000527495#9-0.191188#11-
   2.00723#13+53.8063#17-19.0575#21+0.00832541#27+0.16029#31-
   11.2083#38-5.24492#44+0.000561929#45+0.745482#55+0.58123#57-
   0.0175075#60-0.133877#61-0.425707#63-0.0774075#66-4.20621#68+
   27.9787#69-3.19814#73-11.3455#74+3.25528#75-37.2008#76-
   1.6242#78-1.91935#83-0.478329#84-0.842248#86-1.34194#87-
   0.603847#88+0.000478878#90+0.000330873#91+0.000244214#92+
   0.236073#93-2.79445#98-0.285828#103+3.25104#104&

⇒ coeffT=Sort[Map[Reverse[#]&,
   Partition[Most[Cases[f,_?NumberQ,Infinity]],2]],
   #1[[1]]<#2[[1]]&];
⇒ coeff=Drop[coeffT,{1,1}]
⇐ {{-19.0575,17},{-11.3455,73},{-11.2083,31},{-5.24492,38},
   {-4.20621,66},{-3.19814,69},{-2.79445,93},{-2.00723,11},
   {-1.91935,78},{-1.6242,76},{-1.34194,86},{-0.842248,84},
   {-0.603847,87},{-0.478329,83},{-0.425707,61},{-0.304832,6},
   {-0.285828,98},{-0.191188,9},{-0.133877,60},{-0.0774075,63},
   {-0.0175075,57},{0.000244214,91},{0.000330873,90},
   {0.000478878,88},{0.000527495,8},{0.000561929,44},
   {0.00832541,21},{0.16029,27},{0.236073,92},{0.58123,55},
   {0.745482,45},{3.18472,23.5728},{3.25104,103},{3.25528,74},
   {27.9787,68},{53.8063,13}}

⇒ intersect=Total[coeffT[[1]]]
⇐ 37.7992
```

Now we have somewhat more (37 >33) non-zero coefficients

```
⇒ Length[coeffT]
```

```
⇐ 37
```

But a few of them are quite small,

```
⇒ Select[coeffT,Abs[#[[2]]]>0.001&]//Length
⇐ 36
```

4.2.5 Elastic Net Regression

Mathematica

The Elastic-Net Regression is a combination of the Ridge and the Lasso Regression. In *Mathematica* we employ L2 and L1 regularization simultaneously.

```
⇒ prL=Predict[datatraining,Method → {"LinearRegression",
    "L2Regularization" → 900,"L1Regularization" → 0.06,
    "OptimizationMethod" → "OrthantWiseQuasiNewton"},
    PerformanceGoal → "Quality"]
⇐ PredictorFunction[
    Input type: NumericalVector (Length: 104)
    Method: LinearRegression                    ]
```

Score for training set,

```
⇒ u=Total[Map[#^2&,ytrain-Map[prL[#]&,XtrainT]]]
⇐ 8806.36
```

and

```
⇒ v=Total[Map[#^2&,ytrain-Mean[ytrain]]]
⇐ 32331.8
```

Then the score for the train set

```
⇒ 1-u/v
⇐ 0.727625
```

for test set,

```
⇒ ytest=Flatten[ytest];
⇒ u=Total[Map[#^2&,ytest-Map[prL[#]&,Xtest]]]
⇐ 3731.26
```

and

```
⇒ v=Total[Map[#^2&,ytest-Mean[ytest]]]
⇐ 10375.8
```

Then the score for the test set

```
⇒ 1-u/v
```

```
⇐ 0.640388
```

and the coefficients

```
⇒ f=PredictorInformation[prL,"Function"]
⇒ coeff=Select[Norm/@Most[Cases[f,_?NumberQ,Infinity]],
    Head[#]==Real&]
⇐ {25.5063,0.0439521,2.22507×10^-308,0.848023,0.26222,0.43368,0.26408,
    0.398289,4.52831,0.260777,0.599958,0.051432,0.268433,1.16701,
    0.565828,1.88054,0.604317,518.553,1.29274,14.1373,1.74552,
     ⋮
    0.0387509,0.257668,0.694943,0.0950763,0.347543,0.0770126,
    1.12994,1.00366,0.640493,1.85389,0.478793,1.90365,1.05287}
```

Python

Now we employ *Elastic Net Regression* in *Python*, namely a special variant of it using cross validation technique (ElasticNetCV).

```
import mglearn
X, y = mglearn.datasets.load_extended_boston()
from sklearn.model_selection import train_test_split
X_train, X_test, y_train,
 y_test= train_test_split(X,y,random_state=0)
from sklearn.linear_model import ElasticNetCV
regr=ElasticNetCV(cv=5,max_iter=10000,
 random_state=0).fit(X_train,y_train)
regr.coef_
```

```
⇐ {-1.54547,-1.77251,-1.27239,0.,0.,9.04924,0.,-5.29649,5.07603,
    -0.310374,-0.695627,0.514146,-3.55111,0.,0.,-0.532312,0.161229,
    -1.44735,-1.1207,-1.5445,0.,-2.01746,-1.56299,-1.34282,-1.24166,
     ⋮
    0.,1.29412,1.92353,2.27019,1.3436,-8.46854,0.657055,1.29418,
    -1.13382,-8.34995,1.80211,-1.95155,0.,1.2551,-6.57106,12.9617}
```

The non-zero coefficients are,

```
⇒ coeffs=%;
⇒ Select[coeffs,Abs[#]>0.001&]//Length
⇐ 86
```

```
regr.intercept_
```

```
⇐ 20.5038
```

The scores are

```
print("Training set score: {:.2f}".
 format(regr.score(X_train, y_train)))
print("Test set score: {:.2f}".
 format(regr.score(X_test, y_test)))
```

```
Training set score: 0.89
Test set score: 0.76
```

In this case *Python* seems to be better than *Mathematica*, and for *Python* Lasso provides somewhat higher scores than Elastic Net.

4.2.6 Stitching Images

In image processing, a frequent problem is the *geometrical transformation* of images. To do that, one needs a transformation model with known parameters. Often we have to determine the parameters of the model from the coordinates of corresponding points. As an illustration, let us consider two images, that we want to stitch together, see Fig. 4.18.

⇒ i1= ; i2= ;

Fig. 4.18. The images to be stitched

Let us find corresponding points of the two images, see Fig. 4.19. Using *Mathematica* built in function

```
⇒ z = ImageCorrespondingPoints[i1, i2,MaxFeatures → 15]
  ⇐ {{{227.711,146.822},{258.595,167.002},{201.128,98.6972},
     {195.17,157.643},{197.124,132.347},{230.874,129.912},
     {264.313,140.17},{248.203,132.659},{246.535,97.1216},
     {176.331,112.359}},
     {{92.2366,154.393},{122.395,169.871},{62.6775,110.729},
     {59.3557,169.953},{60.2872,142.867},{93.5135,137.431},
     {125.084,144.26},{110.181,138.568},{107.16,104.919},
     {35.2501,122.952}}}
```

```
⇒ GraphicsGrid[{{HighlightImage[i1,z],HighlightImage[i2,z]}}]
```

⇐ 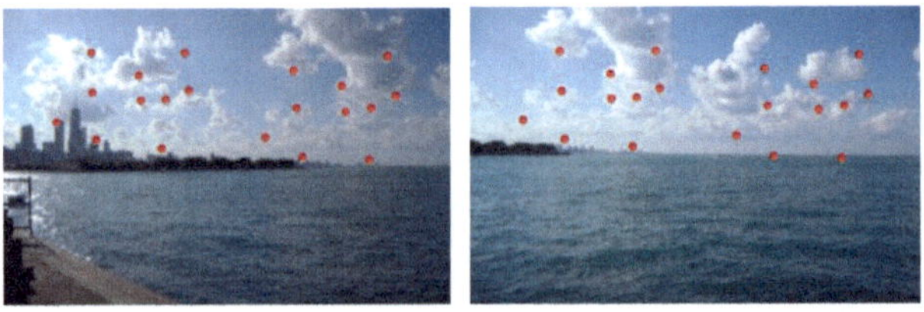

Fig. 4.19. Corresponding points of the two images

We can compute the parameters of a linear geometric transformation on the basis of the corresponding points. The problem is overdetermined and we use the *RANSAC regression* method.

Mathematica

We can find the linear geometric transformation between the two images,

```
⇒ tf=FindGeometricTransform[z[[1]],z[[2]],Method → "RANSAC"][[2]]
```

$$\Leftarrow \texttt{TransformationFunction}\left[\begin{pmatrix} 0.540548 & 0.00736743 & 152.465 \\ -0.135901 & 0.932523 & -0.20188 \\ -0.0016377 & 0.000289486 & 1. \end{pmatrix}\right]$$

Then the second image can be transformed accordingly, see Fig. 4.20.

```
⇒ {w,h}=ImageDimensions[i2];
⇒ i2t=ImagePerspectiveTransformation[i2,tf,DataRange → Full,
    PlotRange → {{0,First@tf[{w,0}]},{0,h}}]
```

⇐

Fig. 4.20. The transformed second image

Then the first and the transformed second image can be stitched, see Fig. 4.21.

```
⇒ ImageCompose[i2t, {i1,1}, Round@({w,h}/2)]
```

⇐

Fig. 4.21. The stitched images

Python

We save the coordinates of the corresponding points of the two images for *Python*

```
⇒ Export["cuki.mtx",z[[2]]]
⇐ cuki.mtx
⇒ Export["caki.mtx",z[[1]]]
```

⇐ caki.mtx

Now the linear geometric transformation can be loaded from the `scikit-image` package (`skimage`)

```
import math
import numpy as np
from skimage import transform as tf
```

Loading the coordinates of the corresponding points

```
src = mmread('cuki.mtx')
dst = mmread('caki.mtx')
```

Computing the parameters of the transformation and checking it,

```
tform3=tf.ProjectiveTransform()
tform3.estimate(src, dst)
```

⇐ True

The parameters are,

```
tform3.params
```

⇐ {0.637741,0.0541895,160.326},
{-0.123859,1.08258,-8.99475},{-0.0016613,0.000519999,1.07164}}

⇐ jj=%;

Then the transformation matrix,

⇒ jj//MatrixForm

$$\Leftarrow \begin{pmatrix} 0.637741 & 0.0541895 & 160.326 \\ -0.123859 & 1.08258 & -8.99475 \\ -0.0016613 & 0.000519999 & 1.07164 \end{pmatrix}$$

The corresponding function

⇒ tlft=LinearFractionalTransform[jj]

$$\Leftarrow \text{TransformationFunction}\left[\begin{pmatrix} 0.637741 & 0.0541895 & 160.326 \\ -0.123859 & 1.08258 & -8.99475 \\ -0.0016613 & 0.000519999 & 1.07164 \end{pmatrix}\right]$$

Now we can transform the second image, see Fig. 4.22.

⇒ i2t=ImagePerspectiveTransformation[i2,tlft,DataRange → Full,
PlotRange → {{0,First@tlft[{w,0}]},{0,h}}]

⇐ 

Fig. 4.22. The transformed second image via Python

The transformed second image will be stitched, see Fig. 4.23.

⇒ `ImageCompose[i2t, {i1,1}, Round@({w,h}/2)]`

⇐ 

Fig. 4.23. The stitched images via Python

It can be seen that the values of the transformation matrices computed by the two codes are different. Even after normalization of the *Python* result we get different matrix,

⇒ `%//MatrixForm`

⇐

$$\begin{pmatrix} 0.595107 & 0.0505669 & 149.608 \\ 0.115579 & 1.01021 & 8.39344 \\ 0.00155024 & 0.000485237 & 1. \end{pmatrix}$$

In this case *Mathematica* provides somewhat better quality, see Fig. 4.24

⇒ `{img1,img2}`

⇐

Fig. 4.24. The stitched images via *Mathematica* (left) and *Python* (right)

4.3 Non-Linear Regression Models

Basic theory

Let us assume that we have observations $\{y_1, y_2, \ldots, y_n\}$ from a random variable Y which we are interested in predicting, based on the observed values $\{\{x_{11}, x_{21}, \cdots, x_{n,1}\}, \{x_{12}, x_{22}, \cdots, x_{n,2}\}, \cdots, \{x_{1p}, x_{2p}, \cdots, x_{n,p}\}\}$, from p independent explanatory random variables $\{X_1, X_2, \ldots, X_p\}$. Our initial assumption is that the X_i variables are independent. Now we are looking for the approximation of the data points in a form:

$$Y = \mathcal{F}\,(\beta\, X) + \epsilon$$

where $\mathcal{F}(\,)$ is a nonlinear function of its argument, namely of the parameter β as well as X.

However, sometimes we have nonlinearity only in the X variable, like

$$Y = \beta\mathcal{F}(X) + \epsilon$$

In this case the model is nonlinear, though it is a linear regression problem, since the model is a linear combination of nonlinear functions, it has linear coefficients β, which should be estimated.

The first problem is much more difficult than the second one and many times leads to nonlinear global optimization task for the unknown parameters or coefficents. Here we discuss only two types of the models, namely polynomial regression and a more general one, the support vector regression, which can belong to the first as well as to the second category, depending on the type of kernels used.

4.3.1 Polynomial Regression

This type of regression belongs to the second category, since the function is a linear combination of basic functions, which are polynomials. Let us consider algebraic polynomials.

Generating synthetic data set in *Python*,

```
import numpy as np
X = np.sort(5 * np.random.rand(40, 1), axis=0)
y = np.sin(X).ravel()
y[::5] += 3 * (0.5 - np.random.rand(8))
```

Saving it for *Mathematica*

```
np.savetxt('M:\\dX.txt',X,fmt='%.5e')
```

```
np.savetxt('M:\\yX.txt',y,fmt='%.5e')
```

Importing the data into *Mathematica*,

```
⇒ Xt=Import["M:\\dX.txt","Table"];
⇒ yt=Import["M:\\yX.txt","Table"];
```

Mathematica

Let us visualize the data, see Fig. 4.25.

```
⇒ data=Transpose[{Flatten[Xt],Flatten[yt]}];
⇒ p0=ListPlot[data,Frame → True,PlotStyle → Tiny,Axes → None]
```

⇐

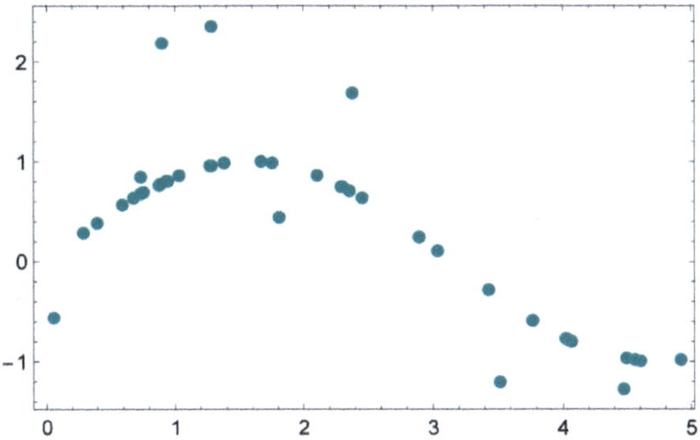

Fig. 4.25 Data points for the polynomial regression problem generated in *Python*.

One may fit first order model,

```
⇒ fitL=Fit[data,{1,x},x]
⇐ 1.3023 -0.449007 x
```

The second order model,

```
⇒ fitP2=Fit[data,1,x,x²,x]
⇐ 0.356727 + 0.722628x - 0.235229x²
```

While the model using third order polynomials,

```
⇒ fitP3=Fit[data,1,x,x²,x³,x]
⇐ -0.617649 + 2.83055x - 1.29828x² + 0.144008x³
```

Fig. 4.26 shows the different polynomials fitted to the data set and the third order model seems to be the best (see red line) one.

```
⇒ Show[{p0,Plot[fitP3,{x,0,40},PlotStyle → {Thin,Red}],
   Plot[fitP2,{x,0,40},PlotStyle → {Thin,Black}],
   Plot[fitL,{x,0,40},PlotStyle → {Thin,Green}]}]//Quiet
```

⇐

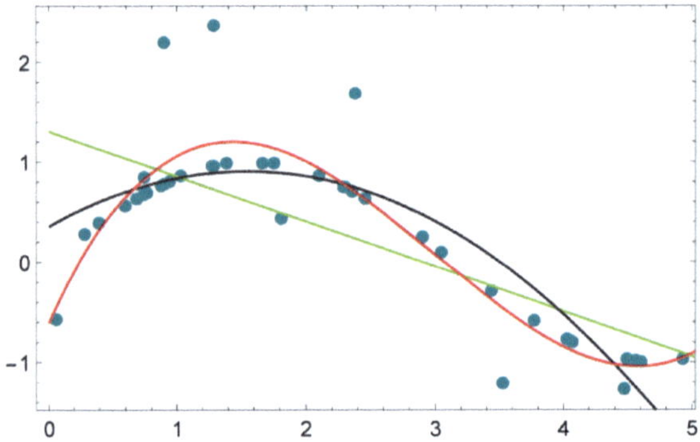

Fig. 4.26 The fitted polynomials, linear model (green), second order model (black), third order model (red).

Python

Now we can use linear regression with Polynomial Features function,

```
from sklearn.preprocessing import PolynomialFeatures
from sklearn.linear_model import LinearRegression
```

Preparing second order regression,

```
regr = LinearRegression()
quadratic = PolynomialFeatures(degree=2)
X_quadratic = quadratic.fit_transform(X)
```

The estimation of the model parameters

```
clf=LinearRegression().fit(X_quadratic,y)
```

Then the interception, the value of the zero order term is,

```
clf.intercept_
```

⇐ 0.356727

The coefficients of the higher order terms are,

```
clf.coef_
```

⇐ {0.,0.722627,-0.235229}

We have the same model as given by *Mathematica*,

⇒ fitP2
⇐ 0.356727 + 0.722628x − 0.235229x^2

Let us predict the function value at $x = 20$

```
clf.predict(quadratic.fit_transform(20))
```

⇐ {-79.2822}

Using *Mathematica*,

⇒ fitP2/.x → 20
⇐ -79.2823

The estimation of third order model is similar

```
regr = LinearRegression()
cubic = PolynomialFeatures(degree=3)
X_cubic = cubic.fit_transform(X)
```

```
clf=LinearRegression().fit(X_cubic,y)
```

```
clf.intercept_
```

⇐ -0.617648

```
clf.coef_
```

⇐ {0.,2.83055,-1.29828,0.144008}

```
clf.predict(cubic.fit_transform(20))
```

⇐ {688.745}

Using *Mathematica*,

```
⇒ fitP3/.x → 20
⇐ 688.746
```

4.3.2 Support Vector Regression (SVR)

The problem of regression is that of finding a function which approximates mapping from an input domain to the real numbers based on a training sample. We refer to the difference between the hypothesis output and its training value as the residual of the output, which indicates the accuracy of the fitting at this point. We must decide how to measure the importance of this accuracy, as small residuals may be inevitable while we wish to avoid large ones. The loss function determines this measure. Each choice of loss function will result in a different overall strategy for performing regression. For example, least squares regression uses the sum of the squares of the residuals (Christianini and Shawe-Taylor 2000).

Although several different approaches are possible, we will provide an analysis for generalization of regression by introducing a threshold test accuracy ε, beyond which we consider a mistake to have been made. We therefore aim to provide a bound on the probability that a randomly drawn test point will have accuracy less than ε.

The linear ε-insensitive loss function $L^{\varepsilon}(x, y, f)$ is defined by

$$L^{\varepsilon}(x,y,f) = \left(\left|y - f(x)\right|\right)_{\varepsilon} = \max\left(0, \left|y - f(x)\right| - \varepsilon\right)$$

where f is a real-valued function on a domain X, $x \in X$ and $y \in \mathbb{R}$. Similarly the quadratic ε-insensitive loss is given by

$$L_2^{\varepsilon}(x,y,f) = \left(\left|y - f(x)\right|\right)_{\varepsilon}^{2}$$

This loss function determines how much a deviation from the true $f(x)$ is penalized; for deviations less than ε, no penalty is considered. Here is how the loss function looks like, see Fig. 4.27. The representation of the linear ε-insensitivity loss function in different coordinate systems. The zero error range in (x, y) system (left), the error variation in (y, L^{ε}) system (right). This latest figure represents the linear variation of the error outside the zero error range.

⇐ 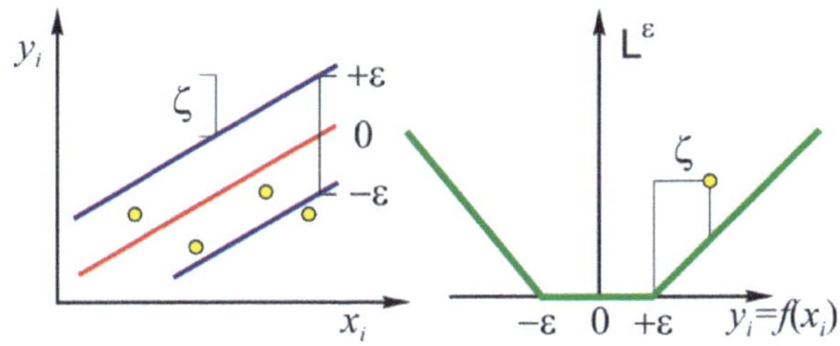

Fig. 4.27. ε-insensitivity loss function.

For example, Fig. 4.28 shows the loss function in case $\varepsilon = 1$,

```
⇒ ε = 1;
⇒ Plot[If[Abs[x]<ε,0,Abs[x]-ε]],{x,-3,3}]
```

⇐ 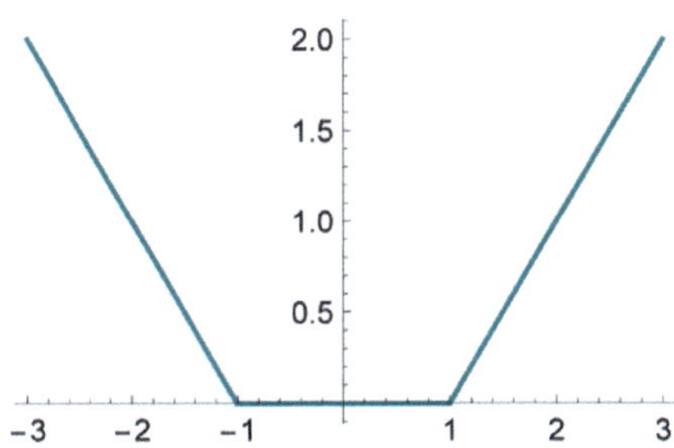

Fig. 4.28 ε-insensitivity loss function.

Here, similar to Support Vector Classifier (SVC), we also can employ different types of kernels, like polynomial kernel,

$$K(u,v) = \left(a + \langle u,v \rangle\right)^d$$

where $d > 0$, is the order of the kernel.

Gaussian kernel belonging to the Radial Bases Function (RBF) family is also a frequently used kernel type,

$$K(u,v) = \mathrm{Exp}(-\beta |u - v|)$$

Let us employ *Python* to approximate our data set using polynomial as well as RBF kernel.

Python

Loading the SVR procedures and parameterizing them (Kak, 2018),

```
from sklearn.svm import SVR
import numpy as np
```

```
svr_poly = SVR(kernel='poly', C=1e2, degree=3)
svr_rbf = SVR(kernel='rbf', C=1e3, gamma=0.3)
```

Estimation of the coefficients of the models, and defining them as predicting functions, then applying them to the input of the data set, we get

```
y_poly =  svr_poly.fit(X, y).predict(X)
```

```
y_rbf= svr_rbf.fit(X, y).predict(X)
```

Provide the results for *Mathematica*

```
y_poly
```

```
⇐ {0.833111,0.832611,0.831791,0.828543,0.826287,0.824534,0.824498,
   0.823624,0.818742,0.81791,0.817593,0.815605,0.815078,0.809572,
   0.789117,0.787873,0.787766,0.776232,0.733654,0.717255,0.705538,
   0.634705,0.576441,0.572947,0.554049,0.546224,0.516203,0.313593,
   0.230771,-0.0328292,-0.10323,-0.313898,-0.560151,-0.589567,
   -0.606949,-1.07855,-1.10354,-1.2002,-1.26313,-1.72156}
⇒ ypoly=%;
```

Let us visualize the result provided by SVR in case of third order polynomial kernel, see Fig. 4.29.

```
⇒ Show[{p0,ListPlot[Transpose[{Flatten[Xt],ypoly}],
   Joined → True,PlotStyle → {Thin,Red}]}]
```

⇐

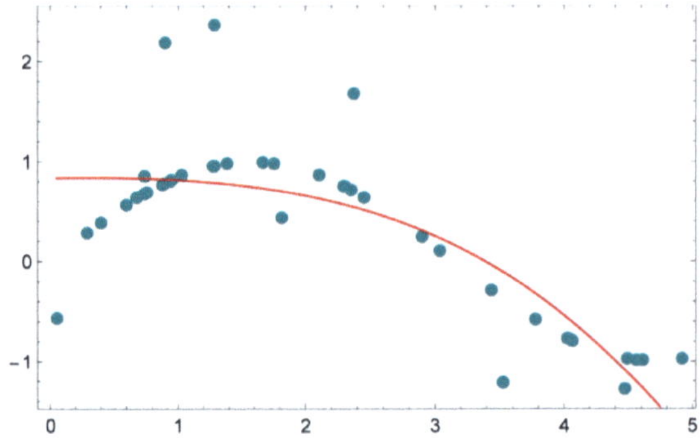

Fig. 4.29 SVR approximation in case of polynomial kernel.

Similarly

```
y_rbf
```

```
⇐ {-0.467642,0.117965,0.332581,0.632064,0.723807,0.772499,0.773344,
   0.792888,0.868541,0.877645,0.880909,0.89914,0.903411,0.937855,
   0.99157,0.993151,0.993282,1.00383,1.01165,1.00725,1.00208,
   0.937755,0.848987,0.84285,0.808327,0.793426,0.733453,0.280319,
   0.10125,-0.364257,-0.460335,-0.688867,-0.871518,-0.888746,
   -0.898533,-1.07061,-1.07526,-1.08928,-1.09505,-1.05904}
⇒ yrbf=%;
```

Fig. 4.30 shows the approximation in case of RBF kernel.

```
⇒ Show[{p0,ListPlot[Transpose[{Flatten[Xt],yrbf}],
   Joined → True,PlotStyle → {Thin,Red}]}]
```

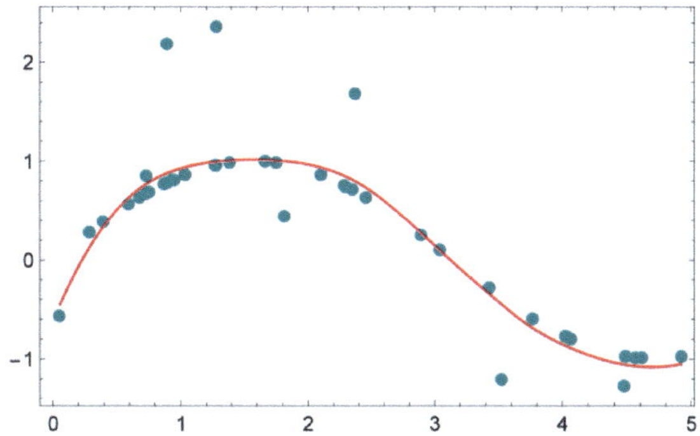

⇐

Fig. 4.30 SVR approximation with RBF kernel.

It can be seen that RBF kernel provides a better approximation than the polynomial kernel.

Mathematica

Mathematica also is able to employ SVR, with an additional advantage. Namely the result, the approximation function itself can be defined in symbolical way, in analytical form. Consequently it can be integrated in any other high level compiled code.

However *Mathematica* has not built in function for SVR, there are more third party codes. Here we employ the procedures developed by *Ruskeepää* (2017).

The first module carries out SVR with any user defined kernel, but with polynomial or Gaussian as basic ones. This module provides the analytical form of the approximation function, as well as the numbering of the support vector points.

```
⇒ supportVectorRegression[data_,kernel_,{param_,c_,ε_},
   scaling_:False]:=Module[{m=Length[data],xx=N[Most/@data],
   yy=N[Last/@data],αα,α,ββ,β,zz,k,obj,val,sol,posα,posβ],sv,b,f},
   αα=Array[α,m];
   ββ=Array[β,m];
   zz=Table[ToExpression["x"<>ToString[i]],{i,Length[xx[[1]]]}];
   Switch[kernel,"polynomial",k[x_,z_]:=(1+x.z)^param,
    "Gaussian",k[x_,z_]:=Exp[-Total[(x-z)^2]/(2param^2]];
   obj=-Total[(αα-ββ) yy]-ε Total[αα+ββ]-
    1/(2c)Total[αα^2 + ββ^2] - 0.5Sum[(α[i]-β[i])(α[j]-
    β[j])k[xx[[i]],xx[[j]]],{i,m},{j,m}];
```

```
{val,sol}=FindMaximum[{obj,Total[αα-ββ]==0,Thread[αα>=0],
 Thread[ββ>=0]},Join[αα,ββ],Method → {"InteriorPoint","
 Scaling" → scaling},AccuracyGoal → 8];
posα=Flatten@Position[αα/.sol,z_/;z>10^-6];
posβ=Flatten@Position[ββ/.sol,z_/;z>10^-6];
sv=Sort@Join[posα,posβ;
b=Mean@Table[yy[[i]]+ε+α[i]/c-Sum[(β[j]-α[j])k[xx[[i]],xx[[j]]],
 {j,m}],{i,posα}];
f=Chop[(Sum[(β[i]-α[i])k[xx[[i]],zz],{i,m}]+b)/.sol,10^-6];
{f,sv}]
```

The second module provides visualization of the approximation curve with the ε-insensitivity region and with the support vector points.

```
⇒ supportVectorRegressionPlot[fit_,data_,sv_,ε_,opts___]:=
  Module[{xdata,ydata},{xdata,ydata}=N[data]ᵀ;
  Show[ListPlot[data,PlotRange → All,
   PlotStyle → {Black,PointSize[Medium]}],
   Graphics[Circle[#,Offset[4]]&/@(data[[sv]])],
   Plot[fit,{x1,Min@xdata,Max@xdata},PlotStyle → Magenta],
   Plot[{fit+ε,fit-ε},{x1,Min@xdata,Max@xdata},
    PlotStyle → {{Darker[Green],Dashing[Small]}}],opts]]
```

Let us employ polynomial kernel, third order model

```
⇒{fit,sv}=supportVectorRegression[data,"polynomial",
  {3,100,0.3}]//Quiet
⇐ { - 0.570647 - 2.49887(1 + 0.683345 x1)^3 - 5.51006 (1 + 0.73845 x1)^3 -
  6.67777(1 + 0.762618 x1)^3 - 10.9886(1 + 0.875769 x1)^3 -
  11.4699(1 + 0.892358 x1)^3 + 69.1679(1 + 0.898521 x1)^3 -
  ⋮
  1.11172(1 + 4.56326 x1)^3 + 0.350979(1 + 4.60986 x1)^3,
  {5,7,8,9,10,11,12,13,14,15,16,17,18,19,
  20,21,26,31,33,34,35,37,38,39}}
```

Fig. 4.31 shows the approximation function,

```
⇒ Show[{p0,supportVectorRegressionPlot[fit,data,sv,0.3],p0}]
```

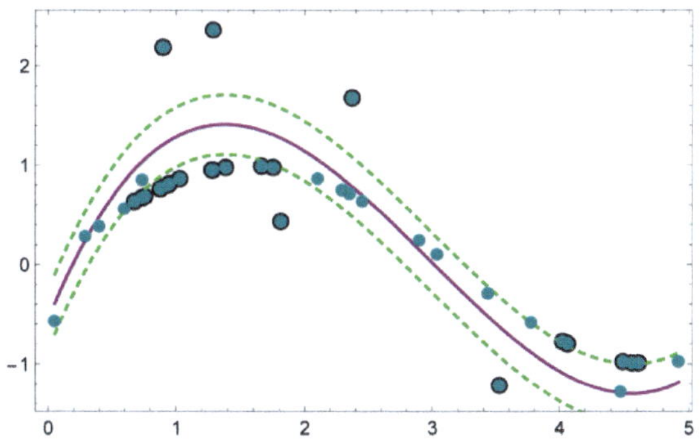

⇐

Fig. 4.31 Approximation with third order polynomial kernel.

The data points, which are inside the insensitivity region do not influence the residual, the loss function value.

Now, let us employ the Gaussian kernel,

```
⇒ {fit,sv}=
   supportVectorRegression[data,"Gaussian",{2,1000,0.3}]//Quiet
```

$$\Leftarrow \{-1.2209 - 66.4909\, e^{-\frac{1}{8}(0.0554427-x1)^2} - 9.6771\, e^{-\frac{1}{8}(0.683345-x1)^2} - 36.3709\, e^{-\frac{1}{8}(0.73845-x1)^2} - 47.0074\, e^{-\frac{1}{8}(0.762618-x1)^2} - 88.4341\, e^{-\frac{1}{8}(0.875769-x1)^2} - 93.3876\, e^{-\frac{1}{8}(0.892358-x1)^2} + 712.9156\, e^{-\frac{1}{8}(0.898521-x1)^2} - 104.9449\, e^{-\frac{1}{8}(0.935335-x1)^2} - \vdots$$

$$35.1427\, e^{-\frac{1}{8}(4.02302-x1)^2} + 34.6850\, e^{-\frac{1}{8}(4.05113-x1)^2} + 34.2259\, e^{-\frac{1}{8}(4.06756-x1)^2},$$

```
{1,5,7,8,9,10,11,12,13,14,15,16,17,
18,19,20,21,26,31,32,33,34,35}}
```

Fig. 4.32 shows the result,

```
⇒ Show[{p0,supportVectorRegressionPlot[fit,data,sv,0.3],p0}]
```

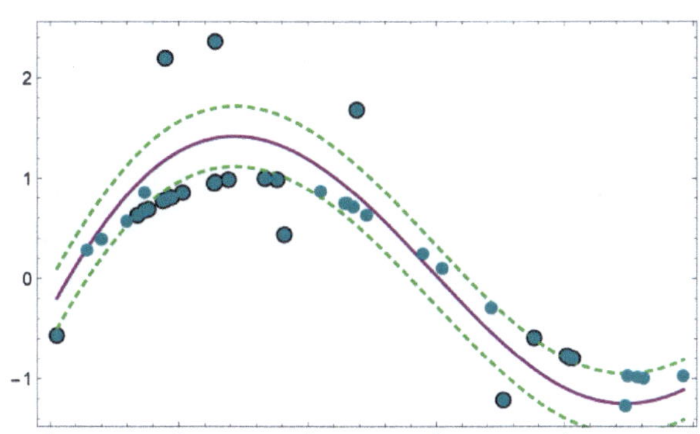

⇐

Fig. 4.32 Approximation of SVR in case of Gaussian kernel.

Mathematica provides nearly the same approximation for the two different kernels, while the *Python* code can approximate the data set properly in case of Gaussian kernel (Liu et al. 2018).

4.3.3 Boundary of the Saturn Ring

This example illustrate, that sometimes it is possible to avoid regression and we can use interpolation as a special case of regression. This can be done by eliminating "wrong" data, the so called outliers. Let us consider a spacecraft view, from Cassini in 2016, showing Saturn' s northern hemisphere. Image via NASA/JPL – Caltech/Space Science Institute.

⇒ `img=`  `;`

Fig. 4.33 Saturn' s northern hemisphere.
Image via NASA/JPL – Caltech/Space Science Institute.

Employing this image we would like to guess the outer boundary of the rings via fitting a curve on it. In order to get some points of the boundary, first we remove the background of the image, see Fig. 4.34.

⇒ `img1=RemoveBackground[img]`

⇐ 

Fig. 4.34 The image without background.

Then we look for some key points of this image, see Fig. 4.35.

```
⇒ corners=ImageCorners[img1,MaxFeatures → 40];
⇒ hu=HighlightImage[img1,corners]
```

⇐ 

Fig. 4.35 The image with key points.

In order to select points on the perimeter, we compute the convex hull of the points,

```
⇒ Needs["ComputationalGeometry`"]
```

The built-in function will provide the numbering of the corner points

```
⇒ convexhull=ConvexHull[corners]
⇐ {34,27,7,20,19,11,15,32,40,37,9,13,23,18,35}
```

The coordinates of these points are

```
⇒ pp=Map[corners[[#]]&,convexhull]
⇐ {{746.5,271.5},{738.5,284.5},{708.5,317.5},{625.5,367.5},
   {511.5,403.5},{232.5,403.5},{130.5,368.5},{44.5,214.5},
   {60.5,188.5},{88.5,158.5},{224.5,88.5},{618.5,87.5},
   {676.5,113.5},{691.5,122.5},{742.5,170.5}}
```

So we selected the perimetrical points, see Fig. 4.36.

```
⇒ hu=HighlightImage[img1,pp]
```

⇐ 

Fig. 4.36 The perimetrical points.

Let us fit an ellipse to the perimetrical points. The general form of an ellipse,

$$Ax^2 + B\,x\,y + Cy^2 + D\,x + E\,y + F = 0$$

Remark

In case of noisy data one should consider a constrain,

$$B^2 - 4AC < 0$$

Employing *Mathematica*,

```
⇒ lin=#1²,#1,#2,2#1#2,#2²&@@@pp;
⇒ lm=LinearModelFit[lin,{1,a,b,c,d},{a,b,c,d}]
⇐ FittedModel[-71352.4-0.246181a+214.937b+507.896c-0.0396957d]
⇒ lm["ParameterTable"]
⇐
```

	Estimate	*Standard Error*	*t-Statistic*	*P-Value*
1	-71352.4	99.228	-719.076	6.65759×10^{-25}
a	-0.246181	0.000555238	-443.379	8.38011×10^{-23}
b	214.937	0.460803	466.439	5.04728×10^{-23}
c	507.896	0.39964	1270.88	2.2389×10^{-27}
d	-0.0396957	0.000394002	-100.75	2.27349×10^{-16}

```
⇒ pa=lm["BestFitParameters"];
⇒ w[x_,y_]:=pa.{1,x^2,x,y,2xy}-y^2;
⇒ cc=ContourPlot[w[x,y]==0,{x,0,800},{y,0,500},
   ContourStyle → {Thick,Red},Epilog → Point[points]];
```

The model equation

```
⇒ TraditionalForm[-w[x,y]==0]
⇒ 0.246181x²+0.0793913xy-214.937x+y²-507.896y+71352.4==0
⇒ Show[{img1,cc}]
```

⇐

Fig. 4.37 The estimated border line of the Rings of Saturn

Alternatively instead of fitting an ellipse to these points we demonstrate a more general technique to estimate the other points of the perimeter as an alternative solution can be the parametric regression, or in our case interpolation. We are looking for the functions

$$x = x(t)$$

$$y = y(t)$$

where $(x(t), y(t))$ are the coordinates of a point of the perimeter. We shall employ the length of the linear lines connecting the neighboring points of the convex hull, see Fig. 4.38.

```
⇒ HighlightImage[img1,PlanarGraphPlot[corners,convexhull]]
```

⇐

Fig. 4.38 The perimetrical points.

Let us compute this parameter values belonging to the different points. Closing the hull,

```
⇒ data=Join[pp,{First[pp]}];
```

Then the parameter values are

```
⇒ n=Length[data];
⇒ t={0};
   Do[t=AppendTo[t,Last[t]+Norm[data[[i+1]]-data[[i]]]],{i,1,n-1}]
```

Assigning these parameter to the x_i and y_i coordinates, we get the lists of (x_i, t_i) and (y_i, t_i),

```
⇒ dataxt=MapThread[{#1,#2[[1]]}&,{t,data}];
⇒ datayt=MapThread[{#1,#2[[2]]}&,{t,data}];
```

Now we employ interpolation instead of regression for $x(t)$ and $y(t)$

```
⇒ xatInt=Interpolation[dataxt,InterpolationOrder → 3];
⇒ yatInt=Interpolation[datayt,InterpolationOrder → 2];
```

Then we can use parametric plot to visualize the outer border line, see Fig.4.39.

```
⇒ par=ParametricPlot[{xatInt[u],yatInt[u]},{u,0,Max[t]},
   PlotStyle → {Thick,Red}];
⇒ Show[{hu,par}]
```

⇐

Fig. 4.39 The estimated border line of the Rings of Saturn.

Interpolation does not require assumption of ellipsis.

4.4 Robust Regression Models

Basic theory

Robust fitting means an estimation technique which is able to estimate accurate model parameters not only despite small- scale noise in the data set but occasionally large scale measurement errors (outliers). *Outlier*'s *definition* is not easy. Perhaps considering the problem from a practical point of the view, we can say that data points, whose appearance in the data set causes dramatic change in the result of the parameter estimation can be labeled as outliers.

Basically there are two different methods to handle outliers:

a) weighting out outliers,
b) discarding outliers.

Weighting outliers means that we do not kick out certain data points labeled as outliers but during the parameter estimation process, we take them with a low weight into consideration in the objective function. Such a technique is the *Local Regression method.*

The other technique will try to identify data points, which make "trouble" during the parameter estimation process. Trouble means that their existence in the data set change the result of the parameter estimation considerably. One of the representatives of this technique is the *RANdom SAmple Consensus* (RANSAC) method.

Both techniques eliminate outliers in a way. However there are softer methods used in case of presence of outliers, too.

The simplest method of estimating parameters in a regression model that are less sensitive to outliers than the least squares estimates, is to use *least absolute deviations*. Even then, gross errors (outliers) can still have a considerable impact on the model, motivating research into even more robust approaches.

In 1973, Huber introduced ML-estimation for regression (Huber 1993). The ML stands for "*maximum likelihood type*". The method is robust to outliers in the response variable (see e.g., Awange and Aduol 1999,2002). This can be considered as a generalization of ordinary least square method, since it can be applied for any error distribution. The error distribution can be estimated by the *Expectation Maximization* method (Awange et al. 2018).

4.4.1 Local Regression (loess)

Let us employ the environmental data set providing relation between ozone concentration, solar radiation, temperature and the strength of wind.

```
⇒ env=Rest[Import["M:\\environmental.txt","Table"]];
⇒ {no,ozone,radiation,temperature,wind}=env//Transpose;
```

Here, we consider the relation between ozone concentration and strength of wind, see Fig.4.40.

```
⇒ data=Transpose[{wind,ozone}];
⇒ pdata=ListPlot[data,AspectRatio → 1,PlotRange → All,
   AxesOrigin → {0,0},PlotStyle → PointSize[0.02],
   Frame → True, FrameLabel → {"wind","ozone concentration"}]
```

⇐

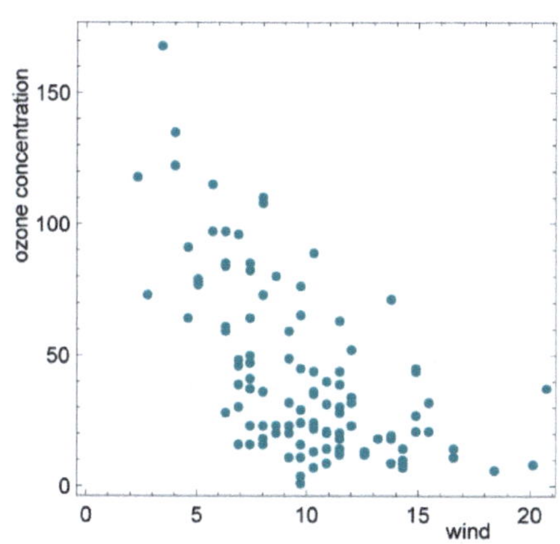

Fig. 4.40 Relation between ozone concentration and the strength of wind.

Mathematica

We approach the data locally via a p order polynomial curve. The normalized [-1, 1] data values will be weighted, see Fig.4.41 below,

```
⇒ Plot[T[u],{u,-1,1},Frame → True,
```

```
⇒ FrameLabel → {"Normalized data value","Local weight"}]
```

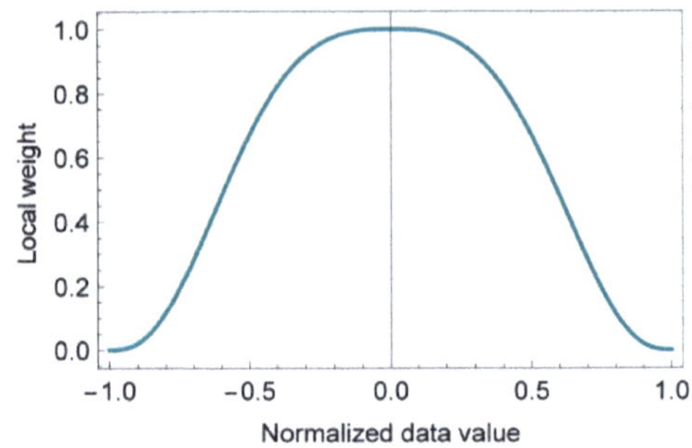

⇐

Fig. 4.41 Local weighted values around the data point in the relative location of (0,0).

The *Mathematica* function carrying out local first or second order regression is,

```
⇒ localRegress[data_,localpols_,α_,λ_]:=
   Module[{xx,ff,a,b,n,x,q,xwei,y},{xx,ff}=dataᵀ;a=Min[xx];
   b=Max[xx];n=Length[xx];x=Range[a,b,(b-a)/(localpols-1)];
   q=Floor[α n];xwei={#,Table[ww[i,#,xx,q],{i,n}]}&/@x;
   Interpolation[{#[[1]],Normal[LinearModelFit[data,y^Range[0,λ],
   y,Weights → #[[2]]+10^-15]]/.y → #[[1]]}&/@xwei]]
```

where the input parameters are:

data	– (x, y) data pairs
localpols	– the number of the applied local points in the interval
α	– smoothing parameter, $\alpha \leq 1$ A small value leads to less smoothing whereas close to 1 results in a strong smoothing
λ	– the degree of the *local polynomials* (either 1 or 2)

Let us employ the method with first and second order polynomial,

```
⇒ {fit1,fit2}=
   {localRegress[data,20,0.9,1],localRegress[data,20,0.9,2]};
```

Fig. 4.42 shows the result

```
⇒ GraphicsGrid[{Map[Show[{pdata,Plot[#[x],{x,2,21},
   PlotStyle → {Thin,Red}]//Quiet}]&,{fit1,fit2}]}]
```

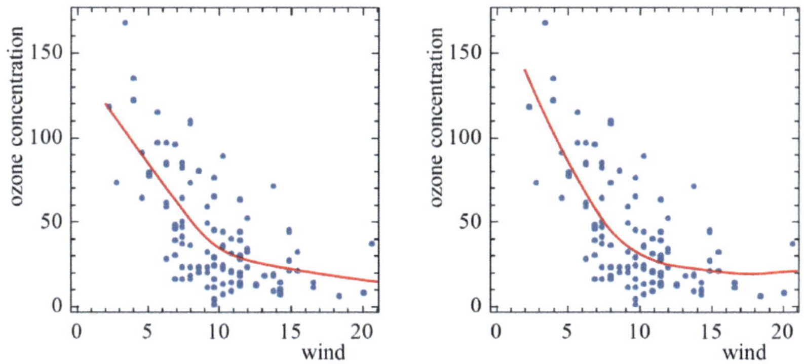

⇐

Fig. 4.42 Local regression with polynomial of first and second degree.

Let us save these data for *Python*

```
⇒ Export["M:\\wind.mtx",{wind}];
⇒ Export["M:\\ozone.mtx",{N[ozone]}];
```

Python

In *scikit* there is not such a "built-in" function. However one can use the *loess.py* function. Let us load some necessary procedures (Mayorov,2018),

```
import numpy as np
import pandas as pd
import scipy
import matplotlib.pyplot as plt
import seaborn as sns
```

The definition of the function is

```
import numpy as np
import pandas as pd
import scipy

def loc_eval(x, b):
    """
    Evaluate `x` using locally-weighted regression parameters.
    Degree of polynomial used in loess is inferred from b. `x`
    is assumed to be a scalar.
    """
    loc_est = 0
    for i in enumerate(b): loc_est+=i[1]*(x**i[0])
    return(loc_est)

    def loess(xvals, yvals, alpha, poly_degree=1):
    """
    Perform locally-weighted regression on xvals & yvals.
    Variables used inside `loess` function:
        n     => number of data points in xvals
```

```
    m      => nbr of LOESS evaluation points
    q      => number of data points used for each
              locally-weighted regression
    v      => x-value locations for evaluating LOESS
    locsDF=> contains local regression details for each
              location v
    evalDF=> contains actual LOESS output for each v
    X      => n-by-(poly_degree+1) design matrix
    W      => n-by-n diagonal weight matrix for each
              local regression
    y      => yvals
    b      => local regression coefficient estimates.
              b = `(X^T*W*X)^T*W*y`. Note that `@`
              replaces `np.dot` in recent numpy versions.
    local_est => response for local regression
"""
# Sort dataset by xvals.
all_data = sorted(zip(xvals, yvals), key=lambda x: x[0])
xvals, yvals = zip(*all_data)

locsDF = pd.DataFrame(
            columns=[
              'loc','x','weights','v','y','raw_dists',
                  'scale_factor','scaled_dists'
              ])
evalDF = pd.DataFrame(
            columns=[
              'loc','est','b','v','g'
              ])

n = len(xvals)
m = n + 1
q = int(np.floor(n * alpha) if alpha <= 1.0 else n)
avg_interval = ((max(xvals)-min(xvals))/len(xvals))
v_lb = max(0,min(xvals)-(.5*avg_interval))
v_ub = (max(xvals)+(.5*avg_interval))
v = enumerate(np.linspace(start=v_lb, stop=v_ub,
    num=m), start=1)

for i in v:
    iterpos = i[0]
    iterval = i[1]
    # Determine q-nearest xvals to iterval.
    iterdists = sorted([(j, np.abs(j-iterval)) \
           for j in xvals], key=lambda x: x[1])
    _, raw_dists = zip(*iterdists)
    # Scale local observations by qth-nearest raw_dist.
    scale_fact = raw_dists[q-1]
    scaled_dists = [(j[0],(j[1]/scale_fact))
         for j in iterdists]
    weights = [(j[0],((1-np.abs(j[1]**3))**3 \
         if j[1]<=1 else 0)) for j in scaled_dists]
```

```
        # Remove xvals from each tuple:
        _, weights      = zip(*sorted(weights,
                          key=lambda x: x[0]))
        _, raw_dists    = zip(*sorted(iterdists,
                          key=lambda x: x[0]))
        _, scaled_dists = zip(*sorted(scaled_dists,
                          key=lambda x: x[0]))

        iterDF1 = pd.DataFrame({
                    'loc'          :iterpos,
                    'x'            :xvals,
                    'v'            :interval,
                    'weights'      :weights,
                    'y'            :yvals,
                    'raw_dists'    :raw_dists,
                    'scale_fact'   :scale_fact,
                    'scaled_dists':scaled_dists
                    })

        locsDF    = pd.concat([locsDF, iterDF1])
        W         = np.diag(weights)
        y         = yvals
        b         = np.linalg.inv(X.T @ W @ X) @ (X.T @ W @ y)
        local_est = loc_eval(interval, b)
        iterDF2   = pd.DataFrame({
                       'loc':[iterpos],
                       'b'  :[b],
                       'v'  :[interval],
                       'g'  :[local_est]
                       })
       evalDF = pd.concat([evalDF, iterDF2])

    # Reset indicies for returned DataFrames.
    locsDF.reset_index(inplace=True)
    locsDF.drop('index', axis=1, inplace=True)
    locsDF['est'] = 0; evalDF['est'] = 0
    locsDF = locsDF[['loc','est','v','x','y','raw_dists',
                     'scale_fact','scaled_dists','weights']]

    # Reset index for evalDF.
    evalDF.reset_index(inplace=True)
    evalDF.drop('index', axis=1, inplace=True)
    evalDF = evalDF[['loc','est', 'v', 'b', 'g']]

    return(locsDF, evalDF)
```

Importing data files

```
import numpy as np
```

```
from numpy import array, matrix
```

```
from scipy.io import mmread, mmwrite
```

```
xv=mmread('M:\\wind.mtx')
```

```
xvals=xv[0]
```

```
yv=mmread('M:\\ozone.mtx')
```

```
yvals=yv[0]
```

Using linear local regression

```
regsDF, evalDF = loess(xvals, yvals, alpha=.9, poly_degree=1)
```

The predicted "wind" values

```
l_x=evalDF['v'].values
l_x
```

```
⇐ {2.21712,2.38438,2.55164,2.71889,2.88615,3.05341,3.22067,3.38793,
   3.55519,3.72245,3.88971,4.05697,4.22423,4.39149,4.55875,4.726,
   4.89326,5.06052,5.22778,5.39504,5.5623,5.72956,5.89682,6.06408,
    ⋮
   18.274,18.4413,18.6085,18.7758,18.943,19.1103,19.2776,19.4448,
   19.6121,19.7793,19.9466,20.1138,20.2811,20.4484,20.6156,20.7829}
```

```
⇒ u=%;
```

The predicted "ozone" values

```
l_y=evalDF['g'].values
l_y
```

```
⇐ {117.46,115.406,113.357,111.315,109.279,107.251,105.23,103.219,
   101.217,99.2251,97.2435,95.2725,93.3119,91.3615,89.4207,87.489,
   85.566,83.6514,81.7446,79.8453,77.953,76.0674,74.1876,72.3127,
    ⋮
   17.8324,17.6265,17.4198,17.2126,17.0051,16.7974,16.5898,16.3824,
   16.1755,15.9692,15.7637,15.5592,15.3557,15.1536,14.9531,14.7544}
```

```
⇒ v=%;
```

Let us visualize the result in *Mathematica*, see Fig.4.43.

```
⇒ p1=ListPlot[Transpose[{u,v}],
   PlotStyle → {Thin,Red},Joined → True];
⇒ Show[{pdata,p1}]
```

⇐

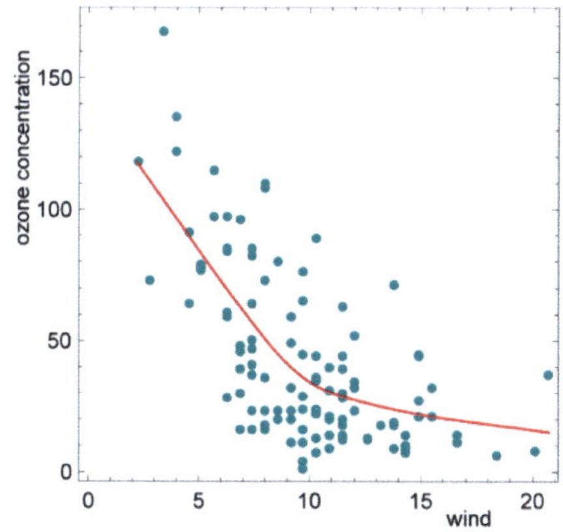

Fig. 4.43 Local regression with local polynomial of first degree.

We can do the same with a polynomial of second degree

```
regsDF, evalDF = loess(xvals, yvals, alpha=.9, poly_degree=2)
```

```
l_x=evalDF['v'].values
l_x
```

```
⇐ {2.21712,2.38438,2.55164,2.71889,2.88615,3.05341,3.22067,3.38793,
   3.55519,3.72245,3.88971,4.05697,4.22423,4.39149,4.55875,4.726,
   4.89326,5.06052,5.22778,5.39504,5.5623,5.72956,5.89682,6.06408,
    ⋮
   18.274,18.4413,18.6085,18.7758,18.943,19.1103,19.2776,19.4448,
   19.6121,19.7793,19.9466,20.1138,20.2811,20.4484,20.6156,20.7829}
⇒ u=%;
```

```
l_y=evalDF['g'].values
l_y
```

```
⇐ {35.951,132.584,129.252,125.958,122.703,119.488,116.315,113.187,
   110.105,107.071,104.085,101.148,98.2602,95.4215,92.6315,89.8898,
   87.1962,84.5504,81.952,79.4004,76.8951,74.4355,72.0205,69.6488
    ⋮
   19.3864,19.41,19.4416,19.4814,19.5295,19.586,19.6512,19.7252,
   19.8083,19.9007,20.0026,20.1143,20.2362,20.3688,20.5126,20.6683
⇒ v=%;
⇒ pl=ListPlot[Transpose[{u,v}],
   PlotStyle → {Thin,Red},Joined → True];
```

The result can be seen in Fig.4.44.

```
⇒ Show[{pdata,pl}]
```

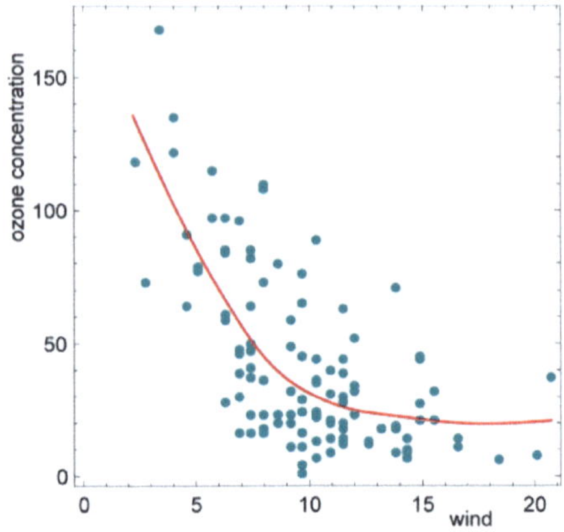

⇐

Fig. 4.44 Local regression with second order local polynomial is more sensitive for points of the less dense data region.

4.4.2 Expectation Maximization

This technique is basically a soft biclustering method, which can be employed for outlier elimination.

Let us consider the following synthetic data set, see Fig.4.45.

⇒ data=Map[#, $\left(\sqrt{10x^{7/9}}\right)$ +
```
  RandomReal[-2.6,1.9]/.x → #&,Range[0.2,30,0.1]];
```
```
⇒ dataAlt=Map[If[RandomInteger[{0,1}]==1&&#[[2]]>10.5,
  {#[[1]],1.5 #[[2]]},#]&,data];
⇒ pL=ListPlot[dataAlt,AxesLabel → {"x","y"},PlotRange → All,
  PlotStyle → PointSize[0.012],AspectRatio → 1,Frame → True]
```

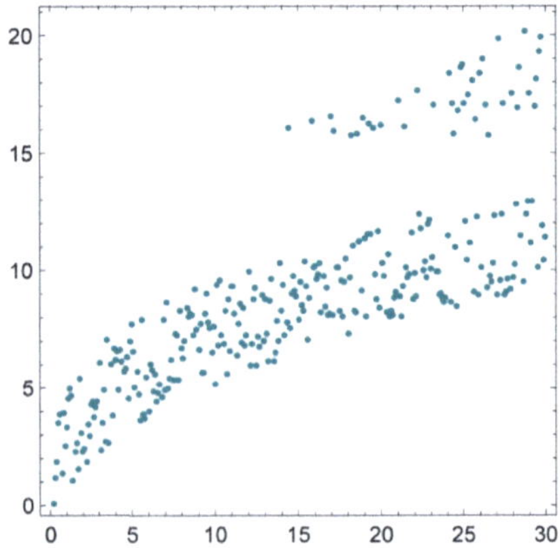

⇐

Fig. 4.45 Synthetic data set with outliers.

As first approach let us employ ordinary least squares method (OLS),

```
⇒ ye=Fit[dataAlt,{1,x,√x},x]
⇐ 1.37543 +1.33926 √x+0.177643 x
```

The error of the approximation

```
⇒ ∈ =Map[#[[2]]-ye/.x → #[[1]]&,dataAlt];
```

Let us save the generated synthetic data for *Python*,

```
⇒ m=Partition[ ∈ ,1];(*data for Python*)
```

The error distribution is not a Gaussian distribution.

```
⇒ pH=Histogram[ ∈ ,Automatic,"PDF"]
```

⇐

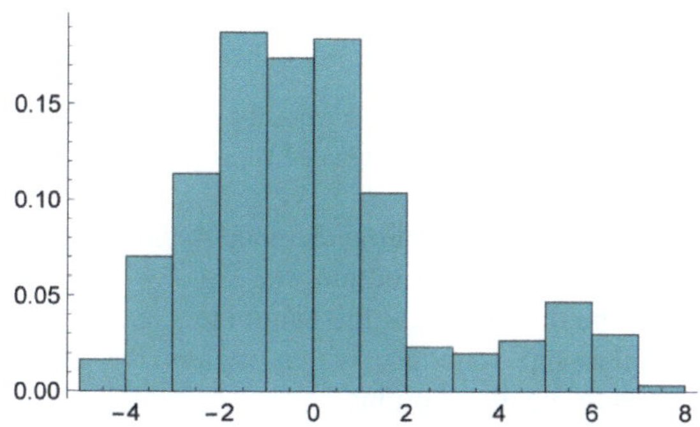

Fig. 4.46 Histogram of the errors of the OLS method.

The second peak indicates outliers. We assume that we have the mixture of two Gaussian distributions. One of the errors of the inliers and one of the outliers data. Then the *Expectation Maximization* algorithm will identify the elements of the two different distributions. The results will be a membership function of values in [0, 1].

Expectation Maximization (EM) Algorithm

Let us consider a two-component Gaussian mixture represented by the mixture model in the following form,

$$N_{12}(x) = \eta_1 N(\mu_1, \sigma_1, x) + \eta_2 N(\mu_2, \sigma_2, x)$$

where

$$N(\mu_i, \sigma_i, x) = \frac{e^{-\frac{(x-\mu_i)^2}{2\sigma_i^2}}}{\sqrt{2\pi}\sigma_i}, \quad i = 1,2$$

and η_i's are the membership weights constrained by

$$\eta_1 + \eta_2 = 1$$

If we know the membership of all x_i elements, then we could compute the parameters (μ, σ) of the two distributions, for example μ_1

$$\mu_1 = \frac{1}{n_1}\sum_i^{n_1} x_{i,1}$$

and σ_1

$$\sigma_1^2 = \frac{1}{n_1}\sum_i^{n_1}\left(x_{i,1} - \mu_1\right)^2$$

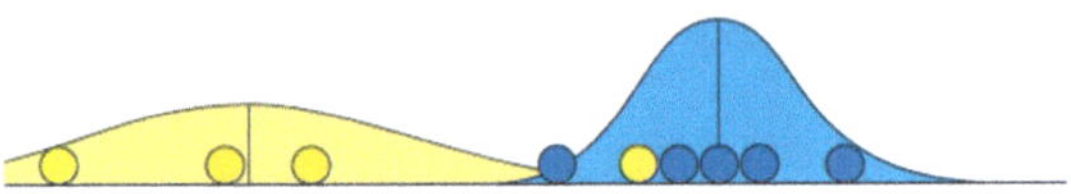

Fig. 4.47 Assuming the membership of the elements, the parameters of the distributions can be estimated.

In this case we know that the yellow elements belong to the first and the blue elements belong to the second distribution, see Fig. 4.47.

Now, let us assume that we know the parameters and would like to guess the value of the membership function, which means the probability of the x_i elements belongings to the distributions.

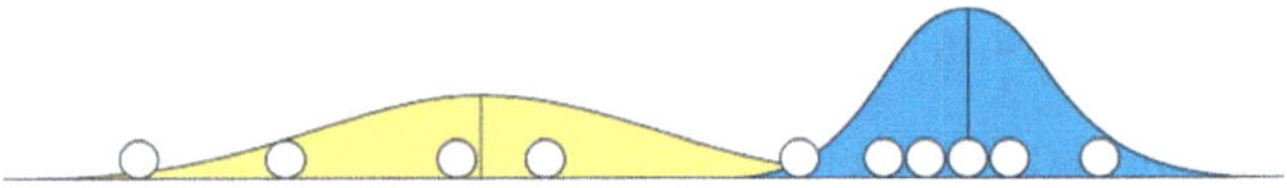

Fig. 4.48 Employing the parameters of the distribution of the probability of the membership of the x_i elements can be estimated.

How likely does a white element belong to the yellow and to the blue distribution? (see Fig.4.48).

The probability that x_i belongs to the first distribution is (Bayesian statistics)

$$P\left(N\left(\mu_1,\sigma_1\right)\middle|x_i\right) = \frac{P\left(x_i\middle|N\left(\mu_1,\sigma_1\right)\right)P\left(N\left(\mu_1,\sigma_1\right)\right)}{P\left(x_i\middle|N\left(\mu_1,\sigma_1\right)\right)P\left(N\left(\mu_1,\sigma_1\right)\right)+P\left(x_i\middle|N\left(\mu_2,\sigma_2\right)\right)P\left(N\left(\mu_2,\sigma_2\right)\right)}$$

where x_i is the observable event and $N(\mu_1,\sigma_1)$ is the hypothesis. In addition

$$P\left(x_i\middle|N\left(\mu_1,\sigma_1\right)\right) = \frac{e^{-\frac{(x-\mu_1)^2}{2\sigma_1^2}}}{\sqrt{2\pi}\sigma_1}$$

The probability that x_i belongs to the second distribution is,

$$P\left(N\left(\mu_2,\sigma_2\right)\middle|x_i\right)=1-P\left(N\left(\mu_1,\sigma_1\right)\middle|x_i\right)$$

So we have a chicken and egg problem. Consequently we can carry out a re-estimation of the parameters, for example,

$$\mu_1=\frac{\sum_i x_i P\left(N\left(\mu_1,\sigma_1\right)\middle|x_i\right)}{\sum_i P\left(N\left(\mu_1,\sigma_1\right)\middle|x_i\right)}$$

and

$$\sigma_1^{\,2}=\frac{\sum_i \left(x_i-\mu_1\right)^2 P\left(N\left(\mu_1,\sigma_1\right)\middle|x_i\right)}{\sum_i P\left(N\left(\mu_1,\sigma_1\right)\middle|x_i\right)}$$

To evaluate the algorithm we start by guessing the parameter values (μ_1,σ_1) and (μ_2,σ_2). The progress of the algorithm can be seen in Fig.4.49.

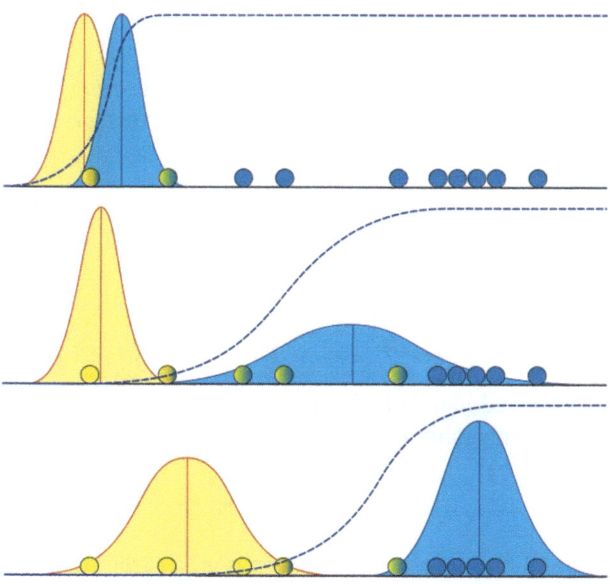

Fig. 4.49 Progress of the EM algorithm.
Figure adopted from https://www.youtube.com/watch?v=REypj2sy_5U

Mathematica

This algorithm is implemented in *Mathematica* see, Fox et al (2013) as a *Mathematica Demonstration* project. The code has been modified and applied here. In the next section we illustrate how this function works.

```
⇒ ExpectationMaximization[samples_,niter_,init_]:=
   Module[{sim,data,updates},
    sim=samples;
    problists[θ_,y_]:=Block[{probs,totalprobs},
     probs=Table[θ[[3,i]]*Map[PDF[Apply[NormalDistribution,
       θ[[i]]],#]&,y],{i,2}];
     totalprobs=Total[probs];
     Map[#/totalprobs&,probs]];
    pi[j_,p_]:=Mean[p[[j]]];
    emu[sim_,j_,p_]:=Total[sim*p[[j]]]/Total[p[[j]]];
    emstd[sim_,j_,u_List,p_]:=Total[(sim-u[[j]])^2*p[[j]]]/
      Total[p[[j]]];
    em[params_,sim_]:=Module[{theprobs,mus,vars,probs},
      theprobs=problists[params,sim];
      mus=Map[emu[sim,#,theprobs]&,{1,2}];
      vars=Map[emstd[sim,#,params[[1;;2,1]],theprobs]&,{1,2}];
      probs=Map[pi[#,theprobs]&,{1,2}];data=theprobs;
      Append[Transpose[{mus,vars^.5}],probs]];
    updates=NestList[em[#,sim]&,init,niter];
    {data,updates}]
```

First we should compute the parameters of the two Gaussian via EM algorithm. Considering the histogram of the model errors, we guess parameters of the two Gaussian are

```
⇒ {Δ,param}=ExpectationMaximization[ ∈ ,100,
   {{6.5,1.6},{-0.5,1.25},{0.5,0.5}}];
```

The result of the parameter values

```
⇒ {{μ1,σ1},{μ2,σ2},{η1,η2}}=Last[param]
⇐ {{5.42244,0.918888},{-0.697125,1.67744},{0.113917,0.886083}}
```

These can be displayed in a table form, see Table 4.2. The first column contains the values of the membership function or mixing parameters,

```
⇒ Grid[{{Panel[TableForm[param[[-1,1;;2]],
   TableHeadings → {param[[-1,3]],{"μ","σ"}}]]}}]
```

Table 4.2 Parameters of the Gaussian mixture after the first iteration

	μ	σ
0.113917	5.42244	0.918888
0.886083	−0.697125	1.67744

Fig. 4.50 shows the density functions of the two clusters.

```
⇒ p3=Plot[{param[[-1,3,1]]*PDF[Apply[NormalDistribution,
    param[[-1,1]]],x],param[[-1,3,2]]*PDF[Apply[NormalDistribution,
    param[[-1,2]]],x]},{x,-4.5,8.2},PlotStyle → {{Red},{Blue}},
    PerformanceGoal → "Speed",PlotRange → All];
⇒ Show[{pdata,p1}]
```

⇐

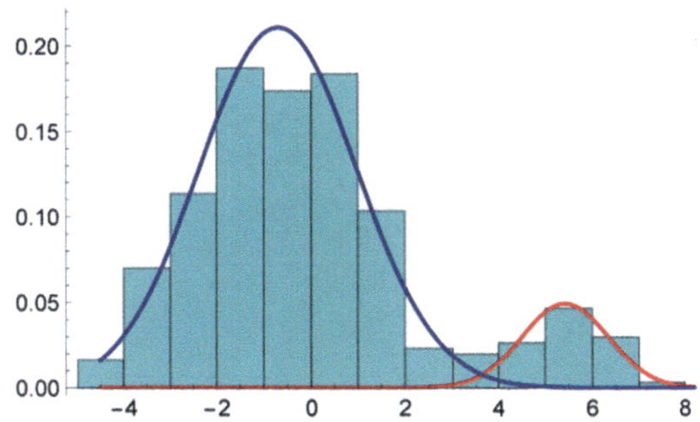

Fig. 4.50 The Probability Density Function (PDF) of the two components and the normalized histogram of the data. The red curve represents the outliers.

Now, we separate the mixture of the data samples into two clusters: cluster of outliers and cluster of inliers. Membership values of the clusters are in the output Δ_i, $i = 1, 2$.

In order to get Boolean (crisp) clustering, let us round the membership values.

```
⇒ S1=Round[Δ[[1]]];S2=Round[Δ[[2]]];
```

Then the elements in the first cluster are

```
⇒ XYZOut=Map[dataAlt[[#]]&,Position[S1,1]//Flatten];
```

The number of the elements

```
⇒ Length[XYZOut]
⇐ 35
```

Similarly, the elements in the second clusters

```
⇒ XYZIn=Map[dataAlt[[#]]&,Position[S2,1]//Flatten];
```

The number of elements

```
⇒ Length[XYZOIn]
⇐ 264
```

Let us visualize the two clusters, the inliers and outliers, see Fig. 4.51.

```
⇒ pOut=ListPlot[XYZOut,AxesLabel → {"x","y"},PlotRange → All,
    PlotStyle → Red,AspectRatio → 1];
⇒ pIn=ListPlot[XYZIn,AxesLabel → {"x","y"},PlotRange → All,
    PlotStyle → Blue,AspectRatio → 1];
```

```
⇒ Show[{pIn,pOut},PlotRange → All,Frame → True]
```

⇐

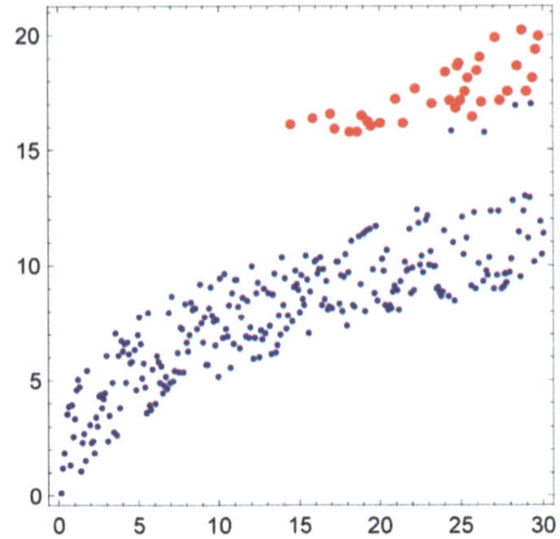

Fig. 4.51 The inliers (blue) and the outliers (red) points after the first iteration.

There are some points which are clustered incorrectly. Let us carry out one more iteration to improve the elimination of the outliers. We consider only the inliers and carry out a new regression,

```
⇒ ye=Fit[XYZIn,{1,x,√x},x]
⇐ 0.542728 +2.31778 √x -0.0603389 x
```

The errors are,

```
⇒ ∈ =Map[#[[2]]-ye/.x → #[[1]]&,XYZIn];
```

Fig. 4.52 shows the histogram of errors,

```
⇒ pH1=Histogram[ ∈ ,Automatic,"PDF"]
```

⇐

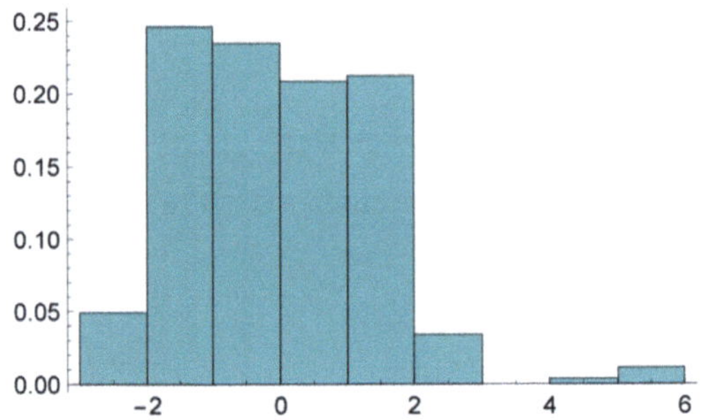

Fig. 4.52 The histogram of errors after the second iteration.

Let us carry out a new Expectation Maximation step,

```
⇒ {Δ,param}=ExpectationMaximization[ ∈ ,100,{{5,0.5},{0,1.25},
    {0.1,0.9}}]//Quiet;
```

The result of the parameter values

```
⇒ {{μ1,σ1},{μ2,σ2},{η1,η2}}=Last[param]
⇐ {{5.41711,0.337091},{-0.0827596,1.26902},{0.0150475,0.984952}}
```

These can be displayed in a table form, see Table 4.3

```
⇒ Grid[{{Panel[TableForm[param[[-1,1;;2]],
    TableHeadings → {param[[-1,3]],{"μ","σ"}}]]}}]
```

Table 4.3 Parameters of the Gaussian mixture after the second iteration

	μ	σ
0.0150475	5.41711	0.337091
0.984952	−0.0827596	1.26902

Then the elements of the two clusters can be computed. The outliers,

```
⇒ S1=Round[Δ[[1]]];S2=Round[Δ[[2]]];
⇒ XYZOut1=Map[XYZIn[[#]]&,Position[S1,1]//Flatten];
⇒ Length[XYZOut]
⇐ 4
```

The inliers

```
⇒ XYZIn1=Map[XYZIn[[#]]&,Position[S2,1]//Flatten];
⇒ Length[XYZIn1]
⇐ 260
```

Fig. 4.53 shows the outliers and inliers,

```
⇒ pOut=ListPlot[XYZOut1,AxesLabel → {"x","y"},PlotRange → All,
    PlotStyle → Red,AspectRatio → 1];
⇒ pIn=ListPlot[XYZIn1,AxesLabel → {"x","y"},PlotRange → All,
    PlotStyle → Green,AspectRatio → 1];
⇒ Show[{pIn,pOut,pIn1,pOut1},PlotRange → All,Frame → True]
```

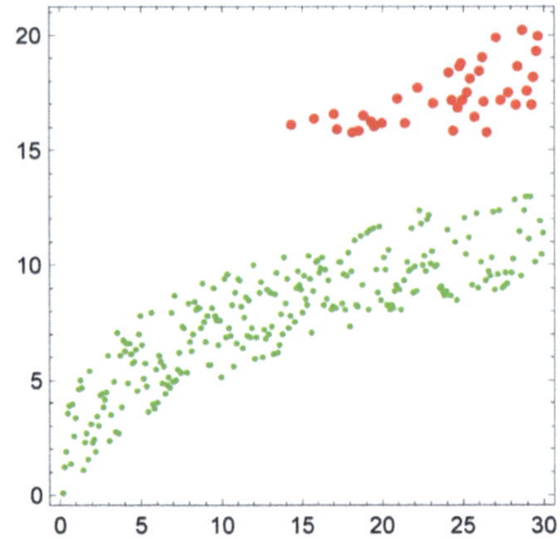

⇐

Fig. 4.53 The histogram of inliers (green) and outliers (red) after the second iteration.

Practically we have now, no outliers. Let us carry out the least squares estimation for the inliers,

```
⇒ ye=Fit[XYZIn1,{1,x,√x},x]
⇐ 0.165426 +2.6727 √x-0.128973 x
```

Fig. 4.54 shows the estimated regression line

```
⇒ Show[{pIn,pOut,pIn1,pOut1,Plot[ye,{x,0,30},
   PlotStyle → {Red,Thin}]},PlotRange → All,Frame → True]
```

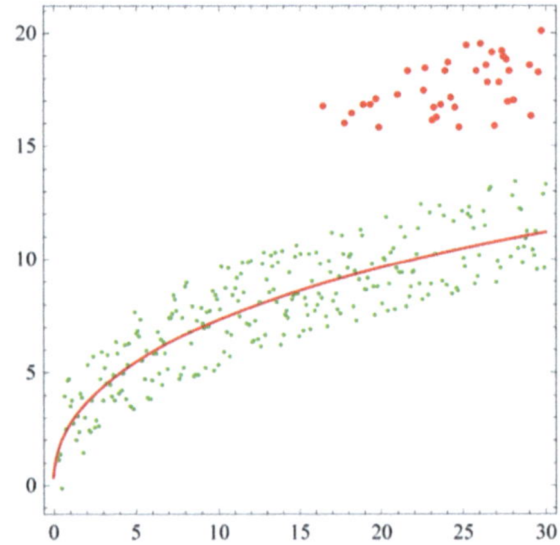

⇐

Fig. 4.54 The regression line after the elimination of outliers via *Expectation Maximization* algorithm.

Python

In *Python* we may employ *Gaussian Mixture* method to cluster data representing a mixture of normal distribution. Let us read the original synthetic data set

```
⇒ Export["G:\\cuki.mtx",m];
```

```
from numpy import array, matrix
from scipy.io import mmread, mmwrite
```

```
caki=mmread('G:\cuki.mtx')
```

Loading the *GaussianMixture* procedure,

```
import numpy as np
from sklearn.mixture import GaussianMixture
```

This method can be applied to multi component mixture, too. Let us separate the mixture,

```
gmm=GaussianMixture(n_components=2).fit(caki)
```

Labels are the crisp membership values of the clusters. The zeros stand for the inliers and one stand for the outliers,

```
labels=gmm.predict(caki)
labels
```

```
⇐ {0,0,0,0,0,0,0,0,0,0,0,0,0,0,0,0,0,0,0,0,0,0,0,0,0,0,0,0,
   0,0,0,0,0,0,0,0,0,0,0,0,0,0,0,0,0,0,0,0,0,0,0,0,0,0,0,0,
   0,0,0,0,0,0,0,0,0,0,0,0,0,0,0,0,0,0,0,0,0,0,0,0,0,0,0,
    ⋮
   1,1,1,1,0,0,1,0,1,0,1,0,0,1,0,1,1,0,0,0,0,0,0,0,1,0,0,1,
   0,0,0,0,1,0,0,0,0,1,0,0,1,0,0,1,0,0,0,1,0,1,1,0,0,0}
```

```
⇒ z=%;
```

Selecting the outlier coordinates

```
⇒ outliers=Select[MapThread[#1 #2&,{dataAlt,z}],#[[1]]!=0&];
```

Fig. 4.55 shows the result,

```
⇒ Show[{pL,ListPlot[outliers,PlotStyle → Red]}]
```

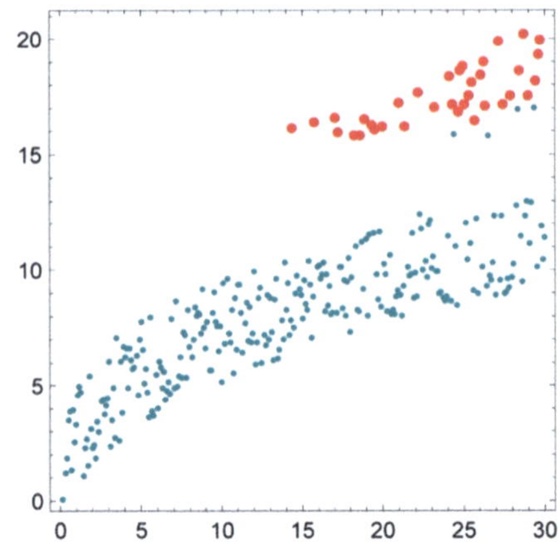

⇐

Fig. 4.55 The separation of the outliers and inliers via *GaussianMixture* procedure.

4.4.3 Maximum Likelihood Estimation

It is known that the least squares is valid if the error distribution follows a Gaussian distribution with zero mean value.

In our case the error is

$$\Delta_i = y_i - \left(a + b\sqrt{x_i} + cx_i\right)$$

The probability density function for a single Gaussian, $N(\mu, \sigma)$

$$PDF = \frac{e^{-\frac{\left(y_i - \left(a + b\sqrt{x_i} + cx_i\right) - \mu\right)^2}{2\sigma^2}}}{\sqrt{2\pi}\sigma}$$

To find the optimal parameters we should maximize the likelihood (ML) function

$$\mathcal{L} = \prod_{i=1}^{\mathbb{N}} PDF = \prod_{i=1}^{\mathbb{N}} \frac{e^{-\frac{\left(y_i - \left(a + b\sqrt{x_i} + cx_i\right) - \mu\right)^2}{2\sigma^2}}}{\sqrt{2\pi}\sigma}$$

or

$$\log \mathcal{L}\left(x_i, \theta\right) = -\sum_{i \in \mathbb{N}} \log\left(N\left(\mu, \sigma, x_i\right)\right) - \log\left(\sqrt{2\pi}\sigma\right)$$

where θ represents the parameters of our model, $\theta = (a, b, c)$.

Otherwise we should minimize

$$\log \mathcal{L}(x_i,\theta) = \sum_{i\in\mathbb{N}} \left(y_i - \left(a + b\sqrt{x_i} + cx_i\right) - \mu\right)^2$$

If $\mu = 0$, then this is the least squares method.

Let us check it in our case. The error of the *i*-th data point is

$$\Delta_i = y_i - \left(a + b\sqrt{x_i} + cx_i\right)$$

We employ the synthetic data generated in the previous section.

The errors as function of the parameters (*a, b, c*) to be estimated,

```
⇒ dL=Map[#[[2]]-(a+b √#[[1]]+c #[[1]])&,dataAlt];
```

For example, the first one is,

```
⇒ dL[[1]]
⇐ 0.0732648 -a-0.447214 b-0.2 c
```

It goes without saying, that from a different point of view, this is an overdetermined linear system for the parameters.

Mathematica

Let us employ the maximum likelihood technique. Assuming normal distribution, the likelihood function to be maximized is,

```
⇒ LL=LogLikelihood[NormalDistribution[0,1],dL];
```

Then

```
⇒ NMaximize[LL,{a,b,c}]
⇐ {-1227.01,{a → 1.37543,b → 1.33926,c → 0.177643}}
```

So we have got the same result as in case of the least squares method, see the beginning of the previous section 4.4.2, OLS solution.

When the error distribution does not satisfies the condition, we may consider that the error distribution can be approximated by a mixture of Gaussian distributions,

$$D = \sum_{i\in\mathbb{N}} N_i(\mu_i,\sigma_i)$$

The minimal case is two components. The likelihood function for a two-component Gaussian mixture can be written as,

$$\log \mathcal{L}(x_i,\theta) = \sum_{i\in\mathbb{N}_1} \log\left(N(\mu_1,\sigma_1,x_i)\right) + \sum_{i\in\mathbb{N}_2} \log\left(N(\mu_2,\sigma_2,x_i)\right)$$

where the likelihood function for one of the components can be developed as follows.

Let us employ the result of the second iteration step of the EM algorithm. Then the error function for the inliers,

```
⇒ dL1=Map[#[[2]]-(a+b √#[[1]] + c #[[1]])&,XYZIn1];
```

For example, the first element is,

```
⇒ dL1[[1]]
⇐ 0.0732648 -a-0.447214 b-0.2 c
```

The estimated parameters of the Gaūssian, see Table 4.3 are.

```
⇒ μ=-0.083;σ=1.269;
```

Then the likelihood function is

```
⇒ LL1=LogLikelihood[NormalDistribution[μ,σ],dL1];
```

Similarly for the outliers,

```
⇒ dL2=Map[#[[2]]-(a+b √#[[1]]+c #[[1]])&,XYZOut1];
⇒ μ = 5.417; σ = 0.337;
⇒ LL2=LogLikelihood[NormalDistribution[μ,σ],dL2];
```

To find the parameters, let us maximized the sum of likelihood functions,

```
⇒ NMaximize[LL1+LL2,{a,b,c}]
⇐ {-431.008,{a → 0.474593,b → 2.47132,c → -0.0922543}}
```

Then the approximation

```
⇒ ye1=a+b√x+c x/.%[[2]]
⇐ 0.474593 +2.47132√x-0.0922543 x
```

Let us display this function with the EM solution, see Fig.4.56.

```
⇒ Show[{pL,Plot[ye1,{x,0,30},PlotStyle → {Purple,Thin}],
    Plot[ye,{x,0,30},PlotStyle → {Red,Thin}]},PlotRange → All,
    Frame → True,Axes → False,AspectRatio → 1]
```

⇐

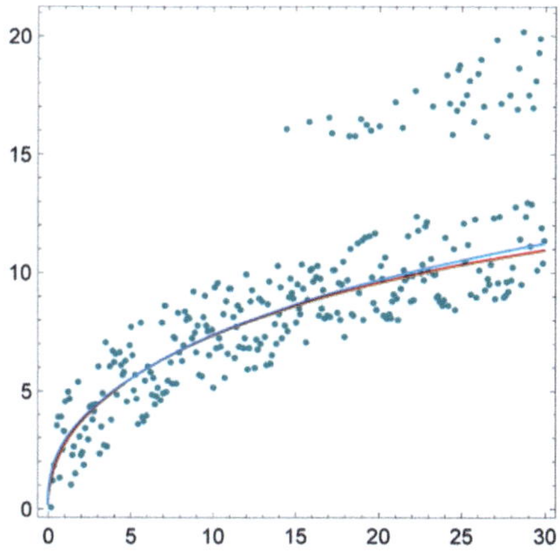

Fig. 4.56 The ML solution based on the EM data and the EM solution.

The ML method for mixture may be employed using inspection to get the mean values and the standard deviation of the error distributions without EM.

Python

Now let us employ ML algorithm in *Python*. We shall use the EM data set, namely, for example the first element.

```
⇒ XYZIn1[[1]]
⇐ {0.2,0.0732648}
```

Let us save the data set for *Python*,

```
⇒ {xx,yy}=Transpose[XYZIn1];
⇒ Export["G:\\zcuki.mtx",{xx}];
⇒ Export["G:\\ycuki.mtx",{yy}];
```

Preparation for reading data

```
from numpy import array, matrix
from scipy.io import mmread, mmwrite
```

Loading the data set

```
xcaki=mmread('G:\zcuki.mtx')
ycaki=mmread('G:\ycuki.mtx')
```

For ML algorithm one needs global optimization method

```
import numpy as np
from scipy.optimize import minimize
import scipy.stats as stats
```

Let us define a Python function for the ML process.

```
def regressLL(params):
 a0 = params[0]
 a1 = params[1]
 a2 = params[2]
 sd = params[3]
 yPred=a0+ a1*xcaki**0.5+a2*xcaki
 logLik = -np.sum(stats.norm.logpdf(ycaki,loc=yPred,scale=sd))
 return(logLik)
```

Guess the initial values of the parameters to be estimated

```
initParams=np.array([1,2,0,1])
```

Employing minimization of the negative ML function via Nelder-Mead method.

```
results = minimize(regressLL,initParams,method='nelder-mead')
```

Then the results

```
results.x
```

⇐ {0.16537,2.67272,-0.128974,1.25777}

⇒ g=%;

⇒ ye=g[[1]]+g[[2]]$\sqrt{x}$+g[[3]]x

⇐ 0.16537+2.67272$\sqrt{x}$-0.128974x

Fig. 4.57 shows the estimated regression function

⇒ Show[{pL,Plot[ye,{x,0,30},
  PlotStyle → {Red,Thin}]},PlotRange → All,Frame → True]

⇐

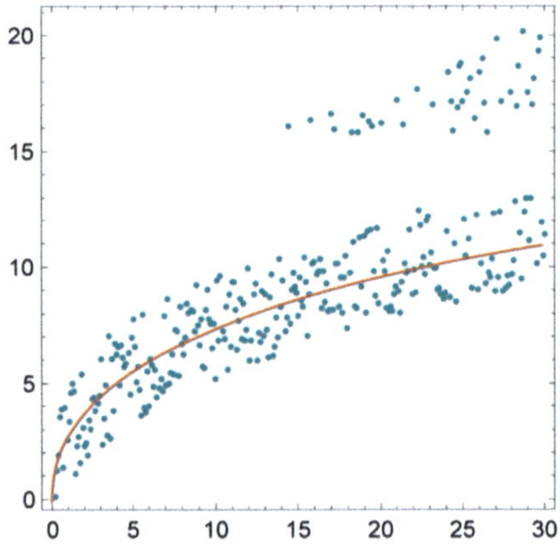

Fig. 4.57 The ML solution based on the EM data using *Python* code.

4.4.4 RANSAC for Linear Models

The RANSAC method has proved to be successful for detecting and eliminating outliers. The basic RANSAC algorithm is as follows,

1) Pick up a model type ($\mathcal{M}$) for fitting
2) Input data as
   `dataQ` - data corrupted with outliers (cardinality (`dataQ`) = n)
   s - number of data elements required per subset
   N - number of subsets to draw from the data
   τ - threshold which defines if data element, $d_i \in$ `dataQ`, agrees with the model $\mathcal{M}$

Remarks

In general s can be the minimal number of the data, which results in a determined system for the unknown parameters of the model.

The number of subsets to draw from the data, N is chosen high enough to ensure that at least one of the subsets of the random examples does not include an outlier (with the probability p, which is usually set to 0.99). Let u represent the probability that any selected data point is an inlier and $v = 1 - u$ the probability of observing an outlier. Then the iterations N can be computed as

$$N = \frac{\log(1 - p)}{\log\left(1 - (1 - v)^s\right)}$$

3) MaximalConsensusSet $\leftarrow \emptyset$
4) Iterate N times:
   a) - ConsensusSet $\leftarrow \emptyset$ (this will contain the inliers)
   b) - Randomly draw a subset containing s elements and estimate the parameters of the model $\mathcal{M}$
   c) - For each data element, $d_i \in$ `dataQ`:
   if (d_i, $\mathcal{M}$, τ) agree, ConsensusSet $\leftarrow d_i$ (the maximum level (threshold) of the local model error, which is a critical parameter of the method)
   d) - if cardinality (maximalConsensusSet) < cardinality(ConsensusSet), maximalConsensusSet $\leftarrow$ ConsensusSet
5) Estimate model parameters using maximalConsensusSet.

To illustrate the method, let us employ the *Housing data set*. From this multivariate data set, we shall consider only a single variable problem (LSTAT) and apply *Python* linear regression model. We concentrate here on the elimination of the outliers.

Loading the data set,

```
import pandas as pd
 df = pd.read_csv('https://raw.githubusercontent.com/rasbt/'
    'python-machine-learning-book-2nd-edition'
    '/master/code/ch10/housing.data.txt',
    header=None,
    sep='\s+')
 df.columns=['CRIM','ZN','INDUS','CHAS','NOX','RM','AGE',
    'DIS','RAD','TAX','PTRATIO','B','LSTAT','MEDV']
 X = df[['LSTAT']].values
 y = df['MEDV'].values
```

where MEDV is the median value of owner occupied homes in $1000, and LSTAT is the percentage of lower status of the population.

Save it for *Mathematica*

```
np.savetxt('G:\\daTaX.txt',X,fmt='%.5e')
```

```
⇒ trainX=Import["G:\\daTaX.txt","Table"];
```

and

```
np.savetxt('G:\\daTay.txt',y,fmt='%.5e')
```

```
⇒ trainy=Import["G:\\daTay.txt","Table"];
```

Then

```
⇒ TrainX=Flatten[trainX];Trainy=Flatten[trainy];
⇒ data=Transpose[{TrainX,Trainy}];
```

Let us visualize the data, see Fig. 4.58.

```
⇒ p0=ListPlot[data,Frame → True,
   FrameLabel → {" percentage of lower status of the population",
   "median value of owner occupied homes in $1000"},
   PlotStyle → PointSize[0.01]]
```

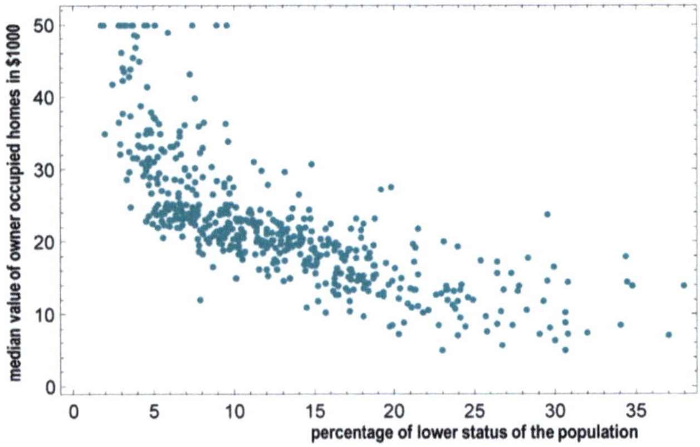

⇐

Fig. 4.58 The single variable problem from Housing data.

RANSAC method always requires a regression method, which should be integrated into the RANSAC algorithm, see steps 4. as well as in the Python code. In this example for the sake of simplicity, we employ linear regression (Straw, 2009).

Python

Let us load the procedures

```
from sklearn.linear_model import LinearRegression
from sklearn.linear_model import RANSACRegressor
```

We should parametrize the RANSAC Regressor function including linear regression as an estimation method. The most sensitive parameter is the threshold value. Using high threshold too many outliers can remains in the training set, using too low threshold one may eliminate also the inliers data points.

```
ransac=RANSACRegressor(LinearRegression(),max_trials=100,
 min_samples=50,loss='absolute_loss',
 residual_threshold=5.0,random_state=0).fit(X,y)
```

The procedure will "masking in" the inliers as active data points

```
inlier_mask=ransac.inlier_mask_
```

Saving inliers for linear regression

```
np.savetxt('G:\\maskedX.txt',X[inlier_mask],fmt='%.5e')
```

```
np.savetxt('G:\\maskedy.txt',y[inlier_mask],fmt='%.5e')
```

Read data into *Mathematica*,

```
⇒ Xinliers=Flatten[Import["G:\\maskedX.txt","Table"]];
⇒ yinliers=Flatten[Import["G:\\maskedy.txt","Table"]];
```

Combining data points and displaying them, see Fig. 4.59.

```
⇒ datainliers=Transpose[{Xinliers,yinliers}];
⇒ pinliers=ListPlot[datainliers,
   Frame → True,PlotStyle → {Red,PointSize[0.01]}];
```

```
⇒ Show[{p0,pinliers}]
```

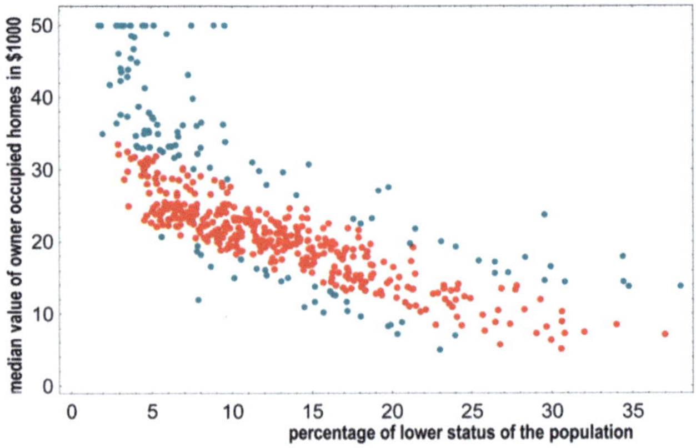

⇐

Fig. 4.59 Separating inliers (red) from outliers (blue) via RANSAC algorithm for Linear Regression.

Now we can fit a line to the inlier points via *Mathematica,* see Fig. 4.60.

Mathematica

```
⇒ lm=LinearModelFit[datainliers,x,x];
⇒ model=lm//Normal
⇐ 29.8819 -0.746351 x

⇒ p1=Plot[model,{x,0,40}];
⇒ Show[{p0,pinliers,p1}]
```

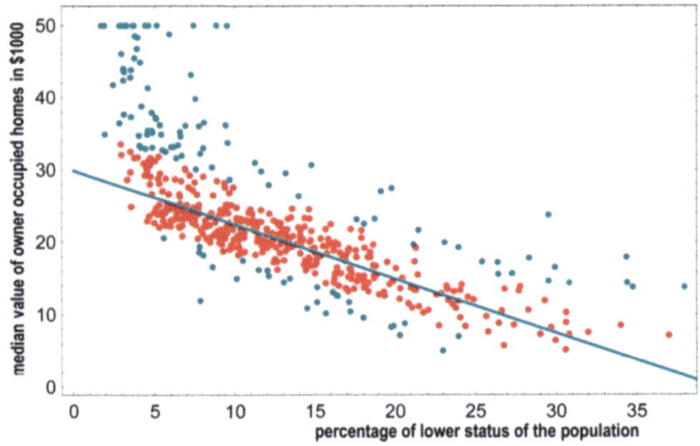

⇐

Fig. 4.60 Fitting line to the inlier points (red).

4.4.5 *Fitting Lidar Cloud of Points to a Slope*

Outdoor laser scanning measurements have been carried out in a hilly park of Budapest, Hungary see Fig. 4.61. The test area is on a steep slope, covered by dense but low vegetation.

Fig. 4.61 The test area in Budapest.

The measurement range of the scanner is 120 m, the ranging error is ± 2 mm, according to the manufacturer's technical specification.

Fig. 4.62 The scanner at the top of the measured steep slope with the different sizes of white spheres as control points in the background.

The scanning parameters were set to ½ resolution that equals to 3mm/10m point spacing. This measurement resulted in 178.8 million points that was acquired in

5 and half minutes. The test data set was cropped from the point cloud; moreover, further resampling was applied in order to reduce the data size. The final data set is composed in ASCII format, and only the *x*, *y*, *z* coordinates were kept (no intensity values). Let us load the measured data:

```
⇒ XYZ=Import["G:\\output_41_392.dat"];
⇒ n=Length[XYZ]
⇐ 41392
```

Eliminating corrupted data points,

```
⇒ XYZP=Select[XYZ,And[NumberQ[#[[1]]],NumberQ[#[[2]]],
    NumberQ[#[[3]]]]&];
⇒ n=Length[XYZP]
⇐ 33292
```

Fig. 4.63 shows measured cloud of points

```
⇒ pS=ListPointPlot3D[XYZP,PlotStyle → Blue,BoxRatios → {1,1,1}]
```

⇐

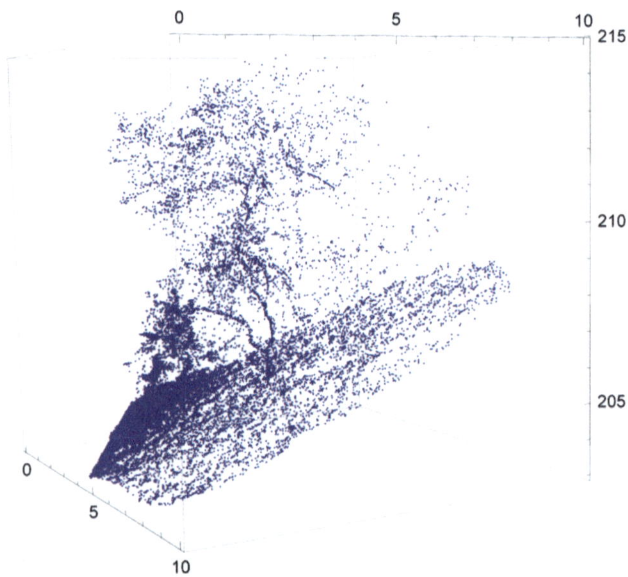

Fig. 4.63 The measured Lidar points.

Organizing data in a format ((*x*, *y*), *z*)

```
⇒ XY=Map[{#[[1]],#[[2]]}&,XYZP];
⇒ Z=Map[#[[3]]&,XYZP];
```

Exporting to *Python*,

```
⇒ Export["G:\\lidarxy.mtx",XY];
⇒ Export["G:\\lidarz.mtx",{Z}];
```

Preparation for reading data

```
import numpy as np
from numpy import array, matrix
from scipy.io import mmread, mmwrite)
```

Reading data into *Python*

```
X=mmread('G:\\lidarxy.mtx')
yy=mmread('G:\\lidarz.mtx')
```

```
y=yy[0]
```

We employ Python code to eliminate outliers, but now with a lower threshold value of 0.2,

```
ransac=RANSACRegressor(LinearRegression(),max_trials=100,
 min_samples=50,loss='absolute_loss',residual_threshold=0.2,
 random_state=0).fit(X,y)
```

The procedure will "masking in" the inliers as active data points,

```
inlier_mask=ransac.inlier_mask_
```

Saving inliers for linear regression

```
np.savetxt('G:\\LmaskedX.txt',X[inlier_mask],fmt='%.5e')
```

```
np.savetxt('G:\\Lmaskedy.txt',y[inlier_mask],fmt='%.5e')
```

Read data into *Mathematica*,

```
⇒ Xinliers=Import["G:\\LmaskedX.txt","Table"];
⇒ yinliers=Flatten[Import["G:\\Lmaskedy.txt","Table"]];
```

The number of inlier points

```
⇒ Length[Xinliers]
⇐ 13267
```

Now we can fit a plane to the inliers via OLS,

```
⇒ z=Fit[XYZPIn,{1,x,y},{x,y}]
⇐ 202.723 +0.0817325 x+0.543778 y
```

Fig. 4.64 shows the fitted plane as slope covered by the vegetation

```
⇒ ppS=Plot3D[z,{x,0,10},{y,0,10},BoxRatios → {1,1,1}];
⇒ Show[{pS,ppS}]
```

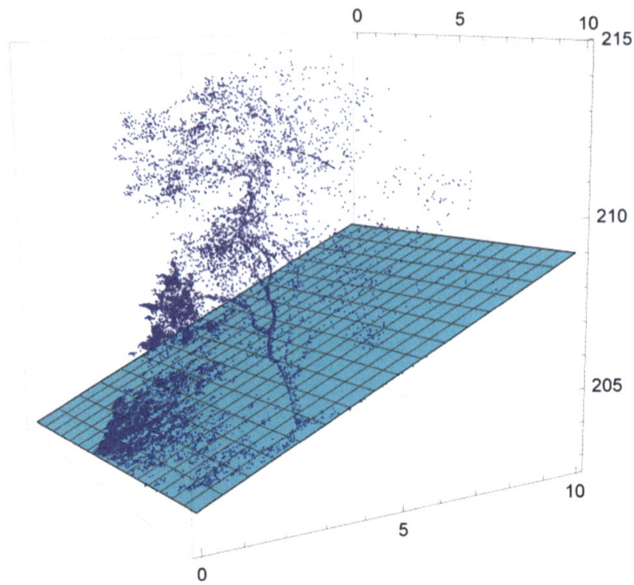

⇐

Fig. 4.64 The slope fitted points without vegetation points as outliers.

4.5 Symbolic Regression Models

Basic theory

In *traditional regression* the model $y = f(w, x)$ is specified and the parameters (w) should be estimated from data (x_i, y_i). In case of *symbolic regression* the structure of the model is not fixed, and the result is a bunch of functions having different complexities and errors (see e.g., Palancz et al., 2015). These functions are generated automatically mostly via genetic algorithm.

The algorithm will search for a set of basic functions (building blocks) and coefficients (weights) in order to minimize the error ε in case of given y_i and x_i pairs.

The standard basic functions are constant, addition, subtraction, multiplication, division, sine, cosine tangent, exponential, power, square root, etc. In order to carry out genetic programming, the individuals (competing functions) should be represented by a binary tree. The leaves of the binary tree are called terminal nodes represented by variables and constants, while the other nodes, the so called non-terminal nodes are represented by functions. Let us see a simple example. Consider

$$\beta_i(\boldsymbol{x}) = x_1 x_2 + 1\frac{1}{2}x_3 .$$

Its binary tree representation can be seen in Fig 4.65,

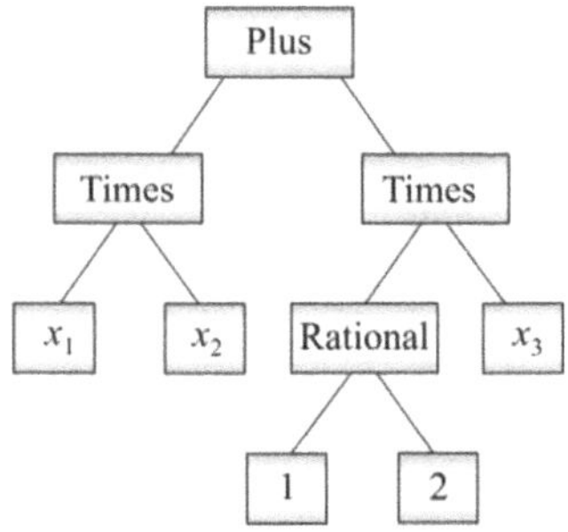

Fig. 4.65 The binary tree representation of a basic function $\beta_i(x)$.

There are two important features of the function represented by a binary tree: complexity and fitness. We define complexity as the number of nodes in a binary tree needed to represent the function, and fitness is basically a loss function characterizing how good the fitting of the function is.

The genetic algorithm tries to minimize this error to improve the fitness of the population consisting of individuals (competing functions) from generation to generation by mutation and cross-over procedure. Mutation is an eligible random change in the structure of the binary tree, which is applied to a randomly chosen sub-tree in the individual. This sub-tree is removed from the individual and replaced by a new randomly created sub-tree. This operation leads to a slightly (or even substantially) different basic function.

The operation "cross-over" representing sexuality can accelerate the improvement of the fitness of a function more effectively than mutation alone can do. It is a random combination of two different basic functions (parents), based on their fitness, in order to create a new generation of functions, more fitter than the original functions. To carry out cross-over, crossing points (non-terminal nodes) in tree of both parents should be randomly selected. Then subtrees belonging to these nodes will be exchanged creating offsprings

Complexity and fitness are conflicting features leading to a multiobjective problem, consequently we have not a single solution, but a set of optimal solutions. Pareto front represents the optimal solutions as they vary over expression complexity and maximum prediction error (see e,g., Palancz and Awange (2012).

4.5.1 Model with Single Variable

Considering a function

$$y(x) = \sin(2x) + x\exp(\cos(x)), \quad x \in [-10, 10].$$

let us generate 500 data points with normally distributed noise:

```
⇒ data=Table[{x, Sin[2x]+x Exp[Cos[x]]+
    RandomVariate[NormalDistribution[0, 1.5]]},
    {x, RandomReal[{-10, 10}, 500]}];
```

Fig. 4.66 shows the synthetic data set,

```
⇒ p0=ListPlot[data]
```

⇐

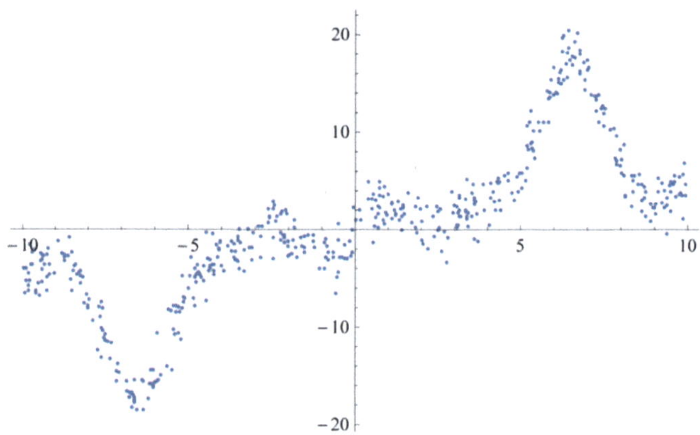

Fig. 4.66 Generated synthetic data set.

Mathematica

We employ symbolic regression generating 10 000 model candidates,

```
⇒ fit = FindFormula[data,x,10000, All,PerformanceGoal → "Quality",
    SpecificityGoal → 1];
⇒ fitN=fit//Normal//Normal;
```

For example the 1st model is

```
⇒ fitN[[1]]
⇐ 0.0460717 + 0.975718x + 1.14309x Cos[x] + 0.570206x Cos[x]^2 →
    <| Score → -1.55681, Error → 2.61224, Complexity → 24 |>
```

Error of this model

```
⇒ fitN[[1]][[2]][[2]]
⇐ 2.61224
```

Complexity of this model

```
⇒ fitN[[1]][[2]][[3]]
⇐ 24
```

The error and complexity of the first 64 models,

```
⇒ Candidates=Map[{#[[2]][[2]],#[[2]][[3]]}&,fitN]
⇐ {{2.61224,24},{4.0778,12},{4.07834,12},{3.98106,13},
   {4.05985,13},{3.96751,14},{3.97264,14},{3.98505,14},
   {3.21578,24},{3.29983,24},{4.53344,13},{2.51369,37},
    ⋮
   {62.2607,1},{62.2691,1},{62.3413,1},{62.236,3},{63.3603,5},
   {62.8032,6},{64.3349,6},{61.5377,9},{61.54,9},{209.445,3}}
```

```
⇒ Length[Candidates]
⇐ 64
```

```
⇒ p2=ListPlot[Candidates,PlotStyle → {PointSize[0.015],Blue},
    AxesLabel → {"Error","Complexity"}];
```

Computing the best models, we shall find the very point of the convex hull of the models, which is closest to the ideal point (0,0).

```
⇒ cH=ConvexHullMesh[Candidates];
```

These models represent the Pareto front, see Fig. 4.67.

```
⇒ ParetoFront=MeshCoordinates[RegionBoundary[cH]]
⇐ {{2.61224,24.},{4.0778,12.},{2.51369,37.},{2.51708,37.},
   {2.52518,37.},{2.53109,37.},{2.53592,37.},{2.53873,37.},
   {2.56256,37.},{2.5822,37.},{2.58571,37.},{2.58605,37.},
   {2.62998,37.},{2.71137,37.},{2.79758,37.},{2.8496,37.},
   {2.87629,37.},{3.01352,37.},{3.14217,37.},{25.3261,1.},
   {62.2566,1.},{62.2607,1.},{62.2691,1.},{62.3413,1.},{209.445,3.}}
```

```
⇒ p1=ListPlot[ParetoFront,PlotStyle → {PointSize[0.015],Red}];
⇒ Show[{p2,p1,cH,p2,p1}]
```

⇐

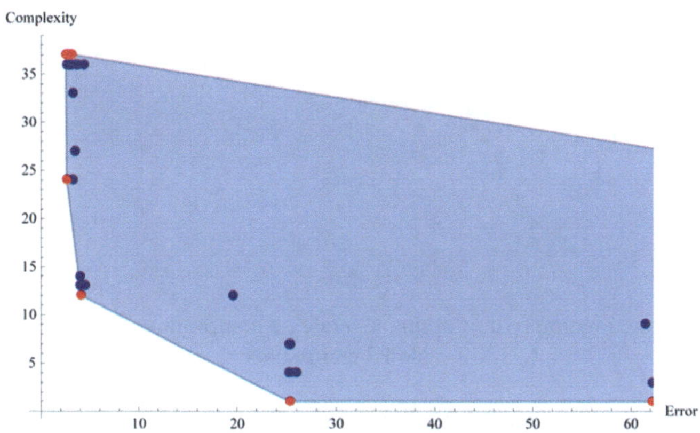

Fig. 4.67 Pareto front of the candidate models (red points).

The best models (red) provide the smallest error among the models having the same complexity. The collection of these models are called *Pareto front* (see e.g., Palancz and Awange 2013; Palancz et al., 2013).

The *ideal point* is the point (0, 0). The user can select the model from the Pareto front, which is closest to this point.

```
⇒ Map[Norm[{{0,0}-#},1]&,ParetoFront]
⇐ {24.,12.,37.,37.,37.,37.,37.,37.,37.,37.,37.,37.,37.,37.,37.,
   37.,37.,37.,37.,25.3261,62.2566,62.2607,62.2691,62.3413,209.445}
```

```
⇒ Norm[{{0,0}-ParetoFront[[2]]},1]
⇐ 12.
```

The optimal model, which is a compromise between complexity and error, is the 6th one,

```
⇒ modelOptimal=
   Select[fitN,{#[[2]][[2]],#[[2]][[3]]}==ParetoFront[[2]]&]
⇐ {1.28998 x+x Cos[x] → <|Score → -1.62928,
   Error → 4.0778,Complexity → 12|>}
```

```
⇒ fitOpt=(modelOptimal//Normal)[[1,1]]
⇐ 1.28998 x+x Cos[x]
```

Fig. 4.68 shows the model

```
⇒ Show[{p0,Plot[fitOpt,{x,-10,10},PlotStyle → {Thin,Red}]}]
```

⇐ 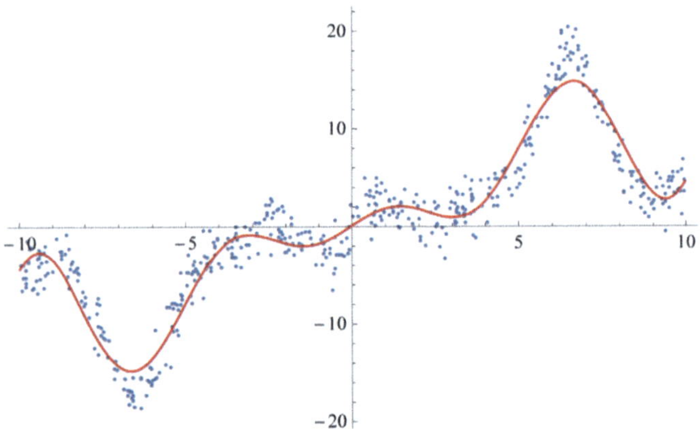

Fig. 4.68 Approximation of the "optimal" model, a compromise between model error and model complexity.

On the other hand, the model having the smallest error,

```
⇒ modelHighAccuracy=
   Select[fitN,{#[[2]][[2]],#[[2]][[3]]}==ParetoFront[[3]]&]
```

```
⇐ {-0.00564615+1.02953 x Cos[x]+9.83944 Sin[0.179446 x]-
    1.60851 Sin[0.447786 x]+1.83557 Sin[2.15024 x] →
    <|Score → -1.84151,Error → 2.51369,Complexity → 37|>}
⇒ fitHigh=(modelHighAccuracy//Normal)[[1,1]]
⇐ -0.00564615+1.02953 x Cos[x]+9.83944 Sin[0.179446 x]-
    1.60851 Sin[0.447786 x]+1.83557 Sin[2.15024 x]
```

Fig. 4.69 shows the model with the smallest error

```
⇒ p1=Plot[fitHigh,{x,-10,10},PlotStyle → {Thin,Red}];
⇒ Show[{p0,p1}]
```

⇐ 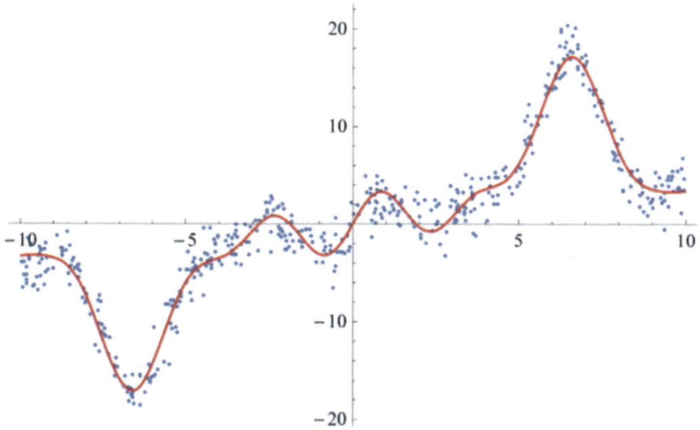

Fig. 4.69 Model provides the smallest error but the higher complexity.

4.5.2 Surface Fitting

This example demonstrates using the *SymbolicRegressor* to fit a symbolic relationship. Let's create some synthetic data based on the relationship:

$$z(x,y) = x^2 - y^2 + y - 1\,.$$

Fig. 4.70 shows the original surface

```
⇒ pi=Plot3D[x²-y²+y-1,x,-1,1,y,-1,1,BoxRatios → 1,1,0.8]
```

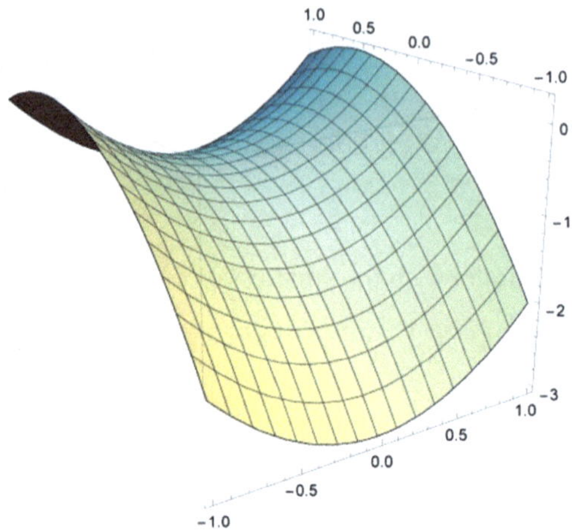

⇐

Fig. 4.70 Surface to be approximated.

Python

Load some basic functions (Stephens, 2016)

```
from numpy import array, matrix
from scipy.io import mmread, mmwrite
from math import exp, sin, cos
function_set = ['add','sub','mul','div','sqrt','log','abs',
                'neg','inv','max','min','sin','cos','exp']
```

Import *Python Symbolic Regressor*

```
from gplearn.genetic import SymbolicRegressor
import numpy as np
from sklearn.utils.random import check_random_state
```

Create a data set on a 10×10 grid in $[-1,1]\times[-1,1]$ using function $z(x,y)=x^2-y^2+y-1$

```
x0 = np.arange(-1, 1, 1/10.)
x1 = np.arange(-1, 1, 1/10.)
x0, x1 = np.meshgrid(x0, x1)
y_truth = x0**2 - x1**2 + x1 - 1
```

```
rng = check_random_state(0)
X_train = rng.uniform(-1,1,100).reshape(50,2)
y_train = X_train[:,0]**2 - X_train[:,1]**2 + X_train[:,1]-1
```

Parametrize the regressor and run it,

```
est_gp = SymbolicRegressor(population_size=5000,
               generations=20, stopping_criteria=
               p_crossover=0.7, p_subtree_mutation=0.1,
               p_hoist_mutation=0.05, p_point_mutation=0.1,
               max_samples=0.9, verbose=1,
               parsimony_coefficient=0.01,
                  random_state=0).fit(X_train, y_train)
```

```
    Population Average    |             Best Individual    |
--- ---------------------- ------------------------------ ----------
Gen Length    Fitness     Length    Fitness    OOB Fitness  Time Left
0   38.13   458.57768152   5     0.32066597  0.55676354     1.61m
1    9.97     1.70232723   5     0.32020176  0.62478715     1.16m
2    7.72     1.94456344   11    0.23953666  0.53314818    57.92s
3    5.41     0.99015681   7     0.23567635  0.71990626    49.56s
4    4.66     0.89444336   11    0.10394641  0.10394641    43.49s
5    5.41     0.94024238   11    0.06080204  0.06080204    38.96s
6    6.78     1.09535926   11    0.00078147  0.00078147    35.26s
```

The optimal model can be printed out in analytical form,

```
print(est_gp._program)
```

```
sub(add(-0.999, X1), mul(sub(X1, X0), add(X0, X1)))
```

which means

```
⇒ (y-0.999)-(y-x)(x+y)//Expand
⇐ -0.999+x²+y-y²
```

This results are quite good, however in case of more complicated functions and in noisy environment *gplearn* does not perform efficiently. In that case, the use of a third party system the *Eureqa* is suggested which has a *Python API* (https://en.wikipedia.org/wiki/Eureqa).

Mathematica

For multivariable problems a third party *Mathematica* function (sym.m) is provided by https://github.com/paulknysh/sym (Knysh, 2018)

```
⇒ Get["M:\\sym.m"]
```

Now we generate 400 noisy data points, see Fig. 4.71.

```
⇒ data=Flatten[Table[x,y,x²-y²+y-1+
    RandomVariate[NormalDistribution[0,0.1]],x,
    RandomReal[-1,1,20],y,RandomReal[-1,1,20]],1];
⇒ Show[{pi,ListPointPlot3D[data,PlotStyle → {Black,Black,Black}]}]
```

⇐

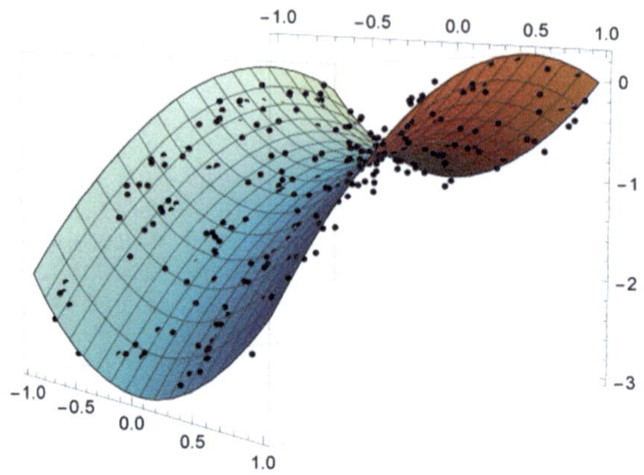

Fig. 4.71 Noisy measurement.

Let us define the error function,

```
⇒ Error[model_]:=Mean@Table[Abs[data[[i,3]]-model/.
    {x → data[[i,1]],y → data[[i,2]]}],{i,Length[data]}]
```

The real but "unknown" continuous model

```
⇒ modi=x²-y²+y-1
⇐ -1+x²+y-y²
```

which provides the smallest error

```
⇒ Error[modi]
⇐ 0.0817619
```

The running time is restricted

```
⇒ time=500;
```

Number of cores for parallel computation

```
⇒ ncores=6;
```

Basic functions,

```
⇒ uops={#²&,Sin[#]&,Cos[#]&,Exp[x]&};
```

Basic operations,

```
⇒ bops={Times,Plus,Subtract,Divide,Power};
```

Basic variables,

```
⇒ vars:=1,x,y,x²,y²;
```

Maximal number of operations used for building an expression

```
⇒ nops=7;
```

The function `Search` provides the hundred best solutions. Best solution means solution with the smallest error.

```
⇒ output=Search[time,ncores,uops,bops,vars,nops];//Quiet
```

The number of the randomly generated functions

```
⇒ output[[2]]
⇐ 612838
```

The first hundred solutions and their errors

```
⇒ output[[1]]//TableForm
```

⇐

0.081762	$-1+x^2+y-y^2$
0.081762	$-1+x^2+y-y^2$
0.127189	$-1+x^4+y-y^4$
0.157028	$-1+x^4+y-y^2$
0.159399	$-1+x^2+y-y^4$
0.222972	$y-y^4-(y^2)^{x^2}$
0.228777	$y-((y^2)^{x^2})^{x^4}$
0.240802	$y-(y^2)^{x^4}$
0.241115	$y-y^2-(y^2)^{(2x^2)}$
0.241115	$y-y^2-(y^4)^{x^2}$
0.276070	$x^2+(-1+y)(1+y^2)$
0.285933	$-(1-x^2)^2+y+xy-y^2$
0.286231	$-1+x^4+y$
0.286231	$-1+x^4+y$
0.286231	$-1+x^4+y$
0.286231	$-1+x^4+y$
0.286231	$-1+x^4+y$
0.286231	$-1+x^4+y$
0.287780	$y-y^2-y^{x8}$
0.288424	$-1+y-y^4$
0.288424	$-1+y-y^4$
0.288424	$-1+y-y^4$
0.288424	$-1+y-y^4$
0.288424	$-1+y-y^4$
0.288424	$-1+y-y^4$
0.288531	$-(1-y)^2-y+x^4y^2$
0.289361	$-1+y-y^6$
0.289502	$-1+y-y^2+x^2y^4$
0.291606	$-1+x^6+y$
0.296542	$y-(y^2)^{x^1}6$

0.296917	$-1+x^{2}-(-y+y^{2})^{2}$
0.299028	$-1+x^{(4(}1+y^{2}))+y$
0.300774	$-1+x^{8}+y$
0.300774	$-1+x^{8}+y$
0.300774	$-1+x^{8}+y$
0.300774	$-1+x^{8}+y$
0.300774	$-1+x^{8}+y$
0.307047	$-(x^{8})^{x^{8}}+y$
0.309605	$x^{4}-(x^{2})^{(2x^{2}y)}$
0.309605	$x^{4}-(x^{4})^{(x^{2}y)}$
0.314205	$-1+x^{4}-x^{8}+y$
0.315808	$y-(-1+x^{2}-y^{4})^{2}$
0.318623	$y-((y^{2})^{-}y^{2})^{y^{2}}$
0.320814	$-1+x^{1}6+y$
0.327563	$-(x^{2})^{x^{4}}+y$
0.333546	$y-(y^{x2})^{x^{4}}$
0.334012	$-1+y-y^{2}+y^{4}$
0.334480	$-(x^{2})^{x^{8}}+y$
0.336557	$(-1+y)(y^{3})^{x^{8}}$
0.338168	$-1+x^{4}+y+x^{4}y^{2}$
0.338454	$y-(y^{2})^{x^{2}}$
0.338454	$y-(y^{2})^{x^{2}}$
0.338454	$y-(y^{2})^{x^{2}}$
0.338454	$y-(y^{2})^{x^{2}}$
0.344800	$y-(1-x^{2}+y^{2})^{2}$
0.347203	$-1+y-x^{2}y^{4}$
0.347222	$-1+(x^{2}/y^{4})^{-}x^{8}y$
0.347675	$-1+y$
0.347675	$-1+y$
0.347675	$-1+y$
0.347675	$(-y+y^{2})/y$
0.347675	$(-y+y^{2})/y$
0.347675	$(-y^{2}+y^{3})/y^{2}$
0.347675	$-1+y$
0.347675	$-1+y$
0.347675	$-1+y$
0.347675	$(-x^{2}+x^{2}y)/x^{2}$
0.347675	$(-y+y^{2})/y$
0.347675	$(-y+y^{2})/y$
0.349789	$-1+y-y^{2}$
0.349789	$-1+y-y^{2}$
0.349789	$-1+y-y^{2}$
0.349789	$-1+y-y^{2}$
0.349789	$-1+y-y^{2}$
0.349789	$-1+y-y^{2}$
0.349955	$-1+(1+x^{2})y-y^{2}$
0.350446	$-1+y-x^{8}y^{4}$
0.352353	$-1+x^{2}+y$
0.352353	$-1+x^{2}+y$
0.352353	$-1+x^{2}+y$

0.352353	$-1+x^2+y$
0.352353	$-1+x^2+y$
0.352353	$-1+x^2+y$
0.352353	$x^2-(y-y^2)/y$
0.353078	$-1+y+x^8y^4$
0.353967	$-1+y-(y^4)^{(1/x^4)}$
0.356288	$-1+(x^4)^{x^5}y$
0.357408	$x^2-(x^2)^{(2x^2y)}+y$
0.359356	$y-y^2-y^{x2}$
0.360486	$-1+y-x^8y^2$
0.360547	$(-1+y)(1+y^4)$
0.362652	$-1+y+x^4y^2$
0.363021	$-1+y+x^4y^8$
0.364334	$-1+x^2+y+xy^2$
0.364334	$-1+x^2+y+xy^2$
0.364987	$(-1+x^2)(1-y)$
0.364987	$(-1+x^2)(1-y)$
0.364987	$(-1+x^2)(1-y)$
0.365459	$-1+y/(1+y^2)$
0.366168	$-1+x^2(x^2+y-y^2)$

The best model

⇒ `model=output[[1]][[1]][[2]]`

⇐ $-1+x^2+y-y^2$

This system cannot be used for serious problems. The best software for symbolic regression for *Mathematica* is the `DataModeler`. (http://www.evolved-analytics.com/)

4.6 Comparison of Regression Methods

To register two chest images means to align them, so that common features overlap and differences, should there be any, between the two are emphasized and readily visible to the naked eye. We refer to the process of aligning two images as image registration. When registering images, we are determining a geometric transformation which aligns one image to fit another. For a number of reasons, simple image subtraction does not work.

Suppose we have two images of a chest region, taken of the same subject, but at different times, say, six months ago and yesterday. We need to align the six month old image, which we will call the source image, with the one acquired yesterday, the target image.

We shall compare here the linear regression approach with the symbolic regression technique.

Let us consider the target image as

⇒ imgT= 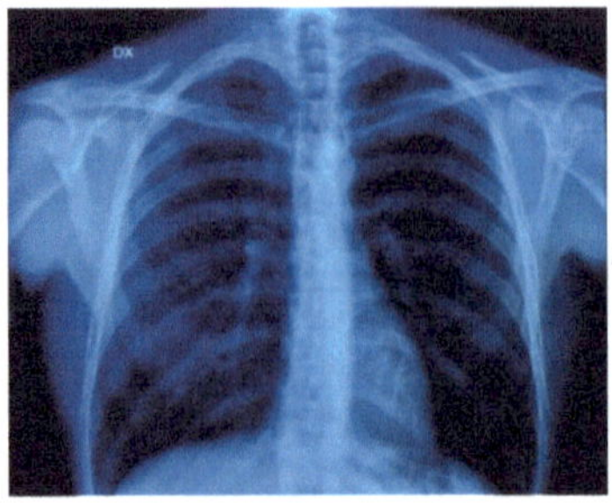 ;

Fig. 4.72 Reference image.

The source image to be aligned to the target image

⇒ imgS= 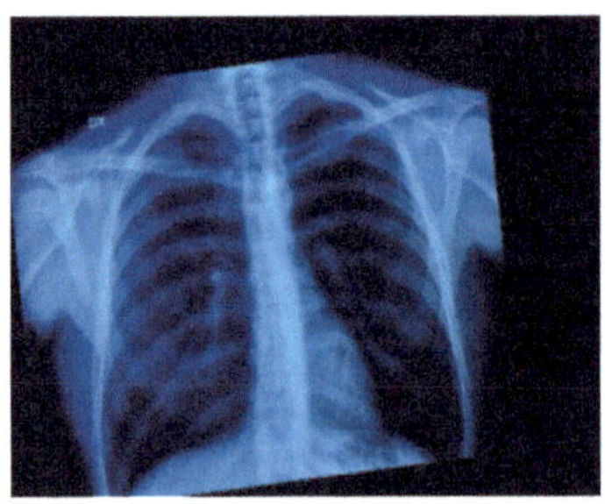 ;

Fig. 4.73 Source image.

Geometric Linear Regression Transform

Automatic registration can be employ with built-in *Mathematica* function,

```
⇒ imgR=ImageAlign[imgT,imgS,Background → Black,
   Method → "Keypoints",TransformationClass → "Affine"]
```

⇐ 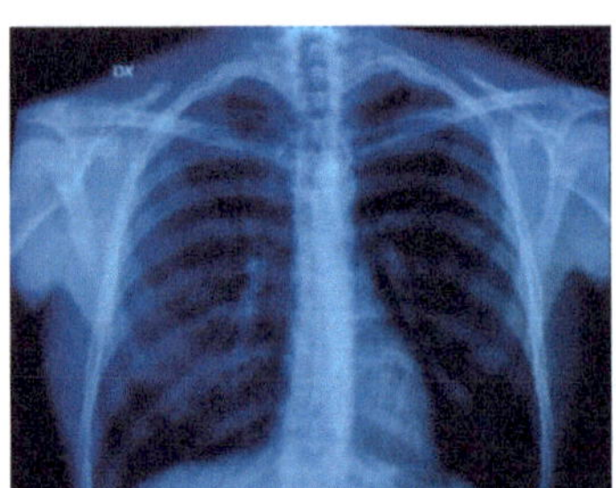

Fig. 4.74 Source image.

Let us compare the target image with the registered image

```
⇒ GraphicsGrid[{{HighlightImage[imgT, imgR]}}]
```

⇐

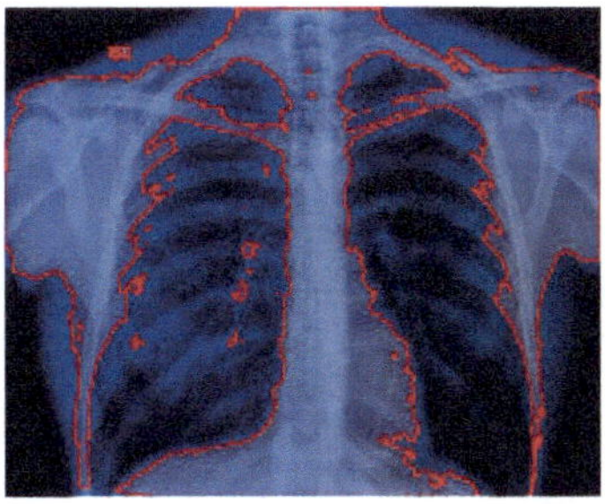

Fig. 4.75 Target image and the registered image (red contour).

A maximal value of the norm of the differences of the intensities of the test and registered images,

```
⇒ ImageDifference[imgT,imgR]//ImageData//Max
⇐ 0.941176
```

As an alternative method, let us try to find the parameters of a geometric transform. This regression technique requires corresponding points, keypoints of the source and target image.

Keypoints are the same thing as interest points. They are spatial locations, or points in the image that define what is interesting or what stands out in the image. The reason why keypoints are special is because no matter how the image changes... whether the image rotates, shrinks/expands, is translated (all of these would be an affine transformation by the way...) or is subject to distortion (i.e. a projective transformation or homography), one should be able to find the same keypoints in this modified image when comparing with the original image.

Let us employ built-in function

```
⇒  z = ImageCorrespondingPoints[imgT, imgS];
```

The number of these corresponding points,

```
⇒ Length[z[[1]]]
⇐ 100
```

We can visualize these points in both images

```
⇒ GraphicsGrid[
    {{HighlightImage[imgT,z[[1]]],HighlightImage[imgS,z[[2]]]}}]
```

⇐ 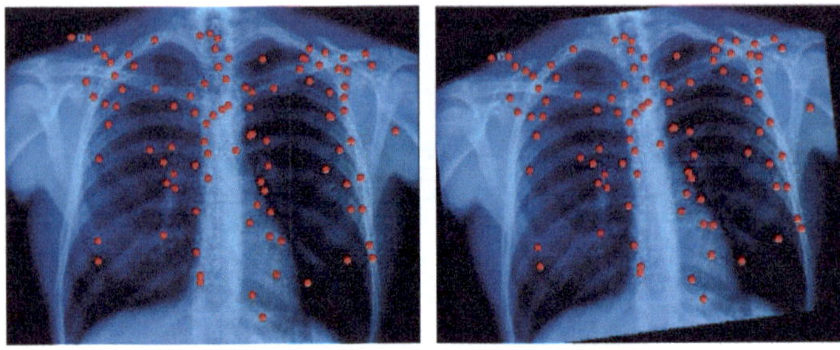

Fig. 4.76 The corresponding keypoints.

We can find the linear geometric transformation between the two images,

```
⇒ {error,tf}=FindGeometricTransform[z[[1]],z[[2]],Method → "RANSAC"]
⇐ {0.397242,TransformationFunction[
```

$$\begin{pmatrix} 1.040321 & 0.108086 & -24.011696 \\ -0.130625 & 1.042260 & 17.502997 \\ 9.41017\times10^{-7} & -5.13586\times10^{-6} & 1. \end{pmatrix}]\}$$

Then the source image can be transformed accordingly,

```
⇒ {w,h}=ImageDimensions[imgS];
⇒ imgR=ImagePerspectiveTransformation[imgS,tf,DataRange → Full]
```

⇐ 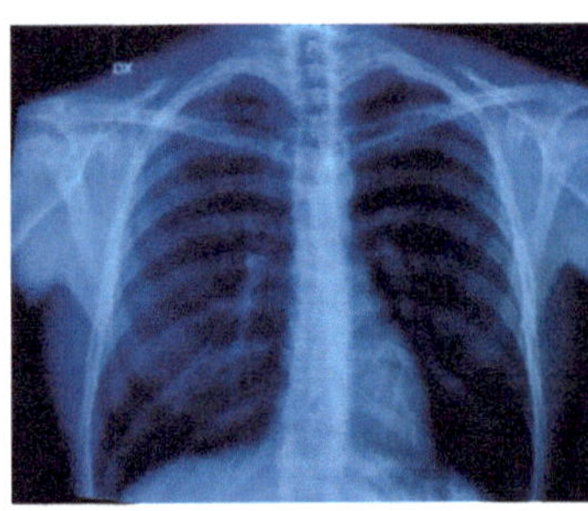

Fig. 4.77 The registered image.

Let us compare the target image with the registered image

```
⇒ GraphicsGrid[{{HighlightImage[imgT, imgR]}}]
```

⇐

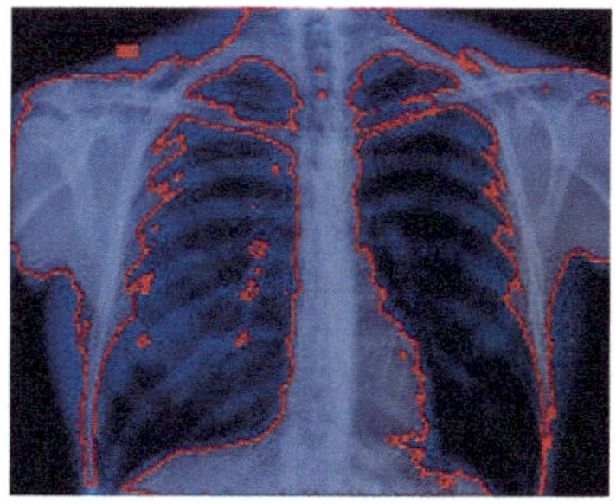

Fig. 4.78 Target image and the registered image (red contour).

A maximal value of the norm of the differences of the intensities of the test and registered images,

```
⇒ ImageDifference[imgT, imgR]//ImageData//Max
⇐ 0.941176
```

Symbolic Regression Transform

As we know, symbolic regression can provide different models (linear and nonlinear) with different complexity. Now we should consider functions for mapping the source image, (η, ξ) into the target or reference image, (x, y) namely

$$(\eta, \xi) \rightarrow (x, y).$$

We define two functions to be determined via nonlinear regression,

$$x = f(\eta, \xi)$$

and

$$y = g(\eta, \xi).$$

Here we can employ again the corresponding keypoints, for example in form (η, ξ, x).

```
⇒ fx=MapThread[({#2,#[[1]]}//Flatten)&,z];
⇒ fy=MapThread[({#2,#[[2]]}//Flatten)&,z];
```

Saving data for Eureqa software

```
⇒ Export["M:\\dataX.dat", fx];
⇒ Export["M:\\dataY.dat", fy];
```

The result for $x = f(\eta, \xi)$ can be seen in Table 4.4. The first model is the Pareto optimum- the closest model to the (0,0) points of the Pareto front.

Table 4.4 The solutions for $x = f(\eta,\xi)$

Models for the function $x = f(\eta,\xi)$	Fit	Complexity	Remark
$1.04\eta + 0.1093\xi - 24.02$	0.042	9	Pareto optimum
$1.04\eta + 0.109\xi - 25.2/\eta - 23.6$	0.041	14	
$1.05\eta + 0.11\xi - 1.5610^{-5}\eta^2 - 24.7$	0.041	15	
$1.04\eta + 0.109\xi + 0.281\sin(93.8\xi) - 23.9$	0.038	19	
$1.04\eta + 0.109\xi + 0.142\sin(\xi) + 0.288\sin(1.07\eta\xi) - 24$	0.038	27	

Considering the Pareto optimum for $y = g(\eta, \xi)$, too

$$x = f(\eta, \xi) = 1.04\ \eta + 0.1083\ \xi - 24.02\,.$$

$$y = g(\eta, \xi) = -0.131\ \eta + 1.04\ \xi - 17.4\,.$$

Now we can carry out the transform on the source image to get the registration image

```
⇒ fg[x_,y_]:={1.04 x+0.1083 y-24.02,-0.131 x+1.04 y+17.4}
  imgR=ImageForwardTransformation[imgS,fg[#[[1]],#[[2]]]&,
  DataRange → Full, Background →  Black]
```

⇐ 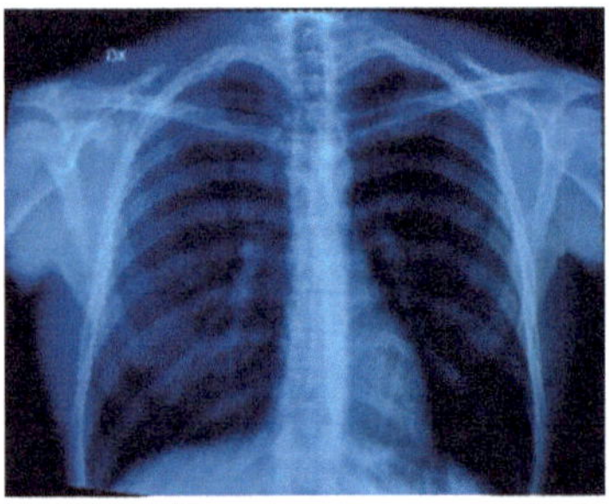

Fig. 4.79 The registered image via symbolic regression.

Let us compare the registered image with the target image,

```
⇒ GraphicsGrid[{{HighlightImage[imgT, imgR]}}]
```

⇐ 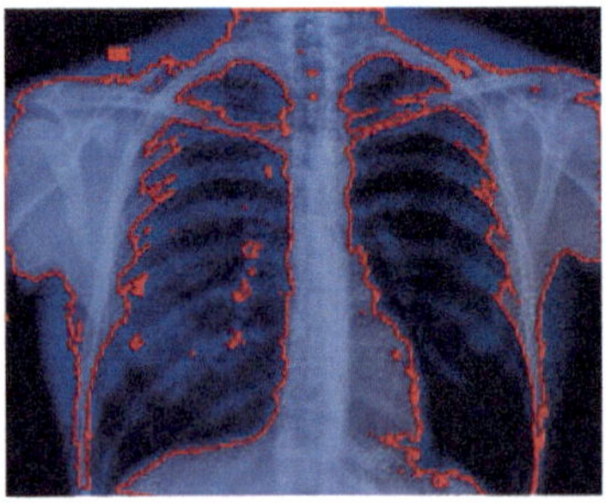

Fig. 4.80 Target image and the registered image (red contour).

A maximal value of the norm of the differences of the intensities of the test and registered images,

```
⇒ ImageDifference[imgT, imgR] // ImageData // Max
⇐ 1.
```

Facit

The symbolic regression takes a considerably longest time than the linear one, although it can provide better result as the linear one (see Table 1.1, solutions with higher complexity than the Pareto solution). Therefore, if the linear regression gives an acceptable approach, then this is the proper choice.

References

Awange JL, Aduol FWO (1999) An evaluation of some robust estimation techniques in the estimation of geodetic parameters. Survey Review 273(35): 146-163, doi: 10.1179/sre.1999.35.273.146.

Awange JL, Aduol FWO (2002) An Evaluation of some Robust Estimation Techniques in the Estimation of Geodetic Parameters - Part II. Survey Review, 283, 36: 380-389, doi: 10.1179/sre.2002.36.283.380.

Awange J, Palancz B, Lewis R H, Völgyesi L (2018): Mathematical Geosciences, Hybrid Symbolic-Numeric Methods, Springer, Heidelberg, New York.

Chen G H, Shah D (2018): Explaining the Success of Nearest Neighbor Methods in Prediction, The essence of knowledge, Boston, Delf

Christianini N, Shawe -Taylor J (2000): Support Vector Machines and other kernel -based, learning methods, Cambridge University Press, Cambridge, New York

Fox A, Smith J, Ford R and Doe J (2013) Expectation Maximization for Gaussian Mixture Distributions, Wolfram Demonstration Project.

Huber P J (1973): Robust Regression Asymptotics, Conjectures and Monte Carlo. Annals of Statistics, 1, 799-821.

Kak A (2018): NonlinearLeastSquares 2.0.0, Python Software Foundation https://pypi.org/project/NonlinearLeastSquares/

Kanevski M, Pozdnoukhov A, Timonin V (2009): Machine Learning for Spatial Environmental Data, CRC Press, Taylor & Francis Group, Boca Raton FL, USA

Knysh P (2018): sym: A Mathematica package for generating symbolic models from data https://github.com/paulknysh/sym

Liu H H, Chang L C, Li C W, Yang C H (2018): Particle Swarm Optimization-Based Support Vector Regression for Tourist Arrivals Forecasting, Computational Intelligence and Neuroscience,Volume 2018, Article ID 6076475, 13 pages, https://doi.org/10.1155/2018/6076475 https://www.hindawi.com/journals/cin/2018/6076475/

Mayorov N (2018) Robust nonlinear regression in scipy, SciPy Cookbook https://scipy-cookbook.readthedocs.io/items/robust_regression.html http://shop.oreilly.com/product/0636920034919.do

Paláncz B, Awange JL (2012) Application of Pareto optimality to linear models with errors-in-all-variables. Journal of Geodesy 86:531–545, doi:10.1007/s00190-011-0536-1.

Paláncz, B, Awange JL (2013) Pareto optimality solution of the multi-objective photogrammetric resection-intersection problem. Earth Science Informatics 6: 1-20, doi: 10.1007/s12145-012-0107-x.

Paláncz, B., Awange JL, Völgyesi L (2013) Pareto optimality solution of the Gauss-Helmert model. Acta Geodetica et Geophysica 48: 293-304, doi: 10.1007/s40328-013-0027-3

Paláncz B. (2014): Algebraic robust estimation employing Expectation Maximization, Wolfram Library Archive, http://library.wolfram.com/infocenter/MathSource/8887/

Paláncz, B, Awange JL, Völgyesi L (2015) Correction of Gravimetric Geoid Using Symbolic Regression. Math Geosciences 47:867–883, doi: 10.1007/s11004-014-9577-3.

Ruskeepää H (2017): Method of Support Vector Regression, http://demonstrations.wolfram.com/MethodOfSupportVectorRegression/

Sargent J S, Stachurski J (2017): Linear Regression in Python, Lectures in Quantitative Economics https://lectures.quantecon.org/py/ols.html

Singh A (2018): A Practical Introduction to K-Nearest Neighbors Algorithm for Regression (with Python code), https://www.analyticsvidhya.com/blog/2018/08/k-nearest-neighbor-introduction-regression-python/

Stephens T (2016): gplearn, Genetic Programming in Python, https://gplearn.readthedocs.io/en/stable/

Straw A (2009) RANSAC, SciPy Cookbook https://scipy-cookbook.readthedocs.io/items/RANSAC.html

VanderPlas J (2016): In Depth: Linear Regression, Python Data Science Handbook

Zielesny A (2016): From Curve Fitting to Machine Learning, Springer, Heidelberg, New York

Chapter 5
Neural Networks

5.1 Single Layer Perceptron

Basic theory

Neural network is a special nonlinear model for classification, clustering as well as regression. A single layer network has *m* input nodes plus a virtual input, called bias. The weighted linear combination of these input values enter into the active node, where it will be transformed by a so called activation function (mostly nonlinear). If this activation function is a threshold function, we call the network (Sullivan 2017) a *Perceptron type* network. The transformed signal will be the output of the network. In case of supervised learning, this output will be compared with the labeled output value. During the training process the weights of the network will be modified in order to minimized the deviation between the actual and the labeled outputs, see Fig. 5.1.

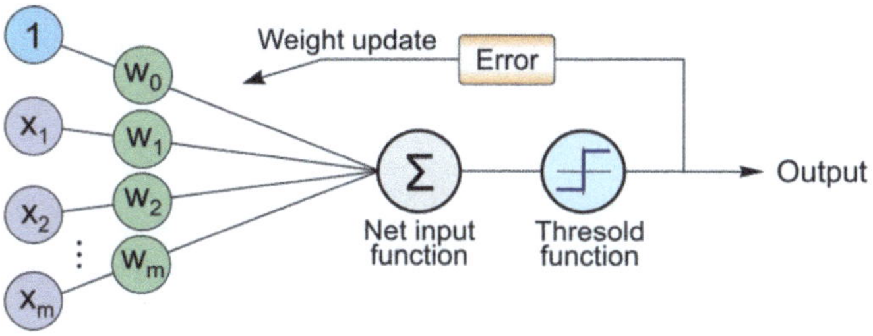

Fig. 5.1 The structure of a single layer Perceptron type neural network.

Starting Python session in *Mathematica,*

J. Awange et al., *Hybrid Imaging and Visualization*,
https://doi.org/10.1007/978-3-030-26153-5_5

```
⇒ session=
    StartExternalSession[<|"System" → "Python",
     "Version" → "3.5.4","Executable" →
      "C:\Users\Ben\AppData\Local\Programs\Python\Python35\
        python.exe"|>]//Quiet
```

```
⇐ ExternalSessionObject[
   SummaryPanel[ System: Python   EvaluationCount: None
                 UUID: 7b76966c2-ebb7-4ee8-bb25-02322d1456a8 ]]
```

5.1.1 Single Layer Perceptron Classifier

Here we employ such a network for classification.

Python

Let us load the *Python* code of a *Perceptron* type single layer network.

```python
import numpy as np

class Perceptron(object):
    """Perceptron classifier.

    Parameters
    ------------
    eta : float
      Learning rate (between 0.0 and 1.0)
    n_iter : int
      Passes over the training dataset.
    random_state : int
      Random number generator seed for random weight
      initialization.

    Attributes
    -----------
    w_ : 1d-array
      Weights after fitting.
    errors_ : list
      Number of misclassifications (updates) in each epoch.

    """
    def __init__(self, eta=0.01, n_iter=50, random_state=1):
        self.eta = eta
        self.n_iter = n_iter
        self.random_state = random_state
```

```
    def fit(self, X, y):
      """Fit training data.

      Parameters
      ----------
      X : {array-like}, shape = [n_samples, n_features]
        Training vectors, where n_samples is the number of
          samples and n_features is the number of features.
      y : array-like, shape = [n_samples]
        Target values.

      Returns
      -------
      self : object
      """
      rgen = np.random.RandomState(self.random_state)
      self.w_ =
        rgen.normal(loc=0.0,scale=0.01,size=1+X.shape[1])
      self.errors_ = []

      for _ in range(self.n_iter):
        errors = 0
        for xi, target in zip(X, y):
          update = self.eta * (target - self.predict(xi))
          self.w_[1:] += update * xi
          self.w_[0] += update
          errors += int(update != 0.0)
        self.errors_.append(errors)
      return self

    def net_input(self, X):
        """Calculate net input"""
        return np.dot(X, self.w_[1:]) + self.w_[0]

    def predict(self, X):
        """Return class label after unit step"""
        return np.where(self.net_input(X) >= 0.0, 1, -1)
```

Data are downloaded from *Python* repository as `mglearn`, and saved for visualization with *Mathematica*,

```
import mglearn
import numpy as np
X, y = mglearn.datasets.make_forge()
np.savetxt('M:\\dataX.txt',X,fmt='%.5e')
```

If the file is short, we do not write the data into a file. So in case of the labels, *y*

```
y
```

⇐ `{1,0,1,0,0,1,1,0,1,1,1,1,0,0,1,1,1,0,0,1,0,0,0,0,1,0}`

⇒ `v=%;`

We relabel (1, 0) into (1, −1) suited to the threshold function in a better way

⇒ `z=Map[If[#==0,-1,1]&,v]`

```
⇐ {1,-1,1,-1,-1,1,1,-1,1,1,1,1,-1,-1,1,1,1,-1,-1,1,-1,-1,-1,-1,1,-1}
```

Import the data set and visualize it, see Fig. 5.2.

```
⇒ u=Import["M:\\dataX.txt","Table"];
```

The data set to be classified can be standardized, which means shifting and rescaling the elements of *list* to have zero mean and unit sample variance.

```
⇒ uu=Standardize[u];
⇒ class1={};class2={};
⇒ MapThread[If[#1==
    -1,AppendTo[class1,#2],AppendTo[class2,#2]]&,{z,uu}];
⇒ p0=ListPlot[{class1,class2},PlotStyle → {Green,Red},Frame → True,
    Axes → None,PlotMarkers → {Automatic,Large},AspectRatio → 1]
```

⇐

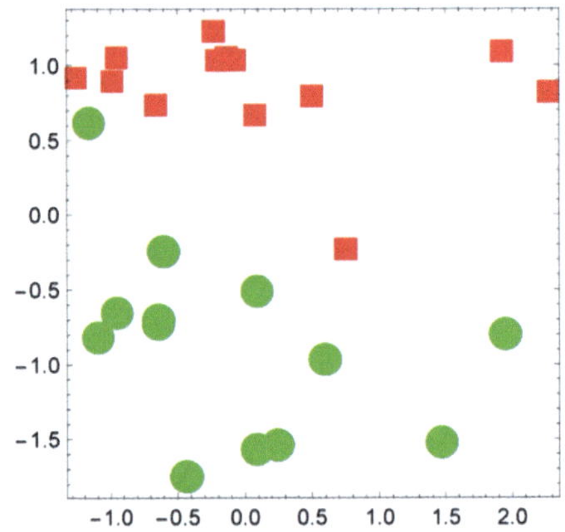

Fig. 5.2 Sample dataset to be classified.

Let us employ *Python* with the relabeled data

```
yy=[1,-1,1,-1,-1,1,1,-1,1,1,1,1,-1,-1,
    1,1,1,-1,-1,1,-1,-1,-1,-1,1,-1]
```

The training process

```
ppn=Perceptron(eta=0.5,n_iter=5000).fit(X,yy)
```

Let us employ the trained network for prediction

```
ppn.predict(X))
```

```
⇐ {1,-1,1,-1,1,1,1,-1,1,1,1,1,-1,-1,1,1,1,-1,-1,1,-1,-1,-1,-1,1,-1}
⇒ zP=%;
```

The error of the classification on the training process

```
⇒ zP-z
```

```
⇐ {1,-1,1,-1,-1,1,1,-1,1,1,1,1,-1,-1,1,1,1,-1,-1,1,-1,-1,-1,-1,1,-1}
```

Only one element is misclassified.

Now we employ *Mathematica* with a "softer" threshold function as activation function, see Fig. 5.3 (Freeman, 1994).

```
⇒ Plot[Tanh[x],{x,-3,3}]
```

⇐

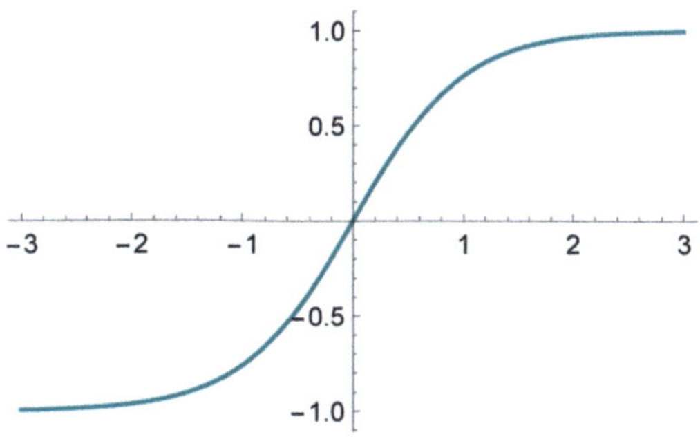

Fig. 5.3 Tanh(x) activation function.

In order to employ this layer we define a special layer,

```
⇒ myL=ElementwiseLayer[Tanh[#]&]
⇐ ElementwiseLayer[ Function: tanh(x)
                    Output:   tensor ]
```

Mathematica

Let us construct a single layer network with two dimensional input,

```
⇒ net=NetInitialize@NetChain[
    {LinearLayer[1,"Input" → 2,"Biases" → -1],myL}]
⇐ NetChain[   Input vector              (size: 2)
            1 LinearLayer vector        (size: 1)
            2 Tanh[x] vector            (size: 1)
              Output vector             (size: 1) ]
```

Preparation of the training input → output data,

```
⇒ trainingData=MapThread[#2 → {#1}&,{z,uu}];
```

Then let us train the network,

```
⇒ trained=NetTrain[net,trainingData,MaxTrainingRounds → 10000]
⇐ NetChain[ Input port:        vector (size: 2)
            Output port:       vector (size: 1)
            Number of layers:  2 ]
⇒ zu=Map[trained[#]&,uu]//Flatten//Round
⇐ {1,-1,1,-1,0,1,1,-1,1,1,1,0,-1,-1,1,1,1,-1,-1,1,-1,-1,-1,-1,1,0}
```

The error of the network is

```
⇒ zu=z
⇐ {0,0,0,0,1,0,0,0,0,0,0,-1,0,0,0,0,0,0,0,0,0,0,0,0,0,1}
```

Now we have three misclassified elements, see Fig. 5.4.

```
⇒ Show[{p0,DensityPlot[trained[{x,y}],{x,-1.4,2.4},{y,-2,1.5},
    PlotPoints → 250,ColorFunction → "CMYKColors"],p0}]
⇐
```

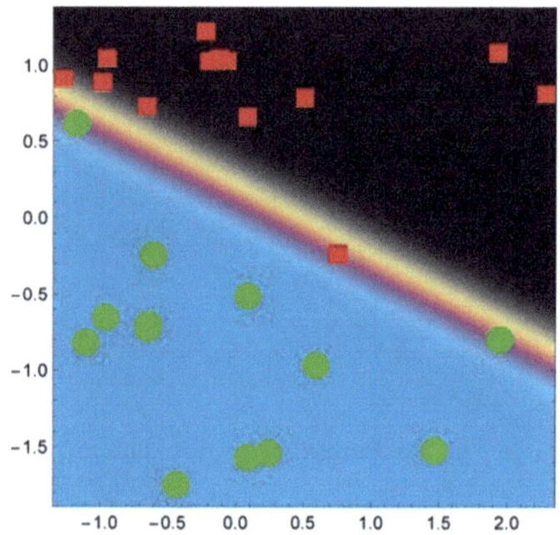

Fig. 5.4 Results of the classification.

With steeper activation function (see Fig 5.5) we can get more robust results.

```
⇒ Plot[Tanh[5 x],{x,-3,3}]
⇐
```

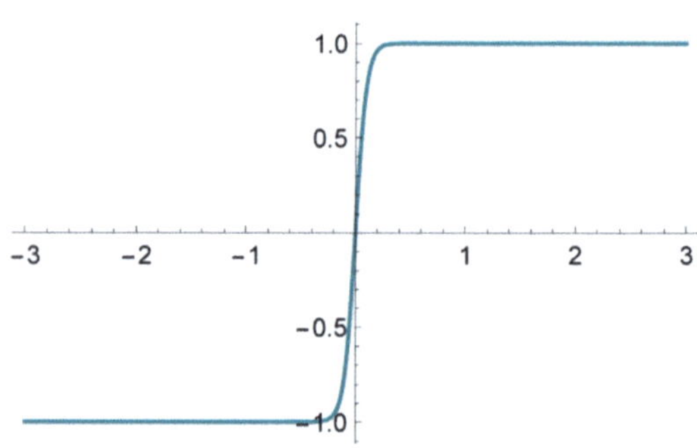

Fig. 5.5 Modified activation function.

```
⇒ myL=ElementwiseLayer[Tanh[5 #]&]
⇐ ElementwiseLayer[ Function: tanh(5x)
                    Output:   tensor ]
⇒ net=NetInitialize@NetChain[
    {LinearLayer[1,"Input" → 2,"Biases" → -1],myL}]
⇐ NetChain[ Input port:        vector (size: 2)
            Output port:       vector (size: 1)
            Number of layers:  2 ]
```

```
⇒ trained1=NetTrain[net,trainingData,MaxTrainingRounds → 10000]
⇐ NetChain[ Input port:         vector (size: 2)
            Output port:        vector (size: 1)
            Number of layers: 2 ]

⇒ zu=Map[trained[#]&,uu]//Flatten//Round
⇐ {1,-1,1,-1,1,1,1,-1,1,1,1,1,-1,-1,1,1,1,-1,-1,1,-1,-1,-1,-1,1,-1}

⇒ zu=-z
⇐ {0,0,0,0,2,0,0,0,0,0,0,0,0,0,0,0,0,0,0,0,0,0,0,0,0,0}

⇒ Show[{p0,DensityPlot[trained1[{x,y}],{x,-1.4,2.4},{y,-2,1.5},
    PlotPoints → 250,ColorFunction → "CMYKColors"],p0}]
⇐
```

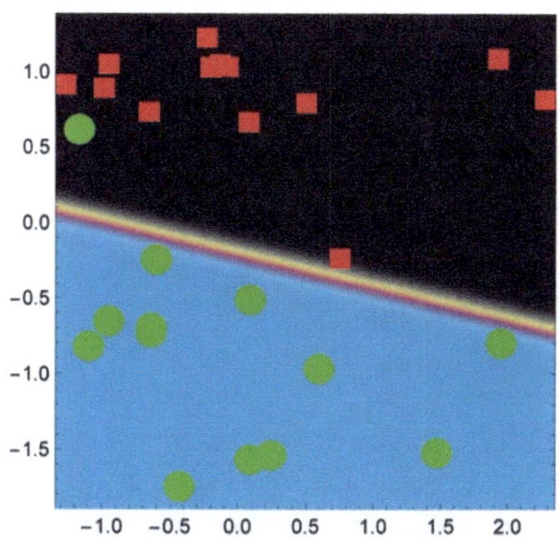

Fig. 5.6 Result of the classification employing steeper tanh($5x$) activation function.

Still, one element is misclassified but we have a considerable margin between the two classes hence the solution is more robust.

5.2 Multi Layer Perceptron

Basic theory

Employing more than one layer, we can considerably improve the efficiency of the nonlinear mapping of the network. For example, linearly non-separable classification problem can be solved using multi layer neural network, see section 5.2.1. However the training algorithm for optimizing the network weights should be different and more complicated than in the case of single layer networks. We should keep it in mind that the network training is a multivariable, global optimization problem, which has more local optimums, and so many variables as many weights in the networks. One of the most popular training methods is the backpropagation technique, which can utilize the feature of the actual activation function (Deshpande 2017).

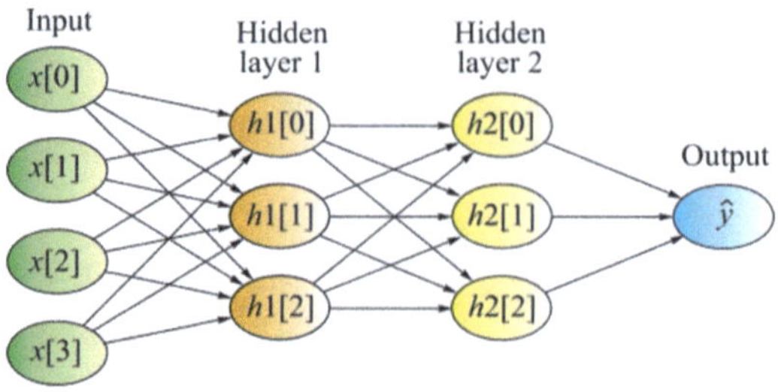

Fig. 5.7 Structure of multi layer neural network.

5.2.1 Multi Layer Perceptron Classifier

Python

Loading some standard procedure

```
from numpy import array, matrix
from scipy.io import mmread, mmwrite
import numpy as np
```

Let us consider the moons data set (an artificial data set from *Python sklearn.dataset library*), which represents a linearly non-separable problem (Chollet 2018).

```
from sklearn.datasets import make_moons
X, y =make_moons(n_samples=100,noise=0.25,random_state=3)
```

Save the coordinate pairs for *Mathematica*,

```
mmwrite('pubi.mtx',X)
```

Import data into *Mathematica*,

```
⇒ trainX=Import["pubi.mtx"];
```

The labels of the data points

```
y
```

```
⇐ {1,1,0,1,1,1,1,0,0,0,1,0,1,1,0,1,1,0,1,0,1,0,0,0,1,
   1,0,1,0,1,0,0,0,0,1,0,1,1,0,0,0,0,0,0,1,1,1,1,1,0,
   1,1,1,1,0,1,1,1,1,1,0,1,1,1,0,1,0,0,0,0,0,1,1,0,1,
   1,0,1,1,0,1,0,1,0,0,0,0,0,0,1,0,0,1,0,0,0,1,1,0,0}
```

Let us visualize the data set

```
⇒ cluster=%;total=MapThread[{#1,#2}&,{trainX,cluster}];
⇒ clust1=Select[total,#[[2]]==0&];
⇒ clust2=Select[total,#[[2]]==1&];
⇒ pclust1=Map[#[[1]]&,clust1];
⇒ pclust2=Map[#[[1]]&,clust2];
```

Fig. 5.8 shows the two clusters,

```
⇒ p0=Show[{ListPlot[pclust1,PlotStyle → {Green,PointSize[0.017]}],
   ListPlot[pclust2,PlotStyle → {Red,PointSize[0.017]}]}]
```

⇐

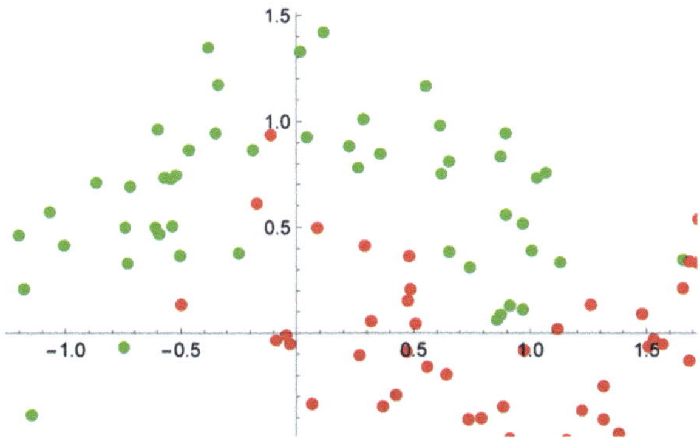

Fig. 5.8 The moons sample dataset to be classified.

Loading the network classifier

```
from sklearn.neural_network import MLPClassifier
import mglearn
```

Employing *Limited memory Broyden-Fletcher-Goldfarb-Shanno* (LBFGS) Saputro and Widyaningsih (2017) method for optimization of the network weights,

```
mlp = MLPClassifier(solver='lbfgs', random_state=0).fit(X,y)
```

Let us predict the membership of the training set

```
cupi=mlp.predict(X)
cupi
```

```
⇐ {1,1,0,1,1,1,1,0,0,0,1,0,1,1,0,1,1,0,1,0,1,0,0,0,1,
   1,0,1,0,1,0,0,0,0,1,0,1,1,0,0,0,0,0,0,1,1,1,1,1,0,
   1,1,1,1,0,1,1,1,1,1,0,1,1,1,0,1,0,0,0,0,0,1,1,0,1,
   1,0,1,1,0,1,0,1,0,0,0,0,0,0,1,0,0,1,0,0,0,1,1,0,0}
```

Comparing the result with the training labels

```
⇒ zu=%;
⇒ zu-cluster
```

```
⇐ {0,0,0,0,0,0,0,0,0,0,0,0,0,0,0,0,0,0,0,0,0,0,0,0,0,0,
   0,0,0,0,0,0,0,0,0,0,0,0,0,0,0,0,0,0,0,0,0,0,0,0,0,0,
   0,0,0,0,0,0,0,0,0,0,0,0,0,0,0,0,0,0,0,0,0,0,0,0,0,0,
   0,0,0,0,0,0,0,0,0,0,0,0,0,0,0,0,0,0,0,0,0,0,0,0,0,0}
```

The classification for the training set is perfect. Let us try to visualize the result with *Python*

```
> import matplotlib.pyplot as plt)
```

```
> mglearn.plots.plot_2d_separator(mlp, X, fill=True, alpha=.3)
```

```
> mglearn.discrete_scatter(X[:,0],X[:,1],y)
```

Running *Python* under *Mathematica* here gives an error message! No worries!

```
> plt.show()
```

This *Python* command provide the figure in a separate window,

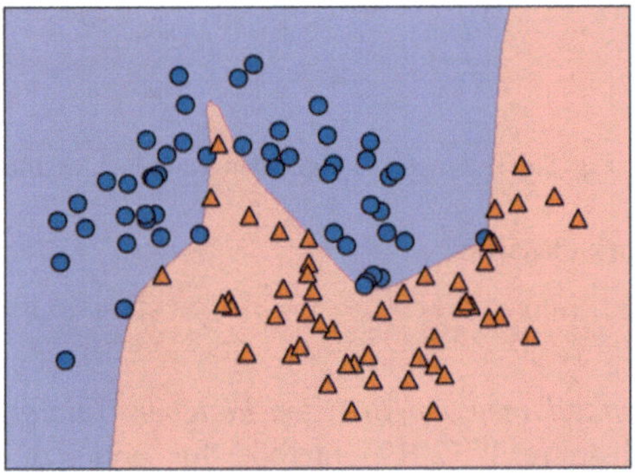

Fig. 5.9 Results of the classification via *Python*.

Mathematica

Let us employ the following activation function, see Fig. 5.10.

```
⇒ Plot[(Tanh[2x]+1)/2,{x,-2,2}]
```

⇐ 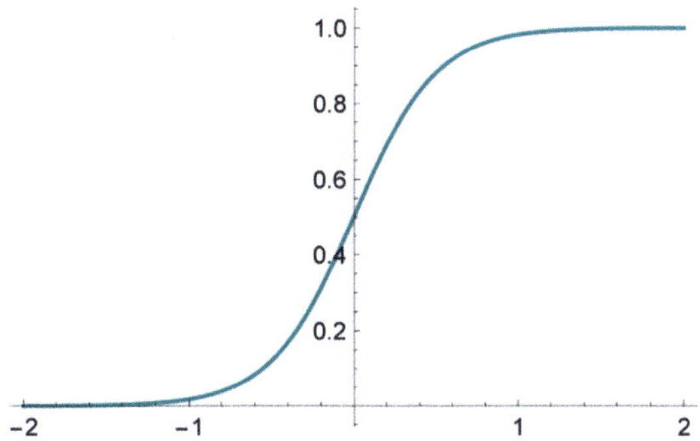

Fig. 5.10 The activation function.

The sigmoid type layer,

```
⇒ myL=ElementwiseLayer[(Tanh[2#]+1)/2&]
⇐ NetChain[ Function:  0.5 tanh(2x)+1
            Output:    tensor ]
```

Let us define the following network where the two linear layers have 15 nodes,

```
⇒ net=NetInitialize@NetChain[{LinearLayer[15,"Input" → 2],myL,
    LinearLayer[15],ElementwiseLayer[Ramp],LinearLayer[1]},
    "Input" → {2}]
⇐ NetChain[    Input              vector(size: 2)
             1 LinearLayer        vector(size: 15)
             2 0.5(1+Tanh[2x])    vector(size: 15)
             3 LinearLayer        vector(size: 15)
             4 Ramp               vector(size: 15)
             5 LinearLayer        vector(size: 1)
               Output             vector(size: 1) ]
```

Visualization of the network, see

```
⇒ NetGraph[net]
```

⇐ NetGraph[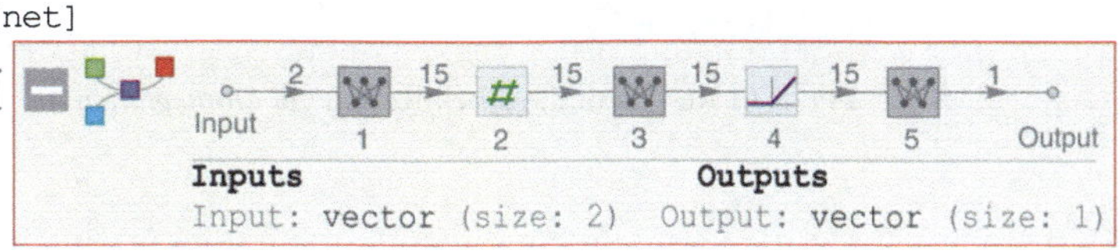]

Preparation of the training set

```
⇒ trainingset=Map[#[[1]] → {#[[2]]}&,total];
⇐ NetChain[ Input port:        vector (size: 2)
            Output port:       vector (size: 1)
            Number of layers:  5 ]
```

The computed labels

```
⇒ uu=Map[#[[1]]&,trainingset];
⇒ zu=Map[NetTrain[#]&,uu]//Flatten//Round
```

```
⇐ {1,1,0,1,1,1,1,0,0,0,1,0,1,1,0,1,1,0,1,0,1,0,0,0,1,
   1,0,1,0,1,0,0,0,0,1,0,1,1,0,0,0,0,0,0,1,1,1,1,1,0,
   1,1,1,1,0,1,1,1,1,1,0,1,1,1,0,1,0,0,0,0,0,1,1,0,1,
   1,0,1,1,0,1,0,1,0,0,0,0,0,0,1,0,0,1,0,0,0,1,1,1,0}
```

The error

```
⇒ zu-cluster
⇐ {0,0,0,0,0,0,0,0,0,0,0,0,0,0,0,0,0,0,0,0,0,0,0,0,0,
   0,0,0,0,0,0,0,0,0,0,0,0,0,0,0,0,0,0,0,0,0,0,0,0,0,
   0,0,0,0,0,0,0,0,0,0,0,0,0,0,0,0,0,0,0,0,0,0,0,0,0,
   0,0,0,0,0,0,0,0,0,0,0,0,0,0,0,0,0,0,0,0,0,0,0,1,0}
```

Let us visualize the result, see Fig. 5.11.

```
⇒ Show[{p0,DensityPlot[trained[{x,y}]//Flatten//Round,{x,-1.4,1.8},
   {y,-0.6,1.5},PlotPoints → 50],p0},AspectRatio → 0.9]
```

⇐

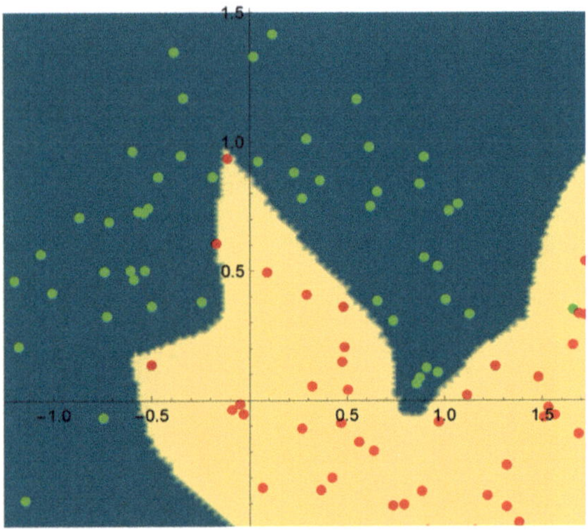

Fig. 5.11 Results of the classification via *Mathematica*.

5.2.2 Multi Layer Perceptron Regressor

Neural network can also be efficiently employed for regression (Melnikov, 2017). Let us illustrate this ability with the regression of a single variable function,

Python

Loading the *Python* regressor

```
from sklearn.neural_network import MLPRegressor
import numpy as np
import matplotlib.pyplot as plt
import random
```

Generating the points for the independent variables

```
x=np.arange(0.,1.,0.01).reshape(-1,1)
```

The function values

```
y=np.sin(0.3*np.pi*x).ravel()+np.cos(3*np.pi*x*x).ravel()+
  np.random.normal(0,0.15,x.shape).ravel()
y
```

```
⇐ {0.90955,1.11354,1.1036,0.94152,0.651125,0.756493,1.08785,0.96283,
   0.90550,1.05755,1.08715,0.89628,1.21879,1.12197,1.35395,1.32278,
   1.06119,1.33492,0.98626,1.16188,0.87097,1.39854,1.12806,0.97950,
    ⋮
   1.51184,1.41921,1.57758,1.29661,1.34853,1.02236,0.81539,0.63867,
   0.519081,0.516352,0.360143,0.064391,0.14143,-0.305298,-0.388398}
```

```
⇒ g=%;
```

Let us visualize the data points, see Fig. 5.12.

```
⇒ ListPlot[g]
```

⇐

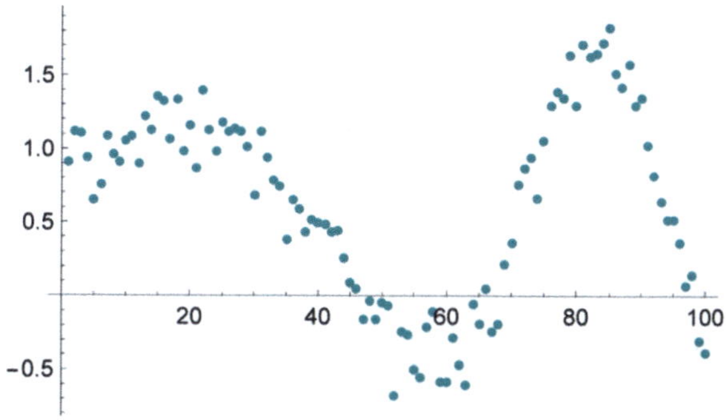

Fig. 5.12 Noisy data of a function to be approximated.

We consider a neural network 12 layers with tanh(x) activation function. Using the same LBFGS optimization we train the net (Raschka 2018).

```
nn=MLPRegressor(hidden_layer_sizes=(12),activation='tanh',
 solver='lbfgs').fit(x,y)
```

Let us employ it for these test points as input values

```
test_x=np.arange(-0.01,1.02,0.01).reshape(-1,1)
test_y=nn.predict(test_x)
```

The visualization

```
fig = plt.figure()
```

```
ax1=fig.add_subplot(111)
```

```
ax1.scatter(x,y,s=5,c='b',marker="o",label='real')
```

The error message does not mean problem.

```
ax1.plot(test_x,test_y,c='r',label='NN Prediction')
```

The results can be seen in Fig. 5.13.

```
plt.show()
```

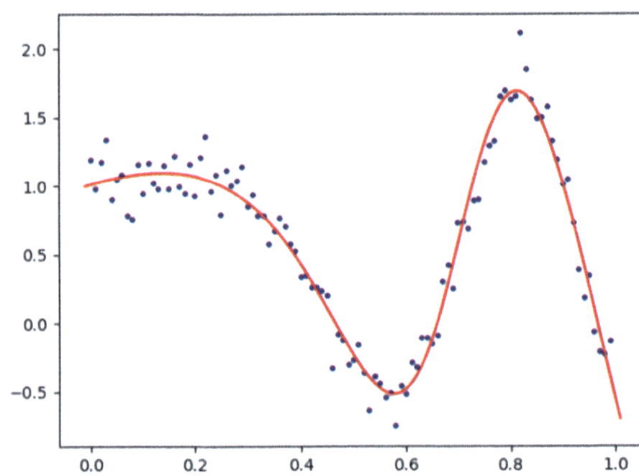

Fig. 5.13 Function approximation via *Python*.

Mathematica

The training data

```
⇒ yp=g;n= Length[yp];
⇒ xp=Table[i 0.01,{i,1,n}];
⇒ data=Transpose[{xp,yp}];
⇒ p0=ListPlot[data,Frame → True,Axes → None]
```

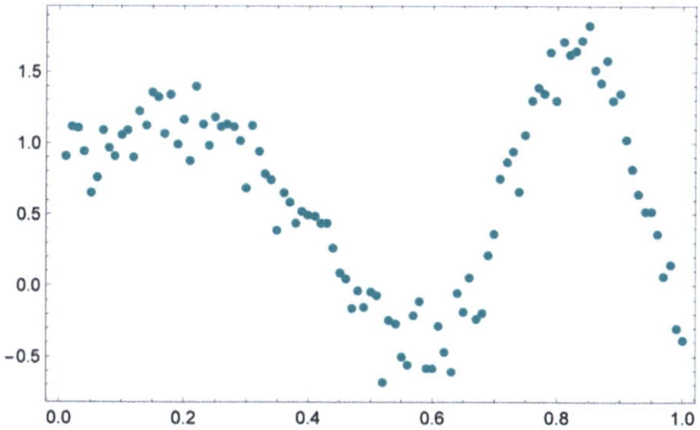

⇐

Fig. 5.14 The training data.

```
⇒ trainingdata=Map[{#[[1]]} → {#[[2]]}&,data];
```

Initializing the network

```
⇒ net=NetInitialize@NetChain[{LinearLayer[12,"Input" → 1],
    ElementwiseLayer[Tanh],LinearLayer[5],
    ElementwiseLayer[Tanh],LinearLayer[1]},"Input" → {1}]
```

⇐ NetChain[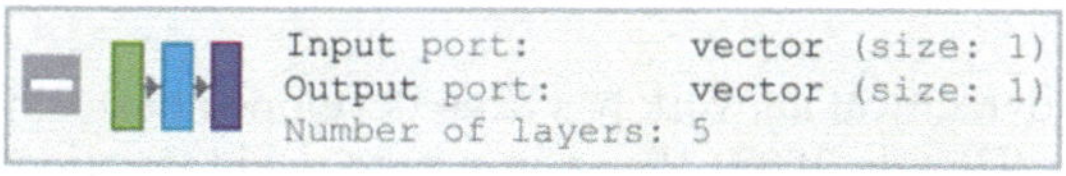]


```
⇒ NetGraph[net]
```

⇐ NetGraph[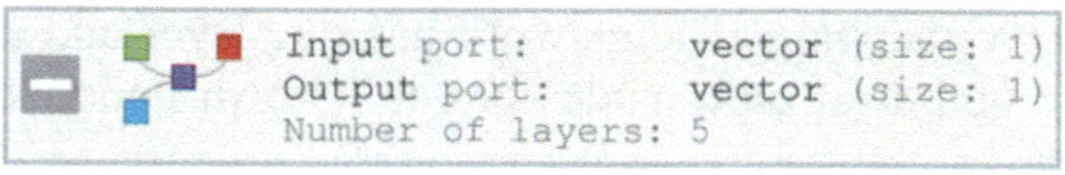]

Training the network

```
⇒ trained=NetTrain[net,trainingdata]
```

⇐ NetChain[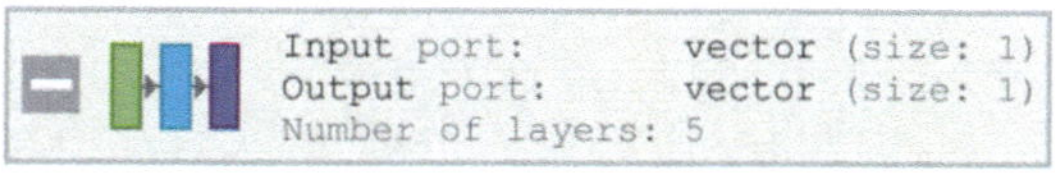]

The result of *Mathematica* can be seen in Fig. 5.15.

```
⇒ Show[{p0,Plot[trained[u],{u,0,1},PlotStyle → Red]},
    Frame → True,Axes → None]
```

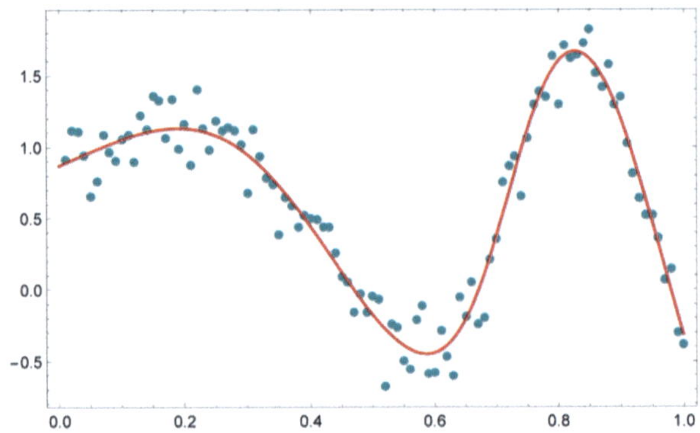

⇐

Fig. 5.15 Function approximation via *Mathematica*.

5.3 Hopfield Network

Basic theory

This most simple dynamical network has been developed basically to solve classification problem (Haykin 2009). The aim was to avoid iteration in the training process. However employing the trained network for classification (recognition) one needs successive mapping. The network has one layer and every node has feedback from the output of other nodes beside its own node, see Fig. 5.16 in case of $n = 2$ nodes

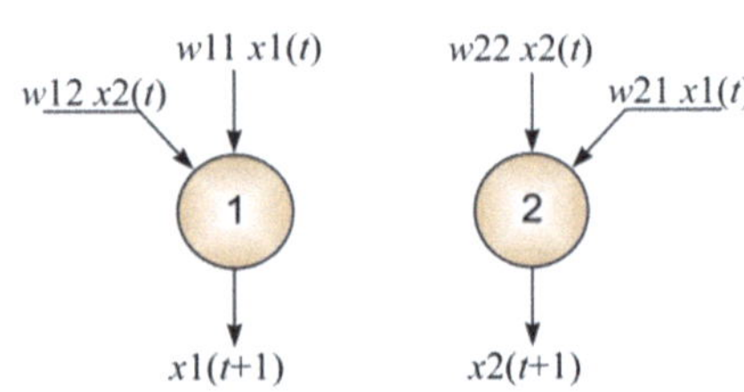

Fig. 5.16 Discrete Hopfield network for $n = 2$ nodes.

The network represents an auto associative memory, which means that its output is associated to its input, in our case

$$\begin{pmatrix} x1(t+1) \\ x2(t+1) \end{pmatrix} = \psi\left(W\begin{pmatrix} x1(t) \\ x2(t) \end{pmatrix}\right)$$

where W is a constant weight matrix represented by the fixed points of the network and ψ is the *Signum* function as activation function. For a fixed point vector

$$x_F = \psi\left(W x_F\right).$$

The vector should be bipolar vectors having elements 1 or −1. Let us consider the weight matrix to be computed without iteration. As an example, let two fixed points be

$$a = \begin{pmatrix} 1 \\ 1 \\ -1 \end{pmatrix} \quad \text{and} \quad b = \begin{pmatrix} 1 \\ -1 \\ 1 \end{pmatrix}.$$

Then, the weight matrix W can be computed as a sum of the outer product of these vectors, namely

$$W = a\,a^T + b\,b^T = \begin{pmatrix} 1 \\ 1 \\ -1 \end{pmatrix}\left(1 \times 1 - 1\right) + \begin{pmatrix} 1 \\ -1 \\ 1 \end{pmatrix}\left(1 - 1 \times 1\right) = \begin{pmatrix} 2 & 0 & 0 \\ 0 & 2 & -2 \\ 0 & -2 & 2 \end{pmatrix}.$$

Applying the network to one of the fixed points, for example

$$\psi\left(Wa\right) = \psi\begin{pmatrix} 2 \\ 4 \\ -4 \end{pmatrix} = \begin{pmatrix} 1 \\ 1 \\ -1 \end{pmatrix} = a.$$

When an input vector is close to one of the fixed points, then the network output will converge to this fixed point vector. However, it may happen that for an input vector, an unknown output is obtained.

5.3.1 Recovery of Digits

The first elementary example illustrates how to recover a digit from its deteriorated digital image. We consider three digits 0, 1 and 2 as equilibrium points of our network and try to recognize imperfect digits, which can be similar to the equilibrium (attractor) digits (Srinivasan et al. 1993).

Python

Load the procedures to handle arrays

```
import numpy as np
from neupy import algorithms
```

A binary image of zero digit in bipolar representation of 5×5 resolution. In *Python* we should use binary, not bipolar representation

```
zero=np.matrix([
    0, 1, 1, 1, 0,
    1, 0, 0, 0, 1,
    1, 0, 0, 0, 1,
    1, 0, 0, 0, 1,
    1, 0, 0, 0, 1,
    0, 1, 1, 1, 0
    ])
```

In vector form

```
zero
```

```
⇐ {{0,1,1,1,0,1,0,0,0,1,1,0,0,0,1,1,0,0,0,1,1,0,0,0,1,0,1,1,1,0}}
⇒ zero=%;
```

Let us visualize it in *Mathematica*, see Fig. 5.17.

```
⇒ Partition[zero//Flatten,5]//MatrixForm
```

$$\Leftarrow \begin{pmatrix} 0 & 1 & 1 & 1 & 0 \\ 1 & 0 & 0 & 0 & 1 \\ 1 & 0 & 0 & 0 & 1 \\ 1 & 0 & 0 & 0 & 1 \\ 1 & 0 & 0 & 0 & 1 \\ 0 & 1 & 1 & 1 & 0 \end{pmatrix}$$

```
⇒ MatrixPlot[%,Mesh → True]
```

⇐

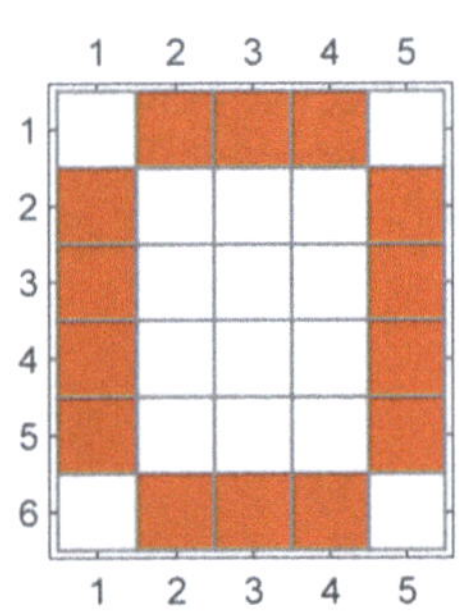

Fig. 5.17 Zero digit, using *Python*.

Similarly, the digits 1 and 2 can be represented by

```
one=np.matrix([
    0, 0, 1, 0, 0,
    0, 1, 1, 0, 0,
    1, 0, 1, 0, 0,
    0, 0, 1, 0, 0,
    0, 0, 1, 0, 0,
    0, 0, 1, 0, 0
    ])
```

```
one
```

⇐ {{0,0,1,0,0,0,1,1,0,0,1,0,1,0,0,0,0,1,0,0,0,0,1,0,0,0,0,1,0,0}}

⇒ one=%;

⇒ Partition[one//Flatten,5]//MatrixForm

$$\Leftarrow \begin{pmatrix} 0 & 0 & 1 & 0 & 0 \\ 0 & 1 & 1 & 0 & 0 \\ 1 & 0 & 1 & 0 & 0 \\ 0 & 0 & 1 & 0 & 0 \\ 0 & 0 & 1 & 0 & 0 \\ 0 & 0 & 1 & 0 & 0 \end{pmatrix}$$

⇒ MatrixPlot[%,Mesh → True]

⇐

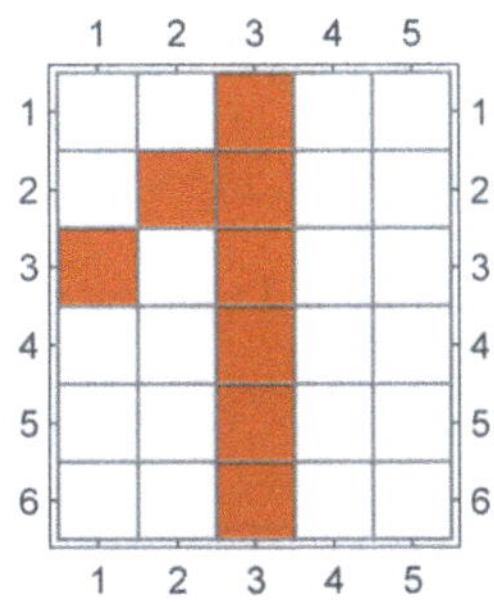

Fig. 5.18 Digit 1 using *Python.*

```
two=np.matrix([
    0, 1, 1, 0, 0,
    1, 0, 0, 1, 0,
    0, 0, 0, 1, 0,
    0, 1, 1, 0, 0,
    1, 0, 0, 0, 0,
    1, 1, 1, 1, 1
    ])
```

```
two
```

⇐ {{0,1,1,0,0,1,0,0,1,0,0,0,0,1,0,0,1,1,0,0,1,0,0,0,0,1,1,1,1,1}}

⇒ two=%;

⇒ Partition[two//Flatten,5]//MatrixForm

$$\Leftarrow \begin{pmatrix} 0 & 1 & 1 & 0 & 0 \\ 1 & 0 & 0 & 1 & 0 \\ 0 & 0 & 0 & 1 & 0 \\ 0 & 1 & 1 & 0 & 0 \\ 1 & 0 & 0 & 0 & 0 \\ 1 & 1 & 1 & 1 & 1 \end{pmatrix}$$

⇒ MatrixPlot[%,Mesh → True]

⇐ 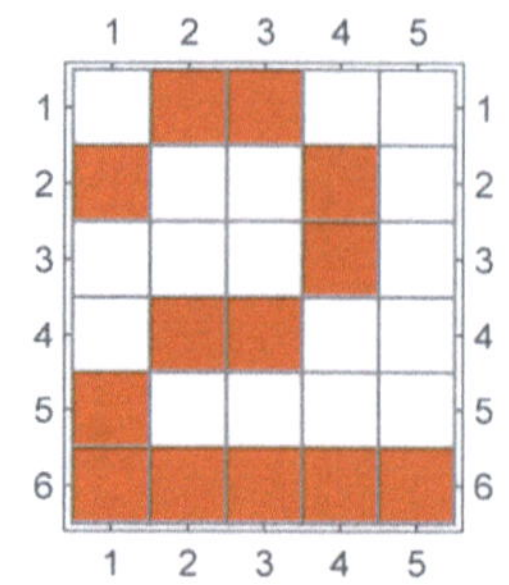

Fig. 5.19 Digit 2 using *Python*.

Preparation of the fixed points data for *Python*

```
data = np.concatenate([zero, one,two], axis=0)
```

Loading discrete Hopfield network and training it

```
dhnet = algorithms.DiscreteHopfieldNetwork(mode='sync')
dhnet.train(data)
```

Let us try to employ our trained network for recognizing an imperfect zero digit, Fig. 5.20.

```
half_zero=np.matrix([
    0, 1, 1, 1, 0,
    1, 0, 0, 0, 1,
    1, 0, 0, 0, 1,
    0, 0, 0, 0, 0,
    0, 0, 0, 0, 0,
    0, 0, 0, 0, 0
    ])
```

```
half_zero
```

```
⇐ {{0,1,1,1,0,1,0,0,0,1,1,0,0,0,1,0,0,0,0,0,0,0,0,0,0,0,0,0,0,0}}
⇒ halfzero=%;
⇒ Partition[halfzero//Flatten,5]//MatrixForm
```

$$\Leftarrow \begin{pmatrix} 0 & 1 & 1 & 1 & 0 \\ 1 & 0 & 0 & 0 & 1 \\ 1 & 0 & 0 & 0 & 1 \\ 0 & 0 & 0 & 0 & 0 \\ 0 & 0 & 0 & 0 & 0 \\ 0 & 0 & 0 & 0 & 0 \end{pmatrix}$$

```
⇒ MatrixPlot[%,Mesh → True]
```

⇐

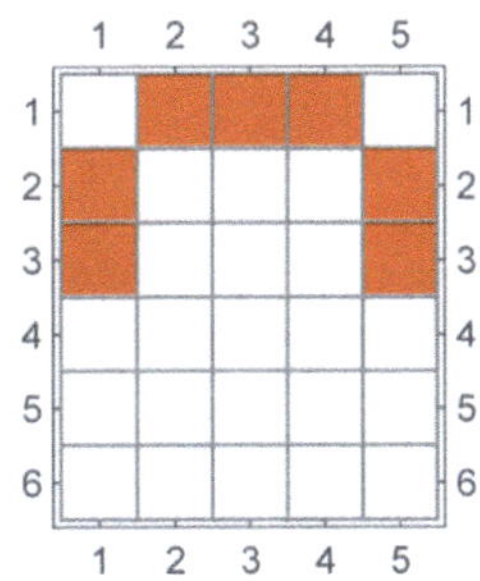

Fig. 5.20 Imperfect digit 0.

Now let us employ our Hopfield net to try to recover the digit

```
result = dhnet.predict(half_zero)
result
```

⇐ {{0,1,1,1,0,1,0,0,0,1,1,0,0,0,1,1,0,0,0,1,1,0,0,0,1,0,1,1,1,0}}

The result is shown by Fig. 5.21.

```
⇒ z=%;
⇒ Partition[z//Flatten,5]//MatrixForm
```

⇐ $\begin{pmatrix} 0 & 1 & 1 & 1 & 0 \\ 1 & 0 & 0 & 0 & 1 \\ 1 & 0 & 0 & 0 & 1 \\ 1 & 0 & 0 & 0 & 1 \\ 1 & 0 & 0 & 0 & 1 \\ 0 & 1 & 1 & 1 & 0 \end{pmatrix}$

```
⇒ MatrixPlot[%,Mesh → True]
```

⇐

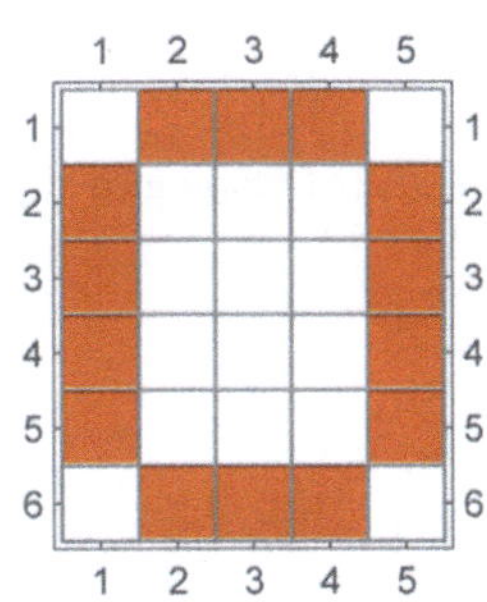

Fig. 5.21 Recovered digit 0.

Now let us try the same with a deteriorated digit 2, see Fig. 5.22.

```python
half_two=np.matrix([
    0, 0, 0, 0, 0,
    0, 0, 0, 0, 0,
    0, 0, 0, 0, 0,
    0, 1, 1, 0, 0,
    1, 0, 0, 0, 0,
    1, 1, 1, 1, 1
    ])
```

```python
half_two
```

```
⇐ {{0,0,0,0,0,0,0,0,0,0,0,0,0,0,0,0,1,1,0,0,1,0,0,0,0,1,1,1,1,1}}
⇒ z=%;
⇒ Partition[z//Flatten,5]//MatrixForm
```

$$\Leftarrow \begin{pmatrix} 0 & 0 & 0 & 0 & 0 \\ 0 & 0 & 0 & 0 & 0 \\ 0 & 0 & 0 & 0 & 0 \\ 0 & 1 & 1 & 0 & 0 \\ 1 & 0 & 0 & 0 & 0 \\ 1 & 1 & 1 & 1 & 1 \end{pmatrix}$$

```
⇒ MatrixPlot[%,Mesh → True]
```

⇐

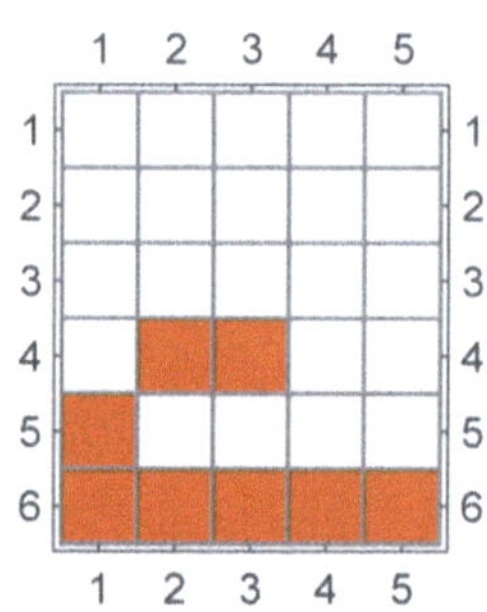

Fig. 5.22 Deteriorated digit 2.

The result can be seen in Fig. 5.23.

```python
result = dhnet.predict(half_two)
result
```

```
⇐ {{0,1,1,0,0,1,0,0,1,0,0,0,0,1,0,0,1,1,0,0,1,0,0,0,0,1,1,1,1,1}}
⇒ z=%;
⇒ Partition[z//Flatten,5]//MatrixForm
```

$$\Leftarrow \begin{pmatrix} 0 & 1 & 1 & 0 & 0 \\ 1 & 0 & 0 & 1 & 0 \\ 0 & 0 & 0 & 1 & 0 \\ 0 & 1 & 1 & 0 & 0 \\ 1 & 0 & 0 & 0 & 0 \\ 1 & 1 & 1 & 1 & 1 \end{pmatrix}$$

```
⇒ MatrixPlot[%,Mesh → True]
```

⇐ 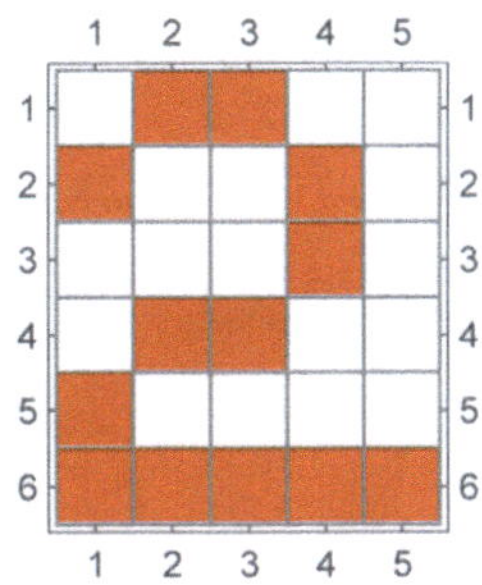

Fig. 5.23 Recovered deteriorated digit 2.

Now let us try a different image, see Fig. 5.24.

```
half_what=np.matrix([
     1, 1, 1, 0, 0,
     0, 0, 0, 1, 0,
     0, 0, 0, 1, 0,
     0, 0, 0, 0, 0,
     0, 0, 0, 0, 0,
     0, 0, 0, 0, 0
     ])
```

```
half_what
```

⇐ {{1,1,1,0,0,0,0,0,1,0,0,0,0,1,0,0,0,0,0,0,0,0,0,0,0,0,0,0,0,0}}

⇒ z=%;

⇒ Partition[z//Flatten,5]//MatrixForm

⇐
$$\begin{pmatrix} 1 & 1 & 1 & 0 & 0 \\ 0 & 0 & 0 & 1 & 0 \\ 0 & 0 & 0 & 1 & 0 \\ 0 & 0 & 0 & 0 & 0 \\ 0 & 0 & 0 & 0 & 0 \\ 0 & 0 & 0 & 0 & 0 \end{pmatrix}$$

⇒ MatrixPlot[%,Mesh → True]

⇐ 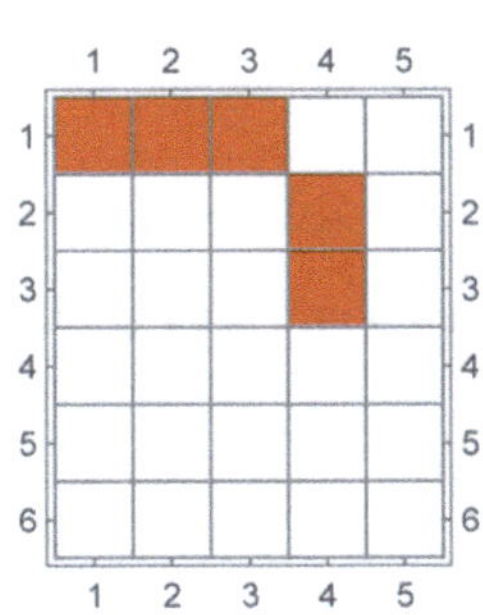

Fig. 5.24 A piece of what (an unknown digit to be recognized) as input.

Employing our net, we get an unknown equilibrium "digit", see Fig. 5.25

```
result = dhnet.predict(half_what)
result
```

⇐ {{0,0,1,0,0,1,1,1,0,0,1,0,1,0,0,0,1,1,0,0,1,0,1,0,0,1,1,1,1,1}}

⇒ z=%;

⇒ Partition[z//Flatten,5]//MatrixForm

$$\Leftarrow \begin{pmatrix} 0 & 0 & 1 & 0 & 0 \\ 1 & 1 & 1 & 0 & 0 \\ 1 & 0 & 1 & 0 & 0 \\ 0 & 1 & 1 & 0 & 0 \\ 1 & 0 & 1 & 0 & 0 \\ 1 & 1 & 1 & 1 & 1 \end{pmatrix}$$

⇒ MatrixPlot[%,Mesh → True]

⇐

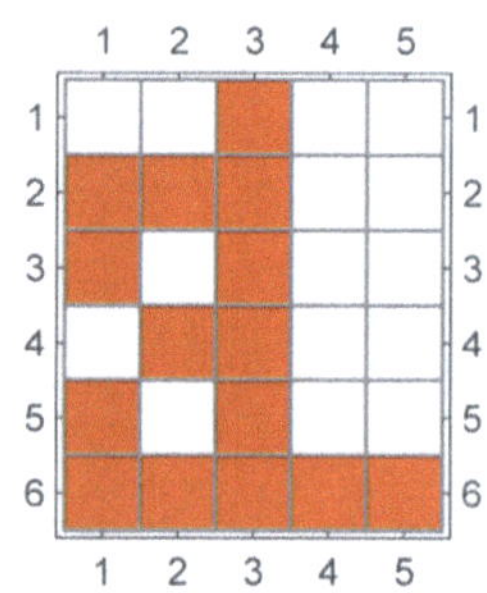

Fig. 5.25 Unknown equilibrium as response.

Unknown fixed point! We can solve this problem using the asynchronous network approach.

```
from neupy import environment
environment.reproducible()
```

In addition let us employ 400 iterations for recognition

```
dhnet.mode='async'
dhnet.n_times=400
```

```
result = dhnet.predict(half_what)
result
```

⇐ {{0,0,1,0,0,0,1,1,0,0,1,0,1,0,0,0,0,1,0,0,0,0,1,0,0,0,0,1,0,0}}

Now, we can find a proper fixed point, see Fig. 5.26.

⇒ z=%;

⇒ Partition[z//Flatten,5]//MatrixForm

$$\Leftarrow \begin{pmatrix} 0 & 0 & 1 & 0 & 0 \\ 0 & 1 & 1 & 0 & 0 \\ 1 & 0 & 1 & 0 & 0 \\ 0 & 0 & 1 & 0 & 0 \\ 0 & 0 & 1 & 0 & 0 \\ 0 & 0 & 1 & 0 & 0 \end{pmatrix}$$

```
⇒ MatrixPlot[%,Mesh → True]
```

⇐

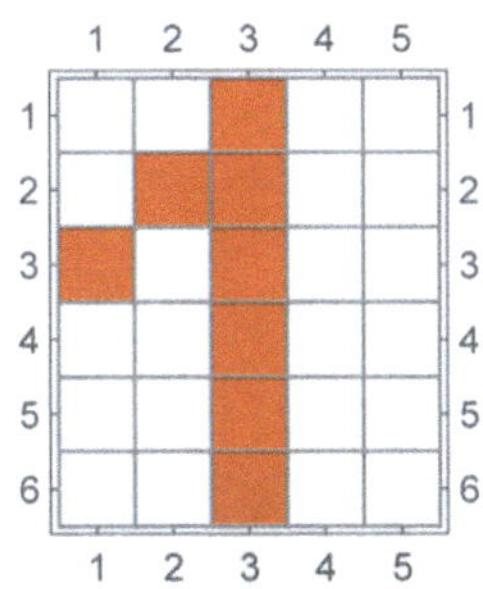

Fig. 5.26 Unknown equilibrium.

Mathematica

Hopfield network is not implemented in the latest version of *Mathematica* 11.3. The following statement can be evaluated in *Mathematica* 10 with *Neural Networks Application Package*. Remark: In the mean time an improved version of this package as NeuralNets has been released, which can be employed with version of Mathematica 11 and higher.

Loading the package

```
<<NeuralNetworks`
```

This function converts the binary input into bipolar input

```
⇒ fun[u_]:=Map[If[#==0,-1,#]&,u]
```

The training set,

```
⇒ x=Map[fun[#]&,{zero,one,two}]
⇐ {{-1,1,1,1,-1,1,-1,-1,-1,1,1,-1,-1,-1,1,
   1,-1,-1,-1,1,1,-1,-1,-1,1,-1,1,1,1,-1},
   {-1,-1,1,-1,-1,-1,1,1,-1,-1,1,-1,1,-1,-1,
   -1,-1,1,-1,-1,-1,-1,1,-1,-1,-1,-1,1,-1,-1},
   {-1,1,1,-1,-1,1,-1,-1,1,-1,-1,-1,-1,1,-1,
   -1,1,1,-1,-1,1,-1,-1,-1,-1,1,1,1,1,1}}
```

The generated network

```
⇒ hop=HopfieldFit[x]
⇐ Hopfield[W,{NetType → Discrete,
   CreationDate → {2018,10,19,10,31,38.4608273}}]
```

Solving the reconstruction problem

```
⇒ halfzeroM=fun[halfzero]
⇐ {-1,1,1,1,-1,1,-1,-1,-1,1,1,-1,-1,-1,1,-1,
   -1,-1,-1,-1,-1,-1,-1,-1,-1,-1,-1,-1,-1,-1}

⇒ hop[halfzeroM]
⇐ {{-1.,1.,1.,1.,-1.,1.,-1.,-1.,-1.,1.,1.,-1.,-1.,-1.,
   1.,1,-1.,-1.,-1.,1,1,-1.,-1.,-1.,1,-1.,1,1,1,-1.}}

⇒ Partition[%//Flatten,5]
⇐ {{-1.,1.,1.,1.,-1.},{1.,-1.,-1.,-1.,1.},{1.,-1.,-1.,-1.,1.},
   {1,-1.,-1.,-1.,1},{1,-1.,-1.,-1.,1},{-1.,1,1,1,-1.}}

⇒ MatrixPlot[%,Mesh → True]
⇐
```

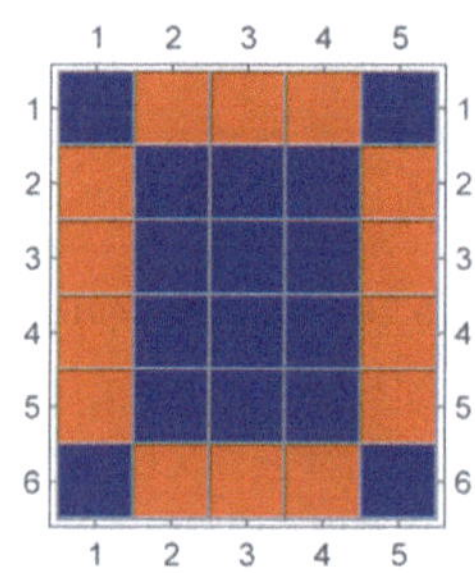

Fig. 5.27 Reconstructed zero.

```
⇒ halftwoM=fun[halftwo]
⇐ {-1,-1,-1,-1,-1,-1,-1,-1,-1,-1,-1,-1,-1,-1,
   -1,-1,1,1,-1,-1,1,-1,-1,-1,-1,1,1,1,1,1}

⇒ hop[halftwoM]
⇐ {{-1.,1,1,-1.,-1.,1,-1.,-1.,1,-1.,-1.,-1.,-1.,1,-1.,
   -1.,1.,1.,-1.,-1.,1.,-1.,-1.,-1.,-1.,1.,1.,1.,1.,1.}}

⇒ Partition[%//Flatten,5]
⇐ {{-1.,1,1,-1.,-1.},{1,-1.,-1.,1,-1.},{-1.,-1.,-1.,1,-1.},
   {-1.,1.,1.,-1.,-1.},{1.,-1.,-1.,-1.,-1.},{1.,1.,1.,1.,1.}}

⇒ MatrixPlot[%,Mesh → True]
⇐
```

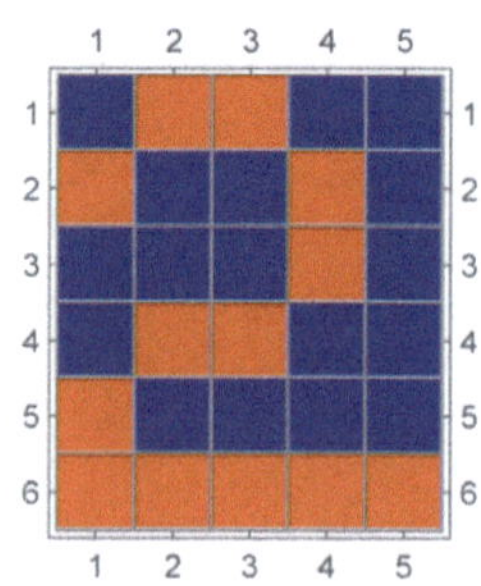

Fig. 5.28 Reconstructed two.

Mathematica Hopfield function can solve the third problem without any modification, however, it provides different fixed points. Instead of digit 1 identified by asynchronous network, now we get digit 2, see Fig.5.29

```
⇒ halfwhatM=fun[halfwhat]
⇐ {1,1,1,-1,-1,-1,-1,-1,1,-1,-1,-1,-1,1,-1,-1,
   -1,-1,-1,-1,-1,-1,-1,-1,-1,-1,-1,-1,-1,-1}
```

```
⇒ hop[halfwhatM]
⇐ {{-1,1.,1.,-1.,-1.,1,-1.,-1.,1.,-1.,-1.,-1.,-1.,1.,
   -1.,-1.,1,1,-1.,-1.,1,-1.,-1.,-1.,-1.,1,1,1,1,1}}
```

```
⇒ Partition[%//Flatten,5]
⇐ {{-1,1.,1.,-1.,-1.},{1,-1.,-1.,1.,-1.},{-1.,-1.,-1.,1.,-1.},
   {-1.,1,1,-1.,-1.},{1,-1.,-1.,-1.,-1.},{1,1,1,1,1}}
```

```
⇒ MatrixPlot[%,Mesh → True]
```

⇐

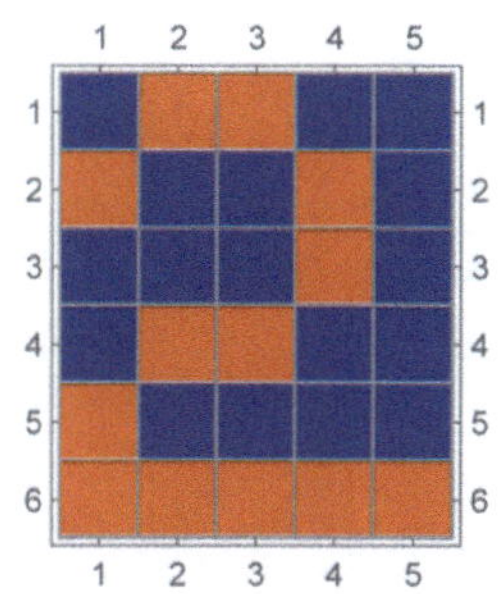

Fig. 5.29 Reconstructed the digit (response of *Mathematica* for input in 5.24).

5.3.2 Reconstruction of Deteriorated Images

We create a Hopfield network with three fixed points. The state vectors have 400 elements representing binarized images of size 20×20. The images are face, sword and cottage, see Fig. 5.30.

Loading images

```
⇒ H1=Import["M:\\Pink\\Head.dat"];
⇒ S1=Import["M:\\Pink\\Sword.dat"];
⇒ P1=Import["M:\\Pink\\Home.dat"];
```

Converting into binary form

```
⇒ H1=H1/.{255 → 1};S1=S1/.{255 → 1};P1=P1/.{255 → 1};
⇒ H1//MatrixForm
```

$$\Leftarrow \begin{pmatrix}
1&1&1&1&1&1&1&0&0&0&0&0&0&1&1&1&1&1&1&1\\
1&1&1&1&1&0&0&0&0&0&0&0&0&0&0&1&1&1&1&1\\
1&1&1&0&0&0&0&0&0&0&0&0&0&0&0&0&0&1&1&1\\
1&1&0&0&0&0&0&0&0&0&0&0&0&0&0&0&0&0&1&1\\
1&1&0&0&0&0&0&0&0&0&0&0&0&0&0&0&0&0&1&1\\
1&0&0&0&0&0&1&1&0&0&0&0&1&1&0&0&0&0&0&1\\
1&0&0&0&0&1&1&1&1&0&0&1&1&1&1&0&0&0&0&1\\
0&0&0&0&0&1&1&1&1&0&0&1&1&1&1&0&0&0&0&0\\
0&0&0&0&0&0&1&1&0&0&0&0&1&1&0&0&0&0&0&0\\
0&0&0&0&0&0&0&0&0&0&0&0&0&0&0&0&0&0&0&0\\
0&0&0&0&0&0&0&0&0&0&0&0&0&0&0&0&0&0&0&0\\
0&0&0&0&0&0&0&0&0&0&0&0&0&0&0&0&0&0&0&0\\
0&0&0&0&1&1&1&1&1&1&1&1&1&1&1&1&1&0&0&0\\
1&0&0&0&0&1&1&1&1&1&1&1&1&1&1&1&0&0&0&1\\
1&0&0&0&0&0&0&1&1&1&1&1&1&1&0&0&0&0&0&1\\
1&1&0&0&0&0&0&0&1&1&1&1&0&0&0&0&0&0&1&1\\
1&1&0&0&0&0&0&0&0&0&0&0&0&0&0&0&0&0&1&1\\
1&1&1&0&0&0&0&0&0&0&0&0&0&0&0&0&0&1&1&1\\
1&1&1&1&1&0&0&0&0&0&0&0&0&0&0&1&1&1&1&1\\
1&1&1&1&1&1&1&0&0&0&0&0&0&1&1&1&1&1&1&1
\end{pmatrix}$$

```
⇒ p1=ListDensityPlot[
    H1//Reverse,ColorFunction → GrayLevel,Mesh → True];
⇒ p2=ListDensityPlot[
    S1//Reverse,ColorFunction → GrayLevel,Mesh → True];
⇒ p3=ListDensityPlot[
    P1//Reverse,ColorFunction → GrayLevel,Mesh → True];
⇒ GraphicsGrid[{{p1,p2,p3}}]
```

⇐ 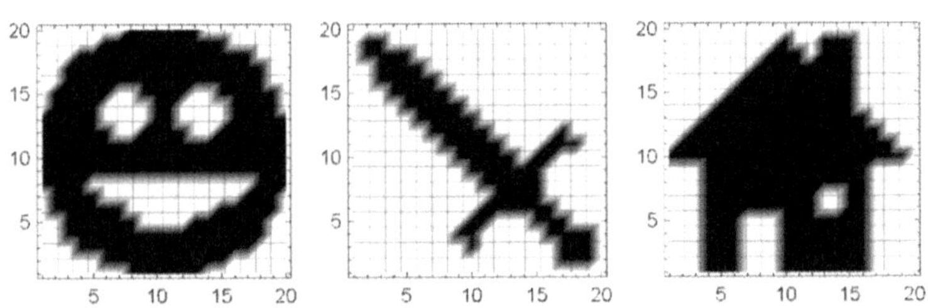

Fig. 5.30 The fixed points of the network: face, sword and cottage.

The deteriorated images are, see Fig. 5.31.

```
⇒ Hd=Import["M:\\Pink\\Head4.dat"];
⇒ Sd=Import["M:\\Pink\\Sword4.dat"];
⇒ Pd=Import["M:\\Pink\\Home4.dat"];
⇒ Hd=Hd/.{255 → 1};Sd=Sd/.{255 → 1};Pd=Pd/.{255 → 1};
⇒ p1d=ListDensityPlot[
    Hd//Reverse,ColorFunction → GrayLevel,Mesh → True];
```

```
⇒ p2d=ListDensityPlot[
    Sd//Reverse,ColorFunction → GrayLevel,Mesh → True];
⇒ p3d=ListDensityPlot[
    Pd//Reverse,ColorFunction → GrayLevel,Mesh → True];
⇒ GraphicsGrid[{{p1d,p2d,p3d}}]
```

⇐

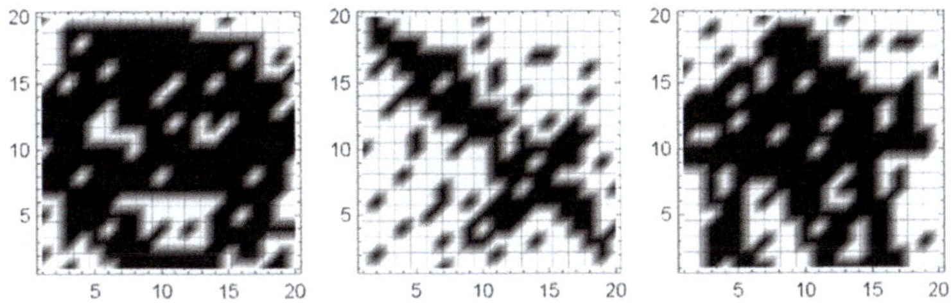

Fig. 5.31 The deteriorated images as input.

Saving data for *Python*

```
⇒ P1n={Flatten[H1]};P2n={Flatten[S1]};P3n={Flatten[P1]};
⇒ Pdn={Flatten[Hd]};P2d={Flatten[Sd]};P3d={Flatten[Pd]};
⇒ Export["cuki1.mtx",P1n];
⇒ Export["cuki2.mtx",P2n];
⇒ Export["cuki3.mtx",P3n];
⇒ Export["cuki11.mtx",Pdn];
⇒ Export["cuki21.mtx",P2d];
⇒ Export["cuki31.mtx",P3d];
```

Python

Loading basic procedure and data set

```python
from numpy import array, matrix
from scipy.io import mmread, mmwrite
```

```python
caki1=mmread('cuki1.mtx')
caki2=mmread('cuki2.mtx')
caki3=mmread('cuki3.mtx')
caki11=mmread('cuki11.mtx')
caki21=mmread('cuki21.mtx')
caki31=mmread('cuki31.mtx')
```

```python
data = np.concatenate([caki1, caki2,caki3], axis=0)
```

Training the network (Shevchuk 2015)

```python
dhnet = algorithms.DiscreteHopfieldNetwork(mode='sync')
dhnet.train(data)
```

Reconstructed the deteriorated images, see Figs. 5.32, 5.34.

```python
t=dhnet.predict(caki11)
mmwrite('cico.mtx',t)
```

```
⇒ z=Import["cico.mtx"];
⇒ pp=Partition[Reverse[Flatten[z]],20];
⇒ ListDensityPlot[pp,ColorFunction → GrayLevel,Mesh → True]
```

⇐ 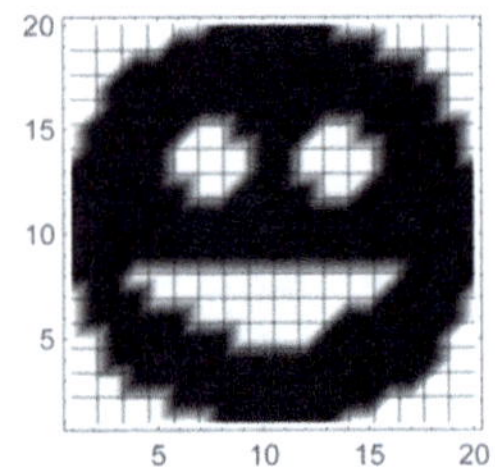

Fig. 5.32 Reconstructed face.

```
t=dhnet.predict(caki21)
mmwrite('cico.mtx',t)
```

```
⇒ z=Import["cico.mtx"];
⇒ pp=Partition[Flatten[z],20]//Transpose//Reverse;
⇒ ListDensityPlot[pp,ColorFunction → GrayLevel,Mesh → True]
```

⇐ 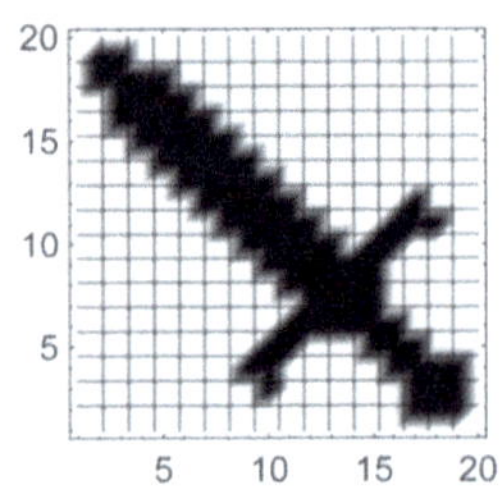

Fig. 5.33 Reconstructed sword.

```
t=dhnet.predict(caki31)
mmwrite('cico.mtx',t)
```

```
⇒ z=Import["cico.mtx"];
⇒ z=%;
⇒ pp=Partition[Flatten[z],20]//Reverse;
⇒ ListDensityPlot[pp,ColorFunction → GrayLevel,Mesh → True]
```

⇐ 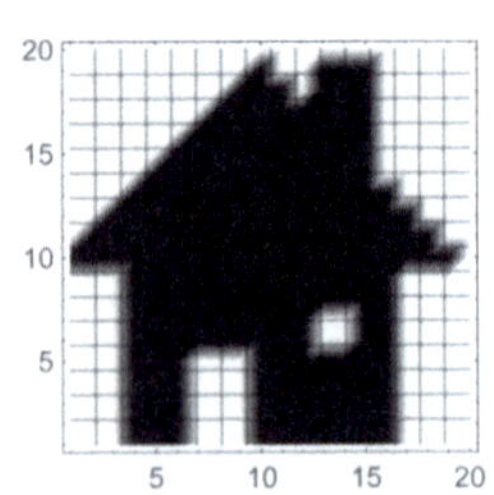

Fig. 5.34 Reconstructed cottage.

Mathematica

In case of *Mathematica*, we need bipolar representation of the pixel values

```
⇒ H1b=H1/.{255 → 1,0 → -1};S1b=S1/.{255 → 1,0 → -1};
⇒ P1b=P1/.{255 → 1,0 → -1};
```

Preparation of the training data

```
⇒ x=Map[Partition[#,20]&,{H1b,S1b,P1b}];
⇒ xv = Map[Flatten, x, {1}];
```

Training Hopfield network

```
⇒ hopD = HopfieldFit[xv]
⇒ Hopfield[W,{NetType → Discrete,
    CreationDate → {2016,10,27,13,6,47.0610803}}]
```

Preparation of deteriorated data

```
⇒ Hdb=Hd/.{255 → 1,0 → -1};Sdb=Sd/.{255 → 1,0 → -1};
⇒ Pdb=Pd/.{255 → 1,0 → -1};
⇒ xd=Map[Partition[#,20]&,{Hdb,Sdb,Pdb}];
```

The reconstructed images

```
⇒ x1v = Map[Flatten, xd, {1}][[1]];
⇒ y1=hopD[x1v];
⇒ ListDensityPlot[Partition[First[y1],20],
    FrameTicks → None,Mesh → True,ColorFunction → GrayLevel]
```

⇐ 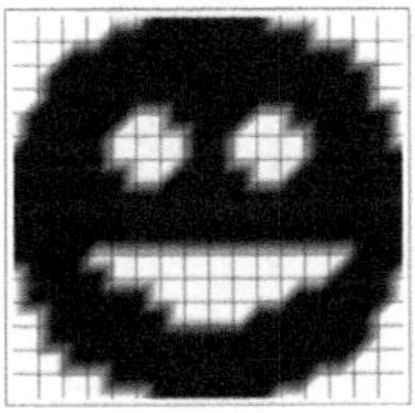

Fig. 5.35 Reconstructed face.

```
⇒ x2v = Map[Flatten, xd, {1}][[2]];
⇒ y2=hopD[x2v];
⇒ ListDensityPlot[Partition[First[y2],20],
    FrameTicks → None,Mesh → True,ColorFunction → GrayLevel]
```

⇐ 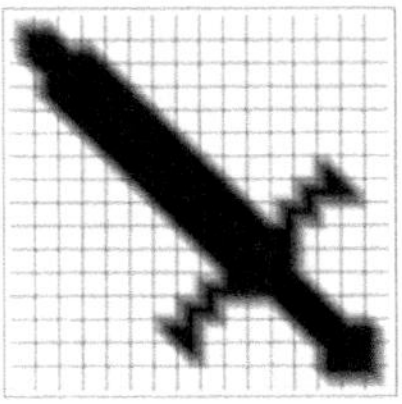

Fig. 5.36 Reconstructed sword.

```
⇒ x3v = Map[Flatten, xd, {1}][[3]];
⇒ y3=hopD[x3v];
⇒ ListDensityPlot[Partition[First[y3],20],
   FrameTicks → None,Mesh → True,ColorFunction → GrayLevel]
```

⇐ 

Fig. 5.37 Reconstructed cottage.

In case of the cottage recognition, the images belonging to one of the iteration steps 1., 26., 52., 78., and 104 can be seen in Fig. 5.38 illustrating the evolution of the recognition process.

```
⇒ y3T=hopD[x3v,Trajectories → True][[1]];
⇒ yy3=Map[Partition[#,20]&,
   {y3T[[1]],y3T[[26]],y3T[[52]],y3T[[78]],y3T[[104]]}];
⇒ yyg=(ListDensityPlot[#1,DisplayFunction → Identity,
   FrameTicks → None,Mesh → True,
   ColorFunction → GrayLevel]&)/@yy3;Show[GraphicsGrid[{yyg}]]
```

⇐ 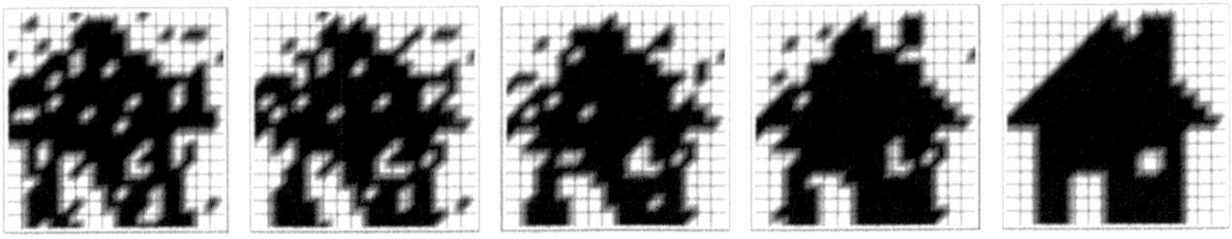

Fig. 5.38 Reconstructed steps of the cottage.

5.4 Unsupervised Network

Basic theory

This type of networks represents an unsupervised learning technique, very similar to clustering. The network tries to find clusters and the centers of these clusters in a data set (Usama et al. 2017). The most simple approach is the so called *competitive learning* method. The basic algorithm is the following:

1) Randomly generate K codebook vectors representing the centers of K clusters, $w_i = 1,2,\ldots K$
2) Randomly choosing a data element, x_k from M total elements and compute its distance from every center,

3) Let us suppose that the winner codebook vector, the closest to the data point is the i-th codebook
4) Then we modify the position of this codebook vector as

$$w_i^{new} = w_i^{old} + \Delta(n)\left(x_k - w_i^{old}\right)$$

see Fig .5.39

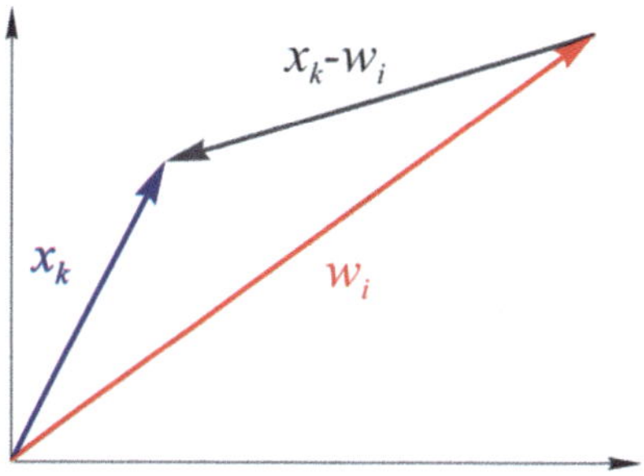

Fig. 5.39 Modification of the winner codebook vector.

It means that this codebook vector will be moved a bit towards the x_k data element, where $\Delta(n)$ is the step-size, which may depend on the actual number of the iteration steps
5) Steps 1) - 4) are repeated M times.
6) We repeat steps 1) - 5) until there is no further change in the position of the code book vectors

Remarks

a) The algorithm minimizes the sum of the distances for all of the codebook vectors (Shevchuk 2017),

$$S\left(w_{1,} w_{2,}\ldots, w_K\right) = \sum_{i=1}^{K}\sum_{k=k_i}^{M_i}\left(x_{i,k} - w_i\right)$$

b) It may happen that there are codebook vectors which were never winners, so there is no data element belonging to them. These are called dead codebooks.

5.4.1 Illustrative Example

Let us consider the following simple data set, and visualize it, see Fig. 5.40.

```
⇒ x={{0.1961`,0.9806`},{-0.1961`,0.9806`},{0.9806`,0.1961`},
   {0.9806`,-0.1961`},{-0.5812`,-0.8137`},{-0.8137`,-0.5812`}};
⇒ p1=ListPlot[x,PlotStyle → {PointSize[0.025],RGBColor[1,0,0]},
   Frame → True,Axes → None]
```

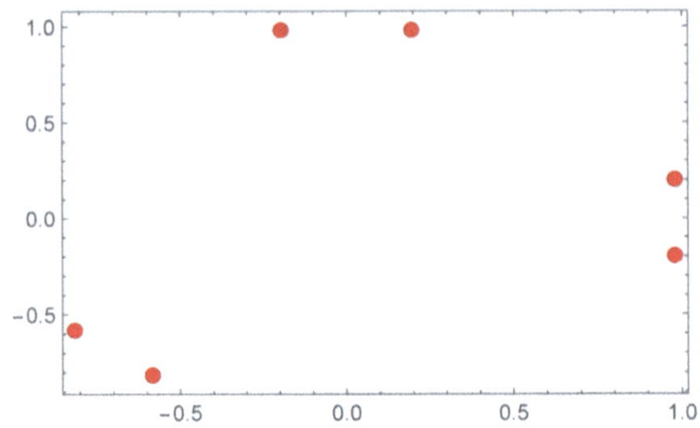

⇐

Fig. 5.40 Simple data set.

Mathematica

Loading the package

```
⇒ <<NeuralNetworks`
```

Let us initialize the network, assuming three clusters

```
⇒ unsup=InitializeUnsupervisedNet[x,3];
```

The training process, see Fig. 5.41.

```
⇒ {unsup,fitrecord}=
    UnsupervisedNetFit[x,unsup,50,ReportFrequency → 5];
```

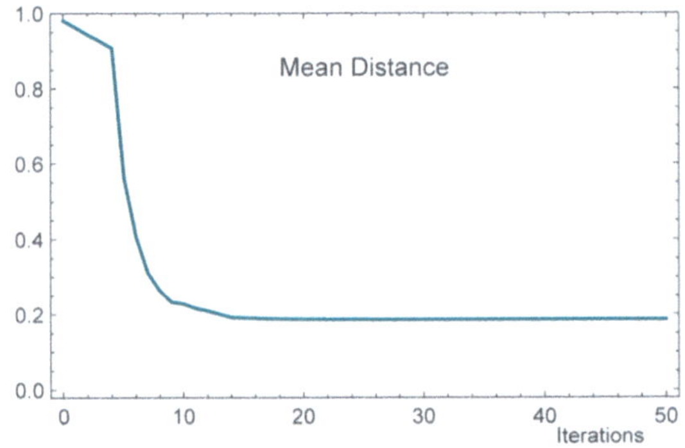

⇐

Fig. 5.41 Convergency of the training process.

During this process the sum of the distances of the elements from their corresponding codebook centers for all cluster will be minimized. In our case the minimum is,

```
⇒ UnsupervisedNetDistance[unsup,x]
⇐ 0.185535
```

Fig. 5.42 shows the trajectories of the moving codebook centers

⇒ `NetPlot[fitrecord,x]`

⇐

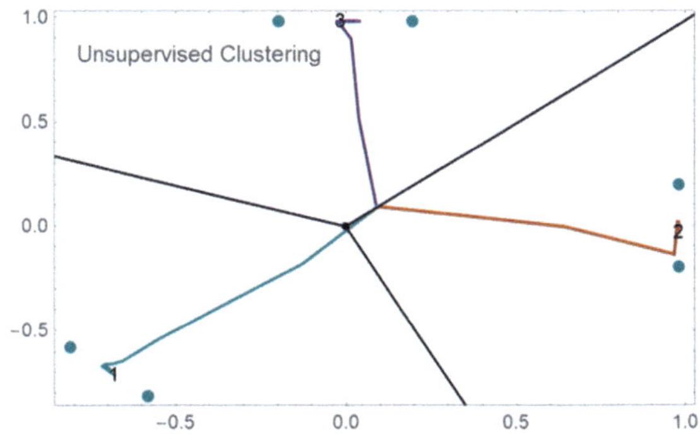

Fig. 5.42 The colored lines are the moving paths of the codebook centers. The black lines represent the linear border of the clusters.

The coordinates of the three centers

⇒ `unsup[[1]]`

⇐ `{{-0.680196,-0.71353},{0.980579,-0.0326197},{-0.0156818,0.980122}}`

Python

Loading procedures and the data

```
import numpy as np
from neupy import algorithms, environment
environment.reproducible()
```

```
input_data = np.array([
    [0.1961, 0.9806],
    [-0.1961, 0.9806],
    [0.9806, 0.1961],
    [0.9806, -0.1961],
    [-0.5812, -0.8137],
    [-0.8137, -0.5812],
    ])
input_data
```

⇐ `{{0.1961,0.9806},{-0.1961,0.9806},{0.9806,0.1961},`
`{0.9806,-0.1961},{-0.5812,-0.8137},{-0.8137,-0.5812}}`

Loading the network algorithm and parametrize the function,

```
sofmnet = algorithms.SOFM(
           n_inputs=2,
     n_outputs=3,
     step=0.5,
     show_epoch=100,
     shuffle_data=True,
     verbose=True,
     learning_radius=0,
     features_grid=(3, 1),
      )
```

```
Main information
[ALGORITHM] SOFM
[OPTION] verbose = True
[OPTION] epoch_end_signal = None
[OPTION] show_epoch = 100
[OPTION] shuffle_data = True
[OPTION] step = 0.5
[OPTION] train_end_signal = None
[OPTION] n_inputs = 2
[OPTION] distance = euclid
[OPTION] features_grid = [3, 1]
[OPTION] grid_type = rect
[OPTION] learning_radius = 0
[OPTION] n_outputs = 3
[OPTION] reduce_radius_after = 100
[OPTION] reduce_std_after = 100
[OPTION] reduce_step_after = 100
[OPTION] std = 1
[OPTION] weight = Normal(mean=0, std=0.01)
```

The training process

```
sofmnet.train(input_data,epochs=100)
```

The coordinates of the codebook vectors, lists of the x and y coordinates

```
sofmnet.weight[0:2, :]
```

⇐ {{-0.0126041,-0.705751,0.9806},{0.9806,-0.689149,-0.0227187}}

Fig. 5.43 shows the codebook vectors

⇒ Show[{p1,ListPlot[Transpose[%],PlotStyle → Blue]}]

⇐

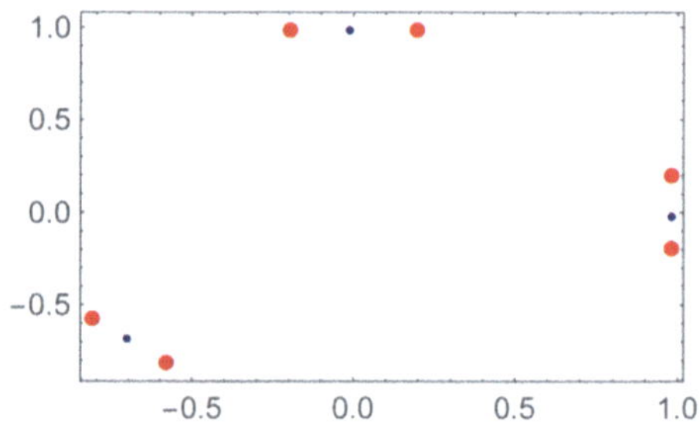

Fig. 5.43 The resulted centers (blue).

The membership functions of the data elements,

```
list1=[]
for data in input_data:
 list1.append(sofmnet.predict(np.reshape(data, (2, 1)).T))
```

```
list1
```

⇐ {{{1,0,0}},{{1,0,0}},{{0,0,1}},{{0,0,1}},{{0,1,0}},{{0,1,0}}}

This means for example, that the first element (see input array -[0.1961, 0.9806]) belongs to the first set (1,0,0).

5.4.2 Iris Data Set

Let us employ unsupervised network for clustering Iris Data Set. We consider two features: the petal width and petal length.

Loading procedures

```
import numpy as np
from sklearn import datasets

from neupy import algorithms, environment
environment.reproducible()
```

The data set

```
iris = datasets.load_iris()
```

Considering two the features

```
X=iris.data[:,[2,3]]
y=iris.target
```

Saving the data for *Mathematica*,

```
np.savetxt('M:\\dataX.txt',X,fmt='%.2e')
```

The labels of the data elements

```
⇐ {0,0,0,0,0,0,0,0,0,0,0,0,0,0,0,0,0,0,0,0,0,0,0,0,0,0,0,0,0,0,
   0,0,0,0,0,0,0,0,0,0,0,0,0,0,0,0,0,0,0,0,1,1,1,1,1,1,1,1,1,1,
   1,1,1,1,1,1,1,1,1,1,1,1,1,1,1,1,1,1,1,1,1,1,1,1,1,1,1,1,1,1,
   1,1,1,1,1,1,1,1,1,1,2,2,2,2,2,2,2,2,2,2,2,2,2,2,2,2,2,2,2,2,
   2,2,2,2,2,2,2,2,2,2,2,2,2,2,2,2,2,2,2,2,2,2,2,2,2,2,2,2,2,2}
⇒ yM=%;
```

Visualizing data

```
⇒ Xtrain=Import["M:\\dataX.txt","Table"];
⇒ datat=MapThread[Join[#1,{#2}]&,{Xtrain,yM}];
⇒ data0=Map[{#[[1]],#[[2]]}&,Select[datat,#[[3]]==0&]];
⇒ data1=Map[{#[[1]],#[[2]]}&,Select[datat,#[[3]]==1&]];
   data2=Map[{#[[1]],#[[2]]}&,Select[datat,#[[3]]==2&]];
⇒ p0=ListPlot[{data0,data1,data2},PlotStyle → {Green,Red,Blue},
   Frame → True,Axes → None,PlotMarkers → {Automatic},
   AspectRatio → 0.9,FrameLabel → {"petallength","petal width"},
   Frame → True]
```

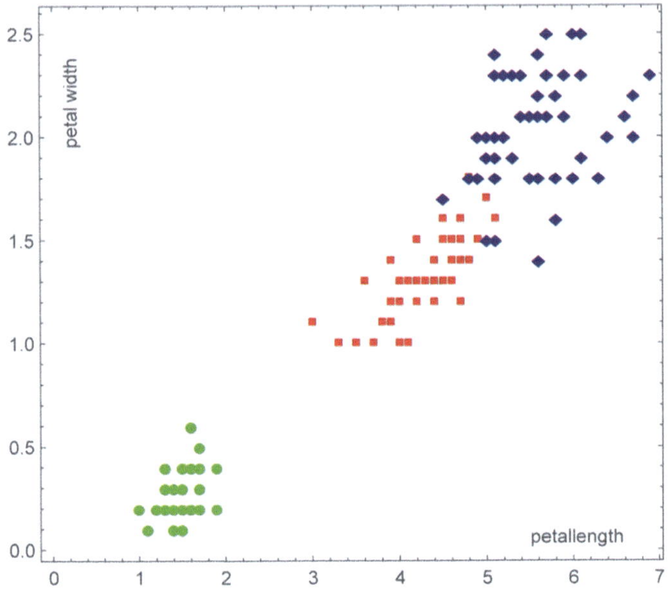

⇐

Fig. 5.44 The dataset to be clustered.

Python

Parametrizing

```
sofmnet = algorithms.SOFM(
        n_inputs=2,
    n_outputs=3,
    step=0.4,
    show_epoch=100,
    shuffle_data=True,
    verbose=True,
    features_grid=(3, 1),
    )
```

```
Main information
[ALGORITHM] SOFM
[OPTION] verbose = True
[OPTION] epoch_end_signal = None
[OPTION] show_epoch = 100
[OPTION] shuffle_data = True
[OPTION] step = 0.4
[OPTION] train_end_signal = None
[OPTION] n_inputs = 2
[OPTION] distance = euclid
[OPTION] features_grid = [3, 1]
[OPTION] grid_type = rect
[OPTION] learning_radius = 0
[OPTION] n_outputs = 3
[OPTION] reduce_radius_after = 100
[OPTION] reduce_std_after = 100
[OPTION] reduce_step_after = 100
[OPTION] std = 1
[OPTION] weight = Normal(mean=0, std=0.01)
```

Training the network

```
sofmnet.train(X,epochs=1500)
```

```
Start training
[TRAINING DATA] shapes: (150, 2)
[TRAINING] Total epochs: 1500
```

Epoch	Train err	Valid err	Time
1	0.46922	-	47 ms
100	0.24516	-	47 ms
200	0.24426	-	47 ms
300	0.23746	-	47 ms
400	0.24027	-	47 ms
500	0.23996	-	47 ms
600	0.23709	-	47 ms
700	0.23738	-	47 ms
800	0.23862	-	47 ms
900	0.23704	-	47 ms
1000	0.23334	-	47 ms
1100	0.23588	-	47 ms
1200	0.23366	-	47 ms
1300	0.23544	-	47 ms
1400	0.23598	-	47 ms
1500	0.23438	-	47 ms

The coordinates of the codebook vectors (x_1,x_2,x_3) and (y_1,y_2,y_3)

```
sofmnet.weight[0:2, :]
```

⇐ {{1.45812,4.26827,5.61301},{0.242962,1.34391,2.03483}}

⇒ c=Transpose[%];

Resulting labels of the elements after training process

```
list1=[]
for data in X:
 list1.append(sofmnet.predict(np.reshape(data, (2, 1)).T))
```

```
list1
```

```
⇐ {{{1,0,0}},{{1,0,0}},{{1,0,0}},{{1,0,0}},{{1,0,0}},{{1,0,0}},
  {{1,0,0}},{{1,0,0}},{{1,0,0}},{{1,0,0}},{{1,0,0}},{{1,0,0}},
  {{1,0,0}},{{1,0,0}},{{1,0,0}},{{1,0,0}},{{1,0,0}},{{1,0,0}},
  ⋮
  {{0,0,1}},{{0,0,1}},{{0,0,1}},{{0,0,1}},{{0,0,1}},
  {{0,0,1}},{{0,0,1}},{{0,0,1}},{{0,0,1}},{{0,0,1}}}
```

⇒ yy=Map[Flatten[#]&,%];

```
⇐ {{1,0,0},{1,0,0},{1,0,0},{1,0,0},{1,0,0},{1,0,0},{1,0,0},
  {1,0,0},{1,0,0},{1,0,0},{1,0,0},{1,0,0},{1,0,0},{1,0,0},
  {1,0,0},{1,0,0},{1,0,0},{1,0,0},{1,0,0},{1,0,0},{1,0,0},
  ⋮
  {0,1,0},{0,0,1},{0,0,1},{0,0,1},{0,0,1},{0,0,1},
  {0,0,1},{0,0,1},{0,0,1},{0,0,1},{0,0,1},{0,0,1}}
```

```
⇒ yyy=Map[#.{0,1,2}&,yy]
⇐ {0,0,0,0,0,0,0,0,0,0,0,0,0,0,0,0,0,0,0,0,0,0,0,0,0,0,0,0,0,0,
   0,0,0,0,0,0,0,0,0,0,0,0,0,0,0,0,0,0,0,0,1,1,1,1,1,1,1,1,1,1,
   1,1,1,1,1,1,1,1,1,1,1,1,1,1,1,1,1,2,1,1,1,1,1,2,1,1,1,1,1,1,
   1,1,1,1,1,1,1,1,1,1,2,2,2,2,2,2,1,2,2,2,2,2,2,2,2,2,2,2,2,1,
   2,2,2,2,2,2,1,2,2,2,2,2,2,2,2,2,2,2,1,2,2,2,2,2,2,2,2,2,2,2}
```

Degree of the misclustered

```
⇒ Norm[yM-yyy]
```

⇐ $\sqrt{6}$

```
⇒ %//N
⇐ 2.44949
```

Visualization of the original and the clustered datasets, see Fig. 5.45/a and Fig. 5.45/b, with the position of the codebook vectors

```
⇒ datat=MapThread[Join[#1,{#2}]&,{Xtrain,yyy}];
⇒ data0=Map[{#[[1]],#[[2]]}&,Select[datat,#[[3]]==0&]];
⇒ data1=Map[{#[[1]],#[[2]]}&,Select[datat,#[[3]]==1&]];
   data2=Map[{#[[1]],#[[2]]}&,Select[datat,#[[3]]==2&]];
⇒ p1=ListPlot[{data0,data1,data2},PlotStyle → {Green,Red,Blue},
   Frame → True,Axes → None,PlotMarkers → {Automatic},
   AspectRatio → 0.9,FrameLabel → {"petallength","petal width"},
   Frame → True];
⇒ p2=ListPlot[
   c,PlotMarkers → "\[SpadeSuit]",PlotStyle → {Black,Large}];
⇒ GraphicsGrid[{{p0,Show[{p1,p2}]}}]
```

⇐

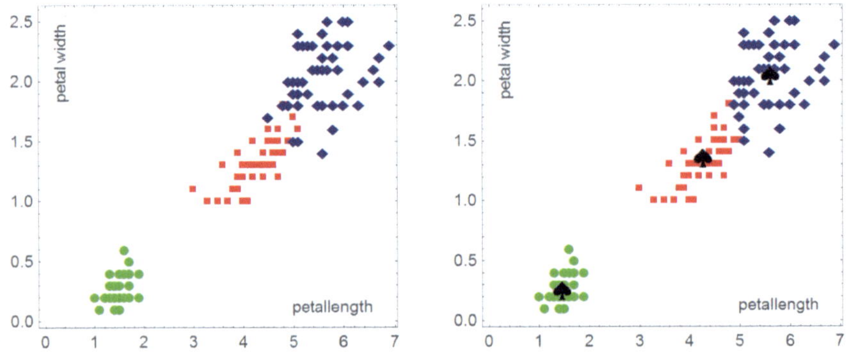

Fig. 5.45 The original (left) and the clustered (right) datasets with the codebook vectors (♠).

5.4.3 Voronoi Mesh

Let us suppose that we would like to partition a set of points, see Fig. 5.46 into ten subsets saving the original topology.

Loading the dataset

⇒ `dataQ=Import["M:\\f_01_05.dat"]/1000;`

```
import numpy as np
from sklearn import datasets

from neupy import algorithms, environment
environment.reproducible()
```

```
from numpy import array, matrix
from scipy.io import mmread, mmwrite
```

⇒ `dataQ2=Map[{#[[1]],#[[2]]}&,dataQ];`

⇒ `p0=ListPlot[dataQ2]`

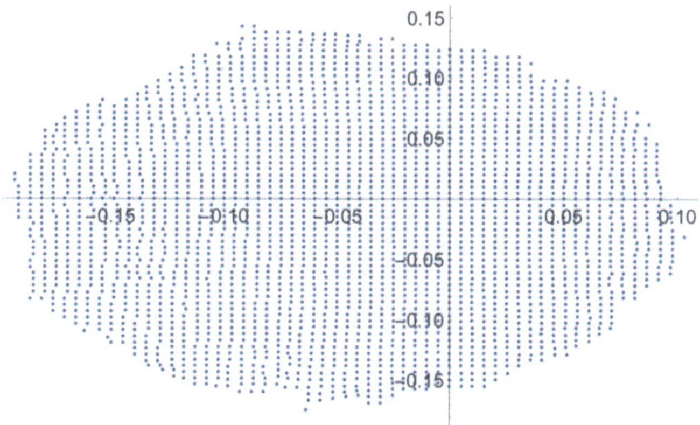

⇐

Fig. 5.46 The dataset to be clustered.

⇒ `Export["cuki.mtx",dataQ2]`

⇒ `cuki.mtx`

Python

Reading the dataset into Python

```
caki=mmread('cuki.mtx')
```

Parametrizing the function

```
sofmnet = algorithms.SOFM(
            n_inputs=2,
     n_outputs=10,
     step=0.4,
     show_epoch=20,
     shuffle_data=True,
     verbose=True,
     features_grid=(5, 2),
     )
```

We have two inputs and ten outputs.

```
Main information
[ALGORITHM] SOFM
[OPTION] verbose = True
[OPTION] epoch_end_signal = None
[OPTION] show_epoch = 20
[OPTION] shuffle_data = True
[OPTION] step = 0.4
[OPTION] train_end_signal = None
[OPTION] n_inputs = 2
[OPTION] distance = euclid
[OPTION] features_grid = [5, 2]
[OPTION] grid_type = rect
[OPTION] learning_radius = 0
[OPTION] n_outputs = 10
[OPTION] reduce_radius_after = 100
[OPTION] reduce_std_after = 100
[OPTION] reduce_step_after = 100
[OPTION] std = 1
[OPTION] weight = Normal(mean=0, std=0.01)
```

Training process

```
sofmnet.train(caki,epochs=50)
```

```
Start training
[TRAINING DATA] shapes: (2670, 2)
[TRAINING] Total epochs: 50
---------------------------------------------------------
|    Epoch      |  Train err   |  Valid err   |    Time     |
---------------------------------------------------------
|             1 |    0.022902 |            - |       1 sec |
|            20 |    0.022270 |            - |     858  ms |
|            40 |    0.021920 |            - |     889  ms |
|            50 |    0.021701 |            - |     858  ms |
---------------------------------------------------------
```

```
list1=[]
for data in caki:
 list1.append(sofmnet.predict(np.reshape(data, (2, 1)).T))
```

Labels of the elements

```
⇐ {{{0,0,0,0,1,0,0,0,0,0}},{{0,0,0,0,1,0,0,0,0,0}},
   {{0,0,0,0,1,0,0,0,0,0}},{{0,0,0,0,1,0,0,0,0,0}},
   {{0,0,0,0,1,0,0,0,0,0}},{{0,0,0,0,1,0,0,0,0,0}},
        OutputSizeLimit`Skeleton [2658],
   {{0,0,0,0,0,0,0,0,1,0}},{{0,0,0,0,0,0,0,0,1,0}},
   {{0,0,0,0,0,0,0,0,1,0}},{{0,0,0,0,0,0,0,0,1,0}},
   {{0,0,0,0,0,0,0,0,1,0}},{{0,0,0,0,0,0,0,0,1,0}}}
 large output   show less   show more   show all   set size limit...
```

```
⇒ yy=Map[Flatten[#]&,%];
⇒ yyy=Map[#.{0,1,2,3,4,5,6,7,8,9}&,yy];
```

The coordinates of the codebook vectors, (x_1,...,x_{10}) and (y_1,...,y_{10})

```
> sofmnet.weight[0:2, :]
```

```
⇐ {{-0.124046,-0.0830807,0.00767717,-0.0265025,-0.15188,
   -0.0204085,0.0463383,-0.10913,0.0576955,-0.0427581},
   {-0.0684907,-0.139941,-0.131987,-0.0525824,0.0128014,
   0.100778,0.0636908,0.0709744,-0.0357107,0.010534}}
```

```
⇒ cc=Transpose[%];
⇒ p1=ListPlot[{cc},PlotStyle → Red];
⇒ p2=Show[{p0,p1}]
```

⇐

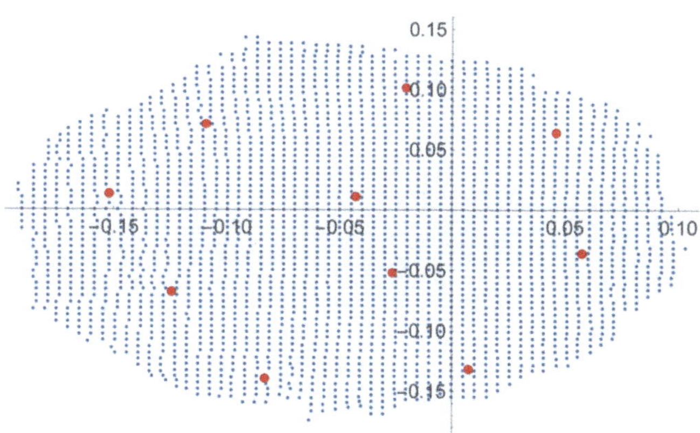

Fig. 5.47 The resulting codebook vectors (red).

Having the coordinates of the codebook vectors, we can identify the points belonging to the different clusters, see Fig. 5.48.

```
⇒ datat=MapThread[Join[#1,{#2}]&,{dataQ2,yyy}];
⇒ index=Range[10]-1;
⇒ fata={};
```

```
⇒ Do[AppendTo[fata,Map[
    {#[[1]],#[[2]]}&,Select[datat,#[[3]]==index[[i+1]]&]]],{i,0,9}]
⇒ p10=ListPlot[fata,Frame → True,Axes → None];
⇒ P=Show[{p0,p1,p10}]
```

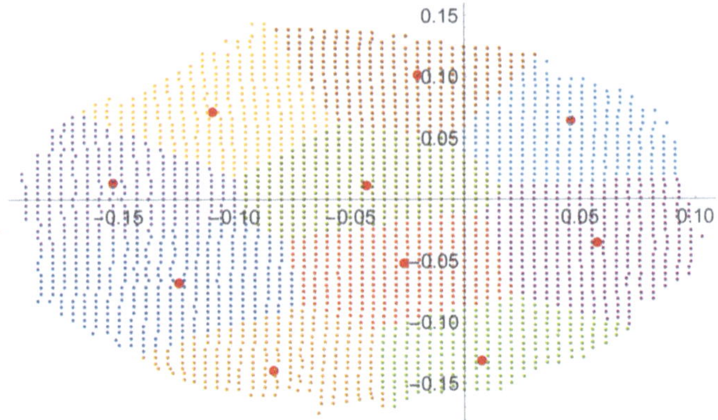

⇐

Fig. 5.48 The ten clusters.

We can get the same subsets via tessellation, namely generating *Voronoi mesh*, see Fig. 5.49.

The partitioning of a plane with *n* points into convex polygons such that each polygon contains exactly one generating point and every point in a given polygon is closer to its generating point than to any other. A Voronoi diagram is sometimes also known as a *Dirichlet tessellation*. The cells are called *Dirichlet regions*, *Thiessen polytopes*, or *Voronoi polygons*.

Mathematica

```
⇒ ki=VoronoiMesh[cc,MeshCellStyle → {{1,All} → White,
    {0,All} → Black},AspectRatio → 0.8];
⇒ Show[{ki,P}]
```

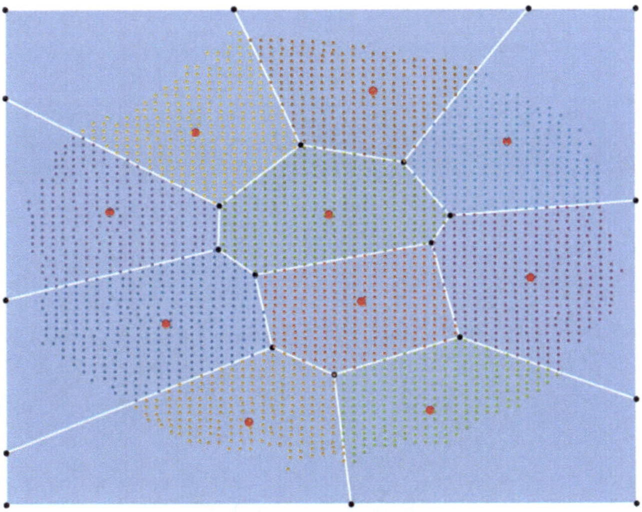

⇐

Fig. 5.49 The resulting centers (blue).

5.4.4 Robust Regression

Unsupervised neural networks can be applied "to eliminate" outliers in order fit a function to clouds of points. Let us consider the following cloud of points, see Fig. 5.50.

```
⇒ Clear[sinc]
⇒ sinc[x_]/;x!=0:=Sin[x]/x
⇒ sinc[x_]/;x==0:=1
⇒ SeedRandom[2];data=Table[{x,sinc[x]+2
     Random[NormalDistribution[0,0.1]]},{x,-10,10,0.05}];
⇒ pdata=ListPlot[data,AspectRatio → 0.8,PlotRange → {-0.7,1.6},
     Epilog → {Green,Map[Point,data]}]
```

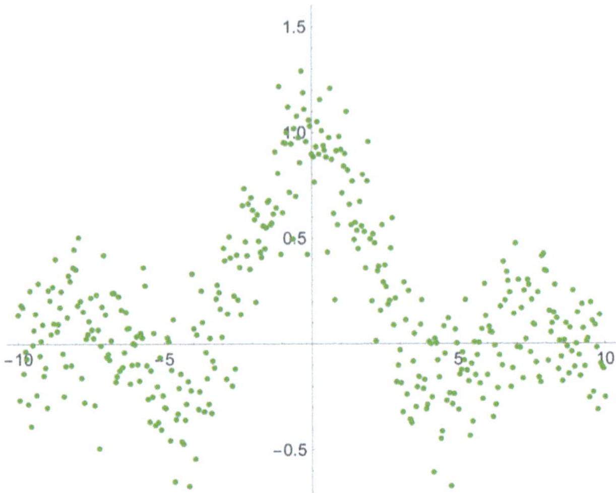

⇐

Fig. 5.50 The set of points.

We would like to approximate these points using non-parametric regression. Let us employ clustering with 16 codebooks.

```
⇒ Export["cuki.mtx",data]
⇐ cuki.mtx
```

Python

```
caki=mmread('cuki.mtx')
```

```
sofmnet = algorithms.SOFM(
            n_inputs=2,
      n_outputs=16,
      step=0.4,
      show_epoch=50,
      shuffle_data=True,
      verbose=True,
      features_grid=(8, 2),
       )
```

```
Main information
[ALGORITHM] SOFM
[OPTION] verbose = True
[OPTION] epoch_end_signal = None
[OPTION] show_epoch = 50
[OPTION] shuffle_data = True
[OPTION] step = 0.4
[OPTION] train_end_signal = None
[OPTION] n_inputs = 2
[OPTION] distance = euclid
[OPTION] features_grid = [8, 2]
[OPTION] grid_type = rect
[OPTION] learning_radius = 0
[OPTION] n_outputs = 16
[OPTION] reduce_radius_after = 100
```

```
[OPTION] reduce_std_after = 100
[OPTION] reduce_step_after = 100
[OPTION] std = 1
[OPTION] weight = Normal(mean=0, std=0.01)
```

```
sofmnet.train(caki,epochs=500)
```

```
Start training
[TRAINING DATA] shapes: (401, 2)
[TRAINING] Total epochs: 500
---------------------------------------------------------
|    Epoch    |  Train err  |  Valid err  |     Time     |
---------------------------------------------------------
|           1 |     0.57089 |           - |       156 ms |
|          50 |     0.25336 |           - |       125 ms |
|         100 |     0.25626 |           - |       140 ms |
|         150 |     0.25190 |           - |       140 ms |
|         200 |     0.24566 |           - |       140 ms |
|         250 |     0.24918 |           - |       140 ms |
|         300 |     0.24723 |           - |       125 ms |
|         350 |     0.24588 |           - |       125 ms |
|         400 |     0.24669 |           - |       125 ms |
|         450 |     0.24495 |           - |       156 ms |
|         500 |     0.24450 |           - |       156 ms |
---------------------------------------------------------
```

The coordinates of the codebook vectors are, see Fig. 5.51.

```
sofmnet.weight[0:2, :]
```

```
⇐ {{7.02132,2.09913,0.730749,3.40938,4.61752,8.1838,-8.37554,
  -1.97268,9.39964,-0.662101,-9.4911,-4.68233,-7.17876,-3.25931,
  5.82016,-5.90641},{0.133833,0.434031,0.803546,-0.0647337,
  -0.192478,0.0401029,0.200271,0.504017,-0.0313606,0.904776,
  -0.032679,-0.261859,0.0245919,0.0110725,-0.0613287,-0.0200913}}
⇒ cc=Transpose[%];
⇒ p1=ListPlot[{cc},PlotStyle → Red];
⇒ Show[{PP,p1}]
```

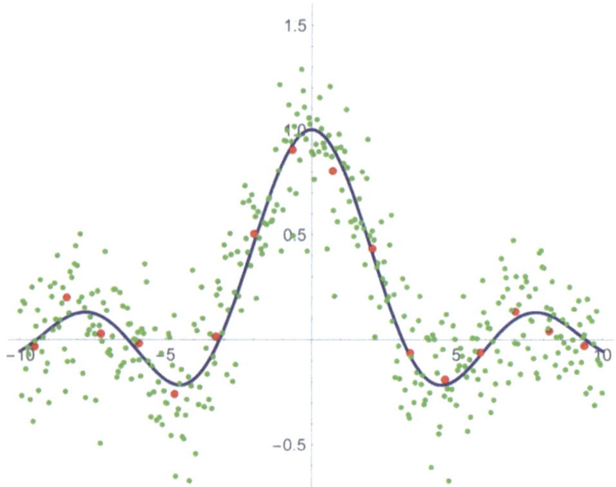

⇐

Fig. 5.51 The original function with the codebook vectors (red).

5.4.5 *Kohonen Map*

In order to improve the efficiency of the clustering, we will modify not only the winner codebook vector, but also the other codebook vectors, which are in its neighborhood, however with different (smaller) weights. To do that, one should define the neighborhood topology of the actual winner codebook vector. This topology of the neighborhood in 2D is called as Kohonen Map, (sometimes Self Organizing Map SOM), see Fig. 5.52. This figure illustrates a topology of 2×3 = 6 codebook vectors.

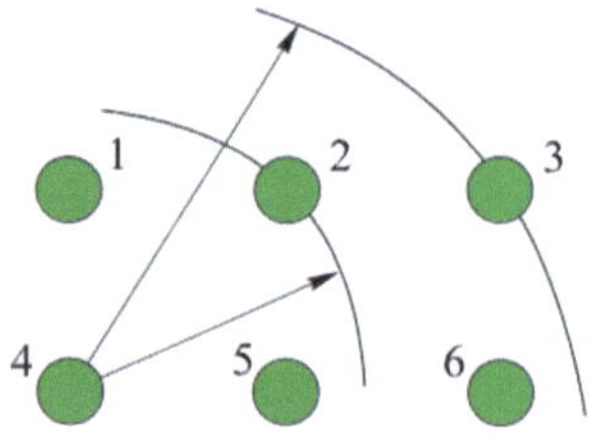

Fig. 5.52 Topology of 2×3 = 6 codebook vectors.

The modification of the codebook vectors belonging to the neighborhood topology is weighted as shown in Fig. 5.53. Weights of the neighboring codebook vectors are represented by a matrix

$$\boldsymbol{\Omega}_{L,J} = \begin{pmatrix} 6 & 4 & 2 & 3 & 4 \\ 4 & 2 & 0 & 1 & 2 \\ 5 & 3 & 1 & 2 & 3 \end{pmatrix}$$

The matrix elements represent the distance from the winner codebook vector. The weights of the codebook vectors can be computed as

$$w_{i,j}{}^{new} = w_{i,j}{}^{old} + \eta(n)\mathrm{e}^{-\xi(n)\Omega_{I,J}}\left(x_k - w_{i,j}{}^{old}\right)$$

where $\mathrm{e}^{-\xi(n)\Omega_{I,J}}$ is a neighborhood function. As an example, let us consider the following set of points, see Fig. 5.53.

```
⇒ angles=Table[Pi/200. i,{i,0,99}];
⇒ x=Map[{Cos[#],Sin[#]}RandomReal[{0.9,1.1}]&,angles];
⇒ p1=ListPlot[x,PlotRange → {{0,1},{0,1}},
    PlotStyle → {PointSize[0.01],RGBColor[0,0,0]},AspectRatio → 1]
```

⇐

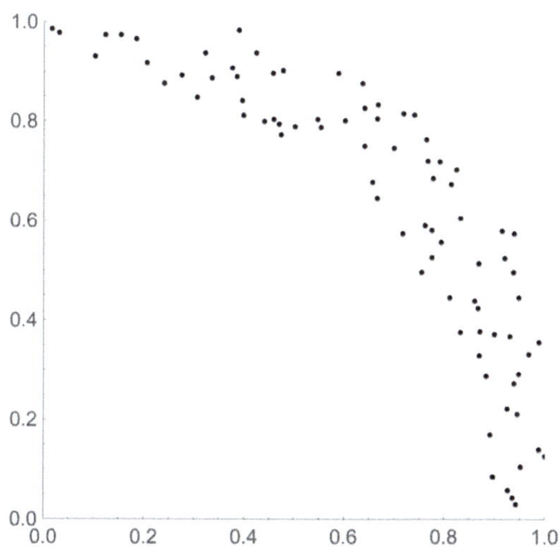

Fig. 5.53 Noisy points of a quadratic circle.

We would like to represent the topology of these points with ten codebook vectors.

Mathematica

First, let us employ standard competing learning technique (without SOM) with six codebook vectors. Initializing and training the network

```
⇒ unsup=InitializeUnsupervisedNet[x,6];
⇒ {unsup,fitrecord}=UnsupervisedNetFit[x,unsup,100,
    ReportFrequency → 1];//Quiet
```

Fig. 5.54 shows the result,

```
⇒ p3=ListPlot[unsup[[1]],PlotRange → {{0,1},{0,1}},
    PlotStyle → {PointSize[0.02],RGBColor[1,0,0]}];
⇒ Show[{p3,p1},AspectRatio → 1,PlotRange → All]
```

⇐

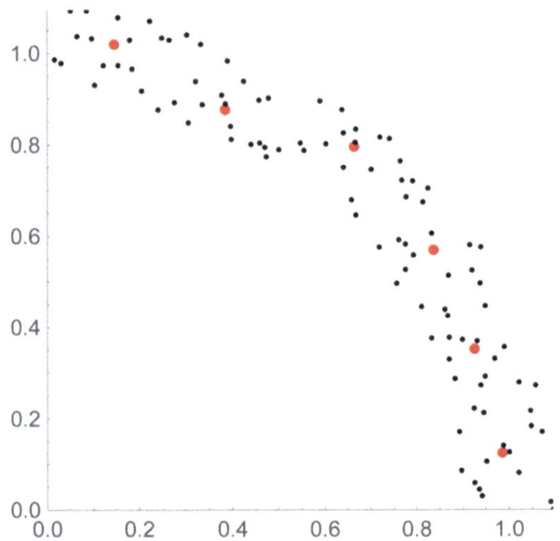

Fig. 5.54 The codebook vectors (red) without SOM.

Now, let us employ competing learning (SOM) with the following symmetric neighborhood definition

$$\begin{pmatrix} 3 \\ 2 \\ 1 \\ 0 \\ 1 \\ 2 \\ 3 \end{pmatrix}$$

where the elements of the vector represents the relative distance from the winner codebook vector

```
⇒ unsup=InitializeUnsupervisedNet[x,6,SOM → {6,1}];
⇒ {som,fitrecord}=UnsupervisedNetFit[x,6,100,SOM → {6,1}]
```

Fig. 5.55 shows the result

```
⇒ p6=NetPlot[som,x];
⇒ p7=ListPlot[som[[1]],PlotRange → {{0,1},{0,1}},
    PlotStyle → {PointSize[0.02],RGBColor[1,0,0]}];
⇒ Show[{p6,p7},AspectRatio → 1]
```

⇐

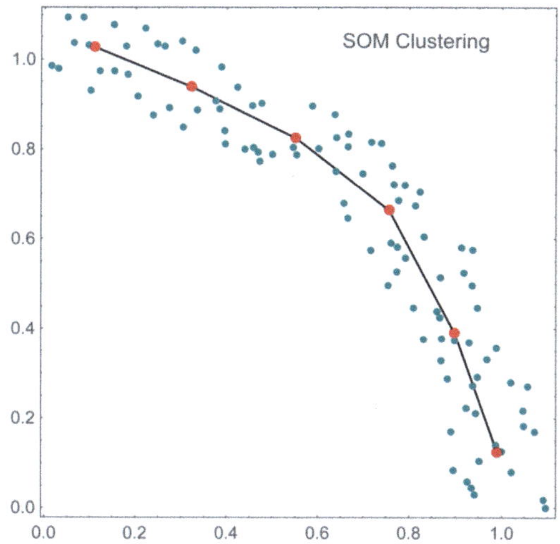

Fig. 5.55 The codebook vectors with SOM technique.

5.4.6 Fitting Sphere to Point Cloud Data

Let us consider a cloud of data resulting from a low resolution sensor causing discontinuity effects. In our case the original object is a sphere having radius $R = 0.152$ m that was placed in the real world position $x = 0$, $y = 0$ with an object distance of $z = 3$ m. The measured data

```
⇒ dataQ=Import["G:\\f_01_05.dat"]/1000;
```

The number of the data points is,

```
⇒ n=Length[dataQ]
⇐ 2670
```

Fig. 5.56 shows the measured points together with the sphere to be estimated

```
⇐ p1=ListPointPlot3D[dataQ,
    PlotStyle → Directive[Blue,PointSize[0.006]],Axes → False,
    Boxed → False,PlotRange → {{-0.2,0.2},{-0.2,0.2},{2.8,3.2}},
    BoxRatios → {1,1,1}];
⇒ a0=a;b0=b;c0=c;d0=d;R0=R;
⇒ p2=Graphics3D[{Specularity[White,30],Orange,Opacity[0.1],
    Sphere[{a,b,c},R ]/.{a → a0,b → b0,c → c0,R → R0}},
    PlotRange → {{-0.2,0.2},{-0.2,0.2},{2.8,3.2}},
    Axes → False,BoxRatios → {1,1,1}];
⇒ Show[{p1,p2}]
```

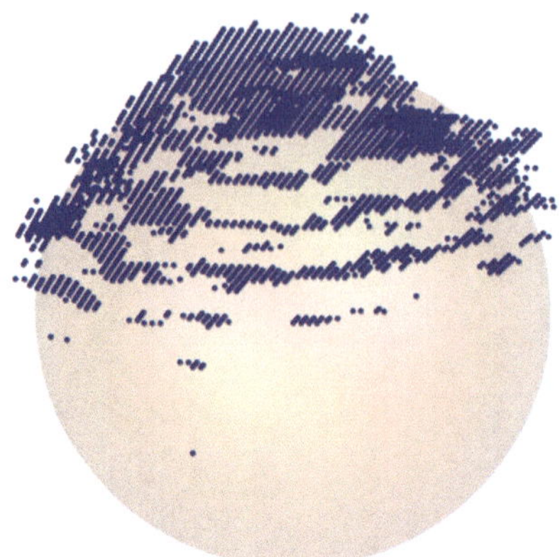

⇐

Fig. 5.56 The measured data with the object to be estimated.

Mathematica

We try to represent the measured points with 25 codebook vectors positioned in 5×5 symmetric SOM

```
⇒ {som,fitrecord}=
    UnsupervisedNetFit[dataQ,25,100,SOM → {5,5}];//Quiet
```

Let us display the code-book vectors with the sphere to be estimated, see Fig. 5.57.

```
⇒ p3=ListPointPlot3D[dataSOM,
    PlotStyle → Directive[Red],BoxRatios → {1,1,1}];
⇒ Show[{p2,p3}]
```

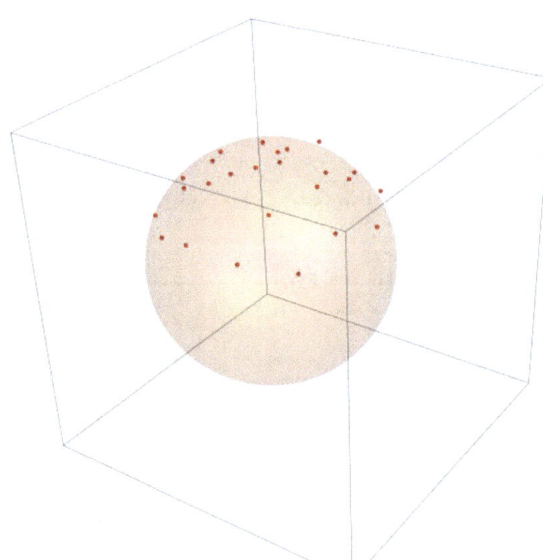

⇐

Fig. 5.57 The codebook vectors with the exact target object.

Now we can use these codebook vectors representing the cloud of the measured data points, to estimate the unknown parameters (*a, b, c, R*) instead of the measured raw data.

This example demonstrates that unsupervised network can also be considered as a dimension reduction method.

5.5 Recurrent Network

Basic theory

The networks we have discussed until now could learn input - output pairs independently on the order they were introduced in the training. However sometimes we need to predict series. For example let us suppose that we have two sequences. An input sequence

$$x = \{1, 2, 3, 2, 1, 3\}$$

and an output sequence

$$y = \{2, 3, 1, 1, 3, 2\};$$

It goes without saying that a standard network is unable to learn this relation, since a normal network has no memory, and for such a network, this task contains contradictions. For example, to learn y_1, $(1 \to 2)$ contradicts with learning $y_5, (3 \to 2)$. To solve the problem, we need a network that is able to remember the previous output. Hence the network needs memory. This is not a static one, like in case of the Hopfield -network's associative memory. The Hopfield network already discussed in the previous section is a special type of the recurrent network where all connections are symmetric and cannot handle sequences (Freeman 1994).

The most simple recurrent network is the Jordan network, see Fig. 5.58.

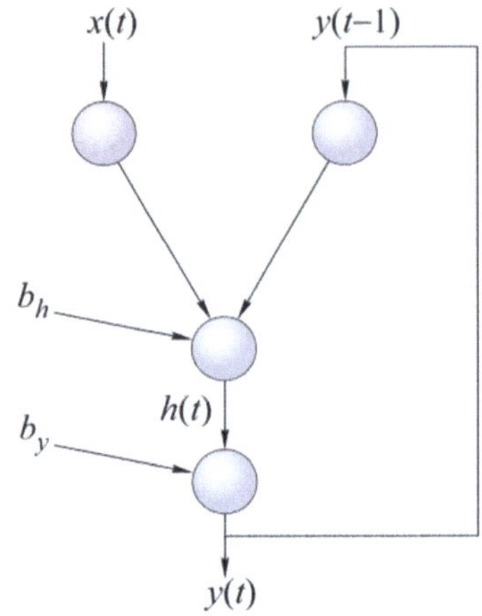

Fig. 5.58 Jordan's network.

It means that the output of the network depends not only on the input, but also on the previous output, too. Since

$$h(t) = \sigma_h(w_x x(t) + w_h y(t-1) + b_h)$$

$$y(t) = \sigma_y(w_y h(t) + b_y)$$

where an activation function, for example

$$\sigma_{(\bullet)}(u) = \frac{1}{1+e^{-u}}$$

therefore

$$y(t) = \sigma_y(w_y \sigma_h(w_x x(t) + w_h y(t-1) + b_h) + b_y) = g(x(t), y(t-1))$$

The output is a nonlinear function (g(•)) representing the network structure and its parameters, with the actual input $x(t)$ as well as of the previous output. This network can be generalized, see Fig. 5.59

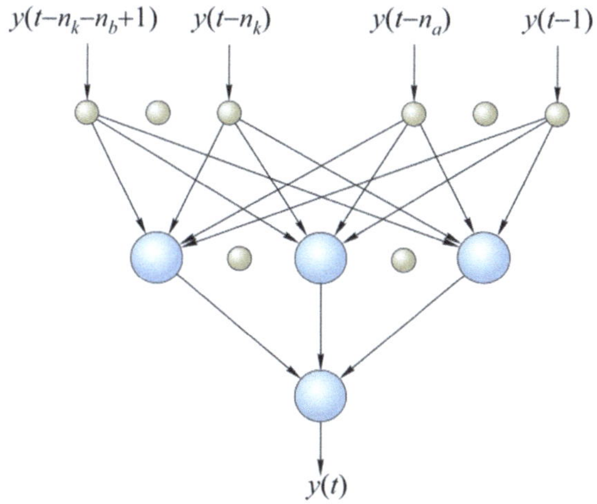

Fig. 5.59 Generalized recurrent network without biases.

Nowadays, there are more sophisticated models for recurrent procedure, Long-Short-Term- Memory (LSTM) and Gated Recurrent Uni (GRU). Let us consider the Elman network, which slightly differ from the Jordan's network concerning its feedback, namely

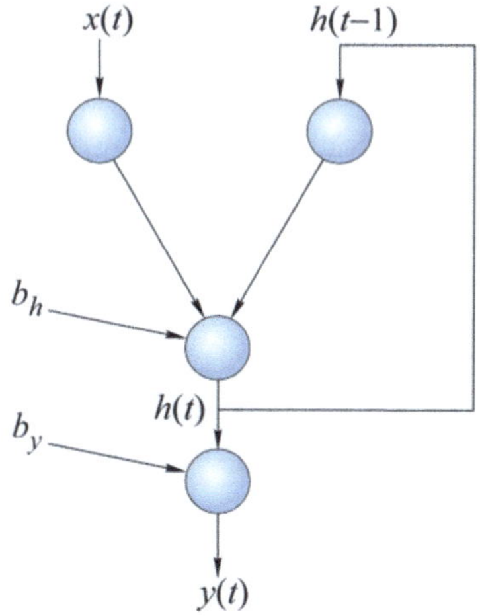

Fig. 5.60 Elman's network.

In this context, we can consider $x(t)$ as input, $h(t)$ as the state and $y(t)$ as the output. The new state in case of the standard Recurrent Neural Network (RNN) can be computed as

$$h(t) = \sigma_h(w_x x(t) + w_h h(t-1) + b_h)$$

The main difference between RNN and LSTM and GRU is, how to compute $h(t)$. The latter two techniques compute the new state in a more complicated way than RNN. These techniques improve very considerable the training process of the network.

5.5.1 Sequence To Sequence

Mathematica

Let us solve the sequence input - sequence output problem introduced above using RNN technique, see Fig. 5.61. The input and output sequences

```
⇒ x={{1},{2},{3},{2},{1},{3}}; y={{2},{3},{1},{1},{3},{2}};
```

We consider the following model

$$y(t) = g(y(t-1), \mathrm{x}(t), \mathrm{x}(t-1))$$

This means that $n_a = 1$, $n_b = 2$ and $n_k = 0$. Our multi-layer network has one hidden layer with 2 nodes containing sigmoid activation function.

Here we employ *Mathematica* 10 with *Neural Networks Application*,

```
⇒ <<NeuralNetworks`
⇒ Clear[model1,fitrecord];
```

Training process can be seen on Fig. 5.61.

```
⇒ {model1,fitrecord}=
   NeuralARXFit[x,y,{1,2,0},FeedForwardNet,{2},1000];//Quiet
```

⇐

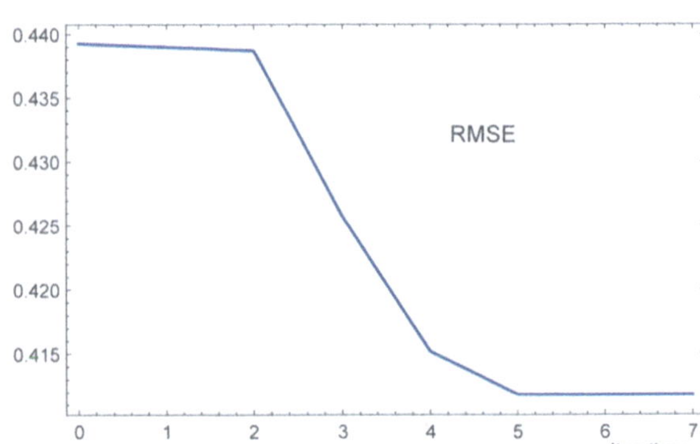

Fig. 5.61 The network error vs. iteration steps.

The network information

```
⇒ NetInformation[model1]
```

```
⇐ NeuralARX model with 1 input signal and 1 output signal.
   The regressor is defined by: na = 1, nb = 2, nk = 0. The
   mapping from regressor to output is defined by a FeedForward
   network created 2018-10-25 at 12:19. The network has 3 inputs
   and 1 output. It consists of 1 hidden layer with 2 neurons
   with activation function of Sigmoid type.
```

Let us employ our network,

```
⇒ NetOut=NetPredict[x,y,model1]
⇐ {{3.04076},{3.},{1.},{1.},{3.},{2.}}
```

The first element of the output sequence is wrong, since its computation is based on the unknown $u(0)$ and $y(0)$ values.

Now let us solve the problem with LSTM model (Rohrer 2017). Further models are in *Mathematica* 11. The input and out sequences are

```
⇒ X=Partition[Flatten[x],1]
⇐ {{1},{2},{3},{2},{1},{3}}
```

```
⇒ Y=Flatten[y]
⇐ {2,3,1,1,3,2}
```

Then the training set

```
⇒ train={X → Y}
⇐ {{{1},{2},{3},{2},{1},{3}} → {2,3,1,1,3,2}}
```

We employ a single LSTM layer with linear output,

```
⇒ netLSTM=NetInitialize@NetChain[{LongShortTermMemoryLayer[1],
   LinearLayer[6]},"Input" → {6,1},"Output" → 6]
⇐ NetChain[    Input                        matrix (size: 6x1)  ]
             1 LongShortTermMemoryLayer     matrix (size: 6x1)
             2 LinearLayer                  matrix (size: 1)
               Output                       matrix (size: 1)
```

The graph of the network

```
⇒ NetGraph[netLSTM]
```

⇐ NetGraph[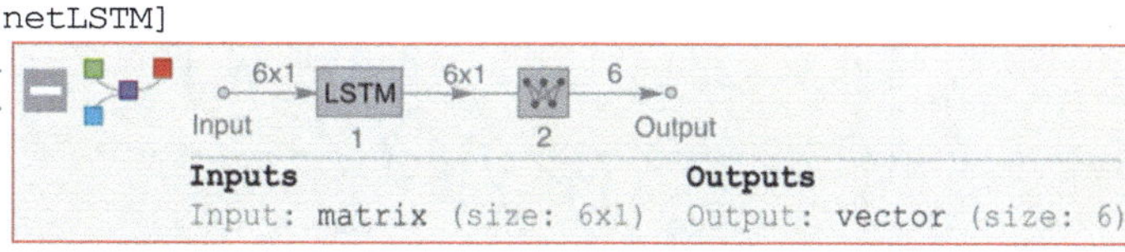]

Training the network

```
⇒ pLSTM=NetTrain[netLSTM,train]//Quiet
⇐ NetChain[ Imput port:          matrix (size: 6x1) ]
            Output port:         vector (size: 6)
            Number of layers:    2
```

Testing the network

⇒ pLSTM[X]
⇐ {2.,3.,1.,0.999999,3.,2.}

LSTM model is more efficient than standard RNN!

5.5.2 *Time Series Prediction*

The problem looks at an Airline Passengers' prediction problem. Given a year and a month, the task is to predict the number of international airline passengers in units of 1, 000. The data ranges from January 1949 to December 1960, or 12 years, with 144 observations. The dataset is freely available from the DataMarket webpage as a CSV download with the filename "international - airline - passengers.csv ".

The example is from: https://machinelearningmastery.com/time-series-prediction-lstm-recurrent-neural-networks-python-keras/ (Brownlee 2016).

Python

Loading the data and visualize it, see Fig. 5.62.

```
import pandas
import matplotlib.pyplot as plt
dataset = pandas.read_csv('M:\\
 international-airline-passengers.csv', usecols=[1],
 engine='python', skipfooter=3)
plt.plot(dataset)
plt.show()
```

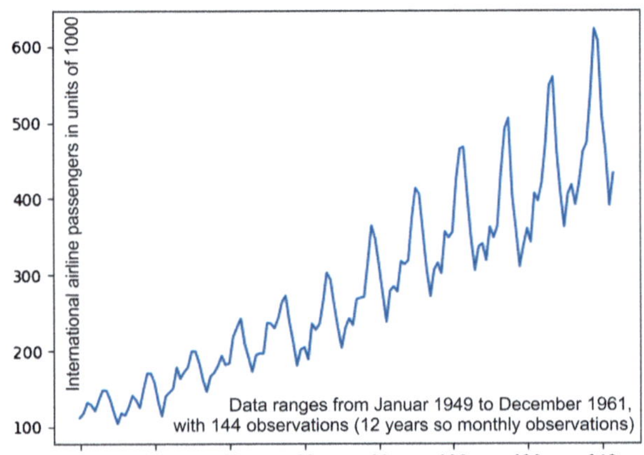

Fig. 5.62 The changes of the number of passengers in unit of 10^3 in time.

We can rephrase the problem as a regression one, i.e., given the number of passengers (in units of thousands) this month, what will be the number of passengers next month? We can write a simple function to convert our single column of data into a two-column dataset: the first column containing this month's $s(t)$ passenger count and the second column containing next month' $s(t+1)$ passenger count, to be predicted. Before we get started, let's first import all of the functions and classes we intend to use. This assumes a working SciPy environment with the Keras deep learning library installed (Chollet 2017).

```
# LSTM for international airline passengers problem with
  regression framing
import numpy
import matplotlib.pyplot as plt
from pandas import read_csv
import math
from keras.models import Sequential
from keras.layers import Dense
from keras.layers import LSTM
from sklearn.preprocessing import MinMaxScaler
from sklearn.metrics import mean_squared_error

# convert an array of values into a dataset matrix
def create_dataset(dataset, look_back=1):
    dataX, dataY = [], []
    for i in range(len(dataset)-look_back-1):
        a = dataset[i:(i+look_back), 0]
        dataX.append(a)
        dataY.append(dataset[i + look_back, 0])
    return numpy.array(dataX), numpy.array(dataY)

# fix random seed for reproducibility
numpy.random.seed(7)

# load the dataset
dataframe =
    read_csv('M:\\international-airline-passengers.csv',
    usecols=[1], engine='python', skipfooter=3)
dataset = dataframe.values
dataset = dataset.astype('float32')

# normalize the dataset
scaler = MinMaxScaler(feature_range=(0, 1))
dataset = scaler.fit_transform(dataset)

# split into train and test sets
train_size = int(len(dataset) * 0.67)
test_size = len(dataset) - train_size
train, test = dataset[0:train_size,:],
    dataset[train_size:len(dataset),:]
```

```
# reshape into X=t and Y=t+1
look_back = 1
trainX, trainY = create_dataset(train, look_back)
testX, testY = create_dataset(test, look_back)
# reshape input to be [samples, time steps, features]
trainX = numpy.reshape(trainX, (trainX.shape[0], 1,
    trainX.shape[1]))
testX = numpy.reshape(testX, (testX.shape[0], 1,
    testX.shape[1]))

# create and fit the LSTM network
model = Sequential()
model.add(LSTM(4, input_shape=(1, look_back)))
model.add(Dense(1))
model.compile(loss='mean_squared_error', optimizer='adam')
model.fit(trainX, trainY, epochs=100, batch_size=1, verbose=0)

# make predictions
trainPredict = model.predict(trainX)
testPredict = model.predict(testX)

# invert predictions
trainPredict = scaler.inverse_transform(trainPredict)
trainY = scaler.inverse_transform([trainY])
testPredict = scaler.inverse_transform(testPredict)
testY = scaler.inverse_transform([testY])

# calculate root mean squared error
trainScore = math.sqrt(mean_squared_error(trainY[0],
    trainPredict[:,0]))
print('Train Score: %.2f RMSE' % (trainScore))
testScore = math.sqrt(mean_squared_error(testY[0],
    testPredict[:,0]))
print('Test Score: %.2f RMSE' % (testScore))

# shift train predictions for plotting
trainPredictPlot = numpy.empty_like(dataset)
trainPredictPlot[:, :] = numpy.nan
trainPredictPlot[look_back:len(trainPredict)+look_back, :] =
    trainPredict

# shift test predictions for plotting
testPredictPlot = numpy.empty_like(dataset)
testPredictPlot[:, :] = numpy.nan
testPredictPlot[len(trainPredict)+(look_back*2)+
    1:len(dataset)-1, :] = testPredict

# plot baseline and predictions
plt.plot(scaler.inverse_transform(dataset))
plt.plot(trainPredictPlot)
plt.plot(testPredictPlot)
plt.show()
```

```
⇐ Train Score: 22.92 RMSE
⇐ Test Score: 47.53 RMSE
```

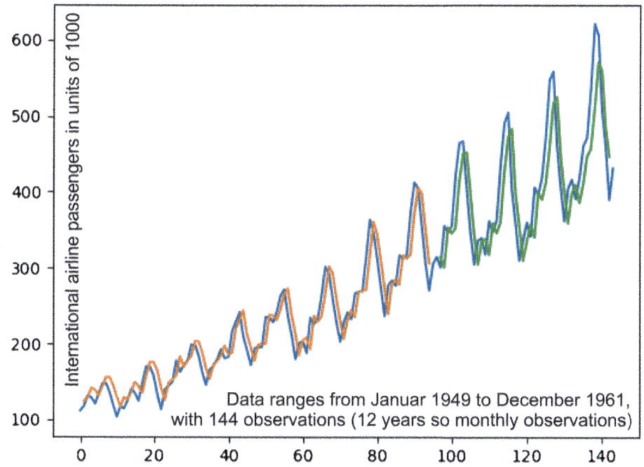

Fig. 5.63 The results of the training of the network (blue).

Fig. 5.63 shows the fitting on the training set (blue on orange) as well as the prediction on the test set (blue on green). For *Mathematica,* the input training and test data are associated with *Mathematica* variables

```
> trainY
⇐ {{118.,132.,129.,121.,135.,148.,148.,136.,119.,104.,118.,115.,
   126.,141.,135.,125.,149.,170.,170.,158.,133.,114.,140.,145.,
   150.,178.,163.,172.,178.,199.,199.,184.,162.,146.,166.,171.,
   180.,193.,181.,183.,218.,230.,242.,209.,191.,172.,194.,196.,
   196.,236.,235.,229.,243.,264.,272.,237.,211.,180.,201.,204.,
   188.,235.,227.,234.,264.,302.,293.,259.,229.,203.,229.,242.,
   233.,267.,269.,270.,315.,364.,347.,312.,274.,237.,278.,284.,
   277.,317.,313.,318.,374.,413.,405.,355.,306.,271.}}
⇒ trainY=Flatten[%];
```

```
> testY
⇐ {{301.,356.,348.,355.,422.,465.,467.,404.,347.,305.,336.,340.,
   318.,362.,348.,363.,435.,491.,505.,404.,359.,310.,337.,360.,
   342.,406.,396.,420.,472.,548.,559.,463.,407.,362.,405.,417.,
   391.,419.,461.,472.,535.,622.,606.,508.,461.,390.}}
⇒ testY=Flatten[%];
```

Mathematica

Let us employ built-in function. The lengths of the datasets are

```
⇒ ntrain=Length[trainY]
⇐ 94
```

```
⇒ ntest=testY//Length
⇐ 46
```

We shift and rescale the elements of the datasets to have zero mean and unit sample variance,

```
⇒ traintestY=Join[trainY,testY];
⇒ traintestYS=Standardize[traintestY];
⇒ trainYS=Take[traintestYS,{1,94}];
⇒ testYS=Take[traintestYS,{95,140}];
```

Our RNN model is

$$y(t+1)=g(y(t),y(t-1))$$

Now creating the training set

```
⇒ yout=Take[trainYS,{3,94}];
  yin1= Take[trainYS,{2,93}];
  yin2= Take[trainYS,{1,92}];
⇒ dataTrain=MapThread[{#2,#3} → #1&,{yout,yin2,yin1}];
```

Employing built in *Predict* function for training

```
⇒ p=Predict[dataTrain,Method → "NeuralNetwork",
   PerformanceGoal → "Quality",TimeGoal → 30]
⇐ PredictorFunction[[ Input type: NumericalVector (Length: 2)
                       Method: NeuralNetwork ]

⇒ PredictorInformation[p]
```

⇐

Predictor Information

Input time	Numerical vector (lenght: 2)
Method	NeuralNetwork
Standard deviation	0.302 ±0.037
Loss	0.0185 ± 0.079
Single evaluation time	1.89 ms/example
Batch evaluation speed	112. example/s
Predictor memory	2.12. MB
Training examples used	92 examples
Training time	35.5 s

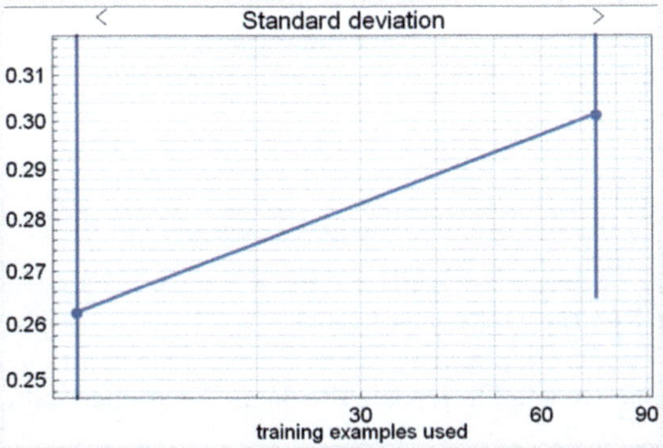


```
⇒ pm=PredictorMeasurements[p,dataTrain]
⇐ PredictorMeasurementsObject[ Predictor: NeuralNetwork
                               Number of test examples: 92 ]
⇒ pm["ComparisonPlot"]
```

⇐ 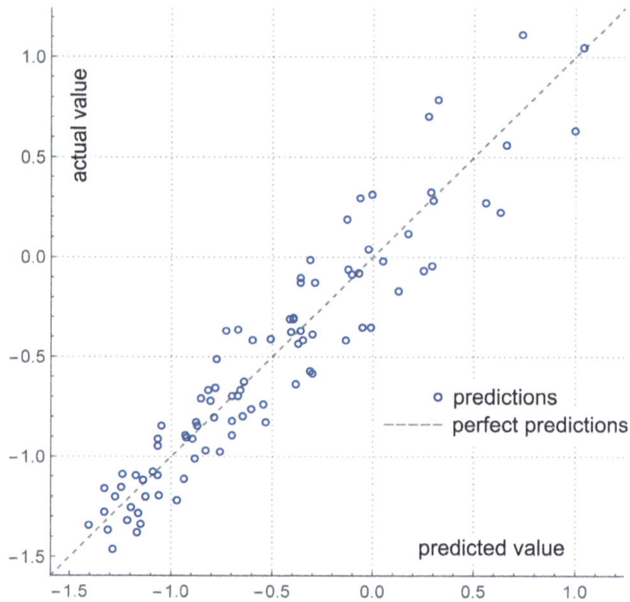

Fig. 5.64 The results of the training.

In order to visualize the results, we need to rescale the data. The trained data

```
⇒ trainedY=MapThread[p[{#1,#2}]&,{yin2,yin1}];
```

The mean value and the variance

```
⇒ me=Mean[traintestY];vari=Variance[traintestY];
```

Then the rescaled predicted values of the training set are

```
⇒ cumi=trainedY*Sqrt[vari]+me;
```

The original values

```
⇒ trainY;
```

Therefore the RMSE of the training values

```
⇒ Sqrt[Total[MapThread[
    (#1-#2)²&,Drop[trainY,1,2],cumi]]/Length[cumi]]
⇐ 23.4732
```

Which is quite close to the values of *Python* (22.92), see above.

Let us plot the output and the predicted values of the training set for visualization

```
⇒ p11=ListPlot[{yout,trainedY},Joined → True];
```

Now, we check the test dataset. Computing the predicted values

```
⇒ youtT=Take[testYS,{3,46}];
  yin1T= Take[testYS,{2,45}];
  yin2T= Take[testYS,{1,44}];
⇒ dataTest=MapThread[{#2,#3} → #1&,{youtT,yin2T,yin1T}];
⇒ testedY=MapThread[p[{#1,#2}]&,{yin2T,yin1T}];
⇒ Xtest=Range[93,136];
⇒ XYtest=Transpose[{Xtest,youtT}];
⇒ XYtrainPredict=Transpose[{Xtest,testedY}];
```

Visualizing the values of the test values and their predicted values

```
⇒ p22=ListPlot[
   {XYtest,XYtrainPredict},Joined → True,PlotStyle → {Blue,Red}];
```

Fig. 5.65 shows the result,

```
⇒ Show[{p11,p22},PlotRange → All,Frame → True,Axes → None]
```

⇐

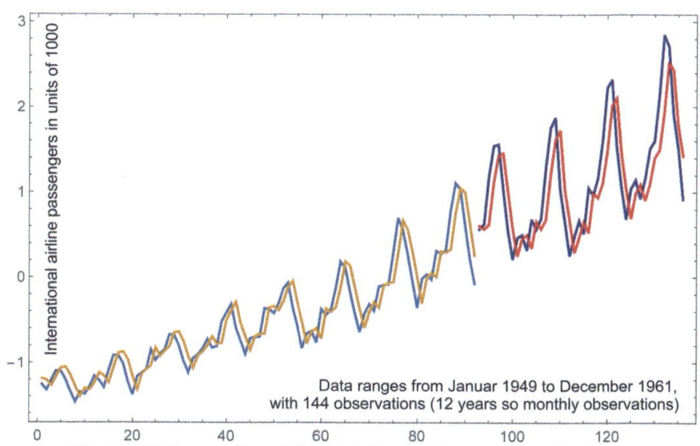

Fig. 5.65 The results of the training and test set prediction.

5.5.3 A Simple Optical Character Recognition

The optical character recognition problem takes an image containing a sequence of characters and returns the list of characters. One simple approach is to pre-process the image to produce images containing only a single character and do classification. This is a fragile approach and completely fails for domains such as cursive handwriting, where the characters run together.

Mathematica

First, generate training and test data, which consists of images of words and the corresponding word string:

```
⇒ imgMap =
    <|" " →  , "a" → a , "b" → b , "c" → c , "d" → d ,
    "e" → e , "f" → f , "g" → g , "h" → h , "i" → i , "j" → j ,
    "k" → k , "l" → l , "m" → m , "n" → n , "o" → o , "p" → p ,
    "q" → q , "r" → r , "s" → s , "t" → t , "u" → u , "v" → v ,
    "w" → w , "x" → x , "y" → y , "z" → z |>;
```

For example, the second element

```
⇒imgMap[[2]]
⇐ a
```

which is an image,

```
⇒ Head[%]
⇐ Image
```

which has a size of 15×9

```
⇒ ImageData[imgMap[[2]]]//Dimensions
⇐ {15,9}
```

The pixel values are

```
⇒ ImageData[imgMap[[2]]]//MatrixForm
```

$$
\Leftarrow \begin{pmatrix}
1.000, & 1.000, & 1.000, & 1.000, & 1.000, & 1.000, & 1.000, & 1.000, & 1.000 \\
1.000, & 1.000, & 1.000, & 1.000, & 1.000, & 1.000, & 1.000, & 1.000, & 1.000 \\
1.000, & 1.000, & 1.000, & 1.000, & 1.000, & 1.000, & 1.000, & 1.000, & 1.000 \\
1.000, & 1.000, & 1.000, & 1.000, & 1.000, & 1.000, & 1.000, & 1.000, & 1.000 \\
1.000, & 1.000, & 1.000, & 1.000, & 1.000, & 1.000, & 1.000, & 1.000, & 1.000 \\
1.000, & 0.890, & 0.498, & 0.000, & 0.000, & 0.420, & 0.929, & 1.000, & 1.000 \\
1.000, & 0.443, & 0.274, & 0.808, & 0.765, & 0.094, & 0.463, & 1.000, & 1.000 \\
1.000, & 1.000, & 1.000, & 0.957, & 0.863, & 0.420, & 0.216, & 0.984, & 1.000 \\
1.000, & 0.839, & 0.120, & 0.000, & 0.298, & 0.074, & 0.216, & 0.984, & 1.000 \\
0.972, & 0.120, & 0.498, & 0.992, & 1.000, & 0.659, & 0.172, & 0.984, & 1.000 \\
0.964, & 0.120, & 0.533, & 0.984, & 0.812, & 0.094, & 0.216, & 0.984, & 1.000 \\
1.000, & 0.678, & 0.000, & 0.000, & 0.047, & 0.572, & 0.216, & 0.984, & 1.000 \\
1.000, & 1.000, & 0.976, & 0.933, & 0.996, & 1.000, & 1.000, & 1.000, & 1.000 \\
1.000, & 1.000, & 1.000, & 1.000, & 1.000, & 1.000, & 1.000, & 1.000, & 1.000 \\
1.000, & 1.000, & 1.000, & 1.000, & 1.000, & 1.000, & 1.000, & 1.000, & 1.000
\end{pmatrix}
$$

```
⇒ MatrixPlot[%]
```

⇐

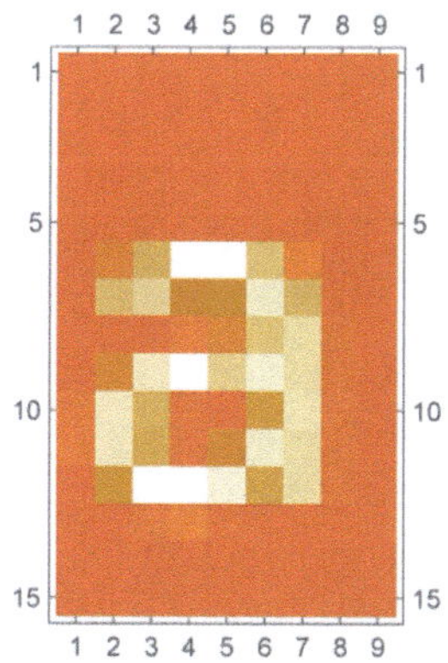

Fig. 5.66 The character as image.

Now we generate words randomly with maximum length of seven characters from English vocabulary of more than 40 thousand words,

```
⇒ Length[WordList[]]
⇐ 40127
```

```
⇒ maxLen=7;
   wordList=ToLowerCase[Select[WordList[],
    StringLength[#]<=maxLen&&LetterQ[#]&]];
   SeedRandom[1234];
   wordList=RandomSample[StringPadRight[wordList,maxLen]];
   dataset=Dataset@Map[<|"Input" → ImageAssemble[Lookup[imgMap,
    Characters[#]]],"Output" → StringTrim[#]|>&,wordList];
```

The number of the elements of the generated input - output data set,

```
⇒ Length[dataset]
⇐ 16022
```

which looks like

```
⇒ Short[dataset]
```

⇐

Input	Output
tool	tool
gingery	gingery
its	its
beacon	beacon
ghee	ghee
source	source
fathead	fathead
bollix	bollix
belated	belated
edge	edge
rancher	rancher
paltry	paltry
viper	viper
backlog	backlog
hit	hit
atrial	atrial
catboat	catboat
algebra	algebra
siphon	siphon
garden	garden

showing 1–20 of 16022

The input is image and the output is a string. The input image size is $15 \times (7 \times 9) = 15 \times 63$

```
⇒ ImageData[dataset[[1]][[1]]]//MatrixPlot
```

⇐

Fig. 5.67 The input string as image.

```
⇒ dataset[[1]][[2]]//Head
⇐ String
```

We define a net that takes an image and then let us treat the width dimension as a sequence dimension. For the network, an input image will be encoded using

```
⇒ enc=NetEncoder[{"Image",{63,15},"Grayscale"}]
⇐ NetEncoder[ Type              Image
              Image size        {63,15}
              Color space       Grayscale
              Color channels    1
              Mean Image        None
              Variance image    None
              Output            3-tensor(size: 1×15×63) ]
```

This will encode the input image into a tensor of size

```
⇒ enc[dataset[[1]][[1]]]//Dimensions
⇐ {1,15,63}
```

Define a net that takes an image and then treats the width dimension as a sequence dimension. A sequence of probability vectors over the width dimension is produced. The probability vector with length of the set of characters including blank will provide a sequence of characters. For example, let us define the following decoder

```
⇒ dec=NetDecoder[{"CTCBeamSearch",{"a","b","c"}}]
⇐ NetDecoder[ Type        CTCBeamSearch
              Labels      {a, b, c}
              Beam size   100           ]
```

Use the decoder on a sequence of probability vectors:

```
⇒ dec[{{0.2,0.4,0.3,0.1},{0.5,0.3,0.1,0.1}}]
⇐ {b,a}
```

The result is the most probable character sequence. The sequences with lower probability can be obtained - first four. The lower the value, the higher is the probability.

```
⇒ dec[{{0.2,0.4,0.3,0.1},{0.5,0.3,0.1,0.1}},
   {"TopNegativeLogLikelihoods",4}]
⇐ {{b,a} → 1.60944,{b} → 1.66073,{a} → 1.77196,{c,a} → 1.89712}
```

In our case all characters are considered,

```
⇒ chars=CharacterRange["a","z"]
⇐ {a,b,c,d,e,f,g,h,i,j,k,l,m,n,o,p,q,r,s,t,u,v,w,x,y,z}
⇒ Length[chars]
⇐ 26
```

and the decoder is

```
⇒ decoder=NetDecoder[{"CTCBeamSearch",chars,"BeamSize" → 50}]
⇐ NetDecoder[ Type        CTCBeamSearch
              Labels      {a,b,c,d,<<18>>,w,x,y,z}
              Beam size   50                        ]
```

Now we can define our network containing GatedRecurrentLayers indicating recurrent type of network,

```
⇒ ocrNet=NetChain[{ConvolutionLayer[20,3],BatchNormalizationLayer[],
    Ramp,PoolingLayer[2],ConvolutionLayer[15,3],
    BatchNormalizationLayer[],Ramp,PoolingLayer[2],FlattenLayer[1],
    TransposeLayer[],GatedRecurrentLayer[19],GatedRecurrentLayer[19],
    NetMapOperator[LinearLayer[Length[chars]+1]],SoftmaxLayer[]},
    "Input" → NetEncoder[{"Image",{63,15},"Grayscale"}],
    "Output" → decoder]
⇐ NetChain[
                                            image
       Input                                3-tensor (size: 1×15×63)
    1  ConvolutionLayer                     3-tensor (size: 20×13×61)
    2  BachNormalizationLayer               3-tensor (size: 20×13×61)
    3  Ramp                                 3-tensor (size: 20×13×61)
    4  PoolingLayer                         3-tensor (size: 20×12×60)
    5  ConvolutionLayer                     3-tensor (size: 15×10×58)
    6  BachNormalizationLayer               3-tensor (size: 15×10×58)
    7  Ramp                                 3-tensor (size: 15×10×58)
    8  PoolingLayer                         3-tensor (size: 15×9×57)
    9  FlattenLayer                         matrix (size: 135×57)
    10 TransposeLayer                       matrix (size: 57×135)
    11 GatedRecurrentLayer                  matrix (size: 57×19)
    12 GatedRecurrentLayer                  matrix (size: 57×19)
    13 NetMapOperator[LinearLayer]          matrix (size: 57×19)
    14 SoftmaxLayer                         matrix (size: 57×27)
       Output                               ctcbeam search
  ]
```

In order to evaluate how the input of the network can fit to the required output, we define a loss function.

```
⇒ loss=CTCLossLayer["Target" → NetEncoder[{"Characters",chars}]]
⇐ CTCLossLayer[ Input:  matrix(size: n1×27)
                Target: string ]
```

Split the dataset into a test set and a training set.

```
⇒ {testData,trainData}=
    TakeDrop[dataset,Ceiling[Length[dataset]/10]];
⇒ Length[trainData]
⇐ 14419
⇒ Length[testData]
⇐ 1603
⇒ trainNet=NetTrain[ocrNet,trainData,LossFunction → loss,
    MaxTrainingRounds → 20,ValidationSet → testData]
```

Evaluate the trained net on images from the test set (see above), the inputs

```
⇒ testIms=Normal@testData[1;;10,"Input"]
⇐ { tool  , gingery, its   , beacon , ghee ,
    source, fathead, bollix, belated, edge }
⇒ trainNet[testIms]
⇐ {{t,o,l},{g,i,n,g,e,r,y},{i,t,s},{b,e,a,c,o,n},
    {g,h,e},{s,o,u,r,c,e},{f,a,t,h,e,a,d},
    {b,o,l,i,x},{b,e,l,a,t,e,d},{e,d,g,e}}
```

Obtain the top 5 decodings for an image, along with the negative log - likelihood of each decoding

```
⇒ trainNet[ somber,
    {"TopNegativeLogLikelihoods", 5}]
⇐ {{s,o,m,b,e,r} → 0.00501868,{s,o,m,u,b,e,r} → 8.53219,
    {u,o,m,b,e,r} → 8.6011,{s,o,m,b,h,e,r} → 8.76271,
    {w,o,m,b,e,r} → 8.92553}
```

or

```
⇒ dataset[[18]][[1]]
⇐ algebra
⇒ trainNet[dataset[[18]][[1]], {"TopNegativeLogLikelihoods", 5}]
⇐ {{a,l,g,e,b,r,a} → 0.00695273,{l,g,e,b,r,a} → 7.21149,
    {t,l,g,e,b,r,a} → 8.16581,{a,l,c,e,b,r,a} → 8.59214,
    {l,g,e,b,r,a} → 7.21149}
```

We may extend this example for licence plate recognition!

5.6 Deep Neural Network

Basic theory

The deep neural networks (DNN) with more hidden layers yielded poorer performance since they represent over fitting and weak learning performances more likely than the traditional shallow networks. Here we discuss some new techniques to improve the performance of the DNN (Banerjee 2018).

5.6.1 Dropout

The most representative solution is the dropout, which trains only some of the randomly selected nodes rather than the entire network. It is very effective, while its implementation is not very complex. Figure 5.68 below explains the concept of the dropout. Some nodes are randomly selected at a certain percentage and their outputs are set to zero in order to deactivate the nodes.

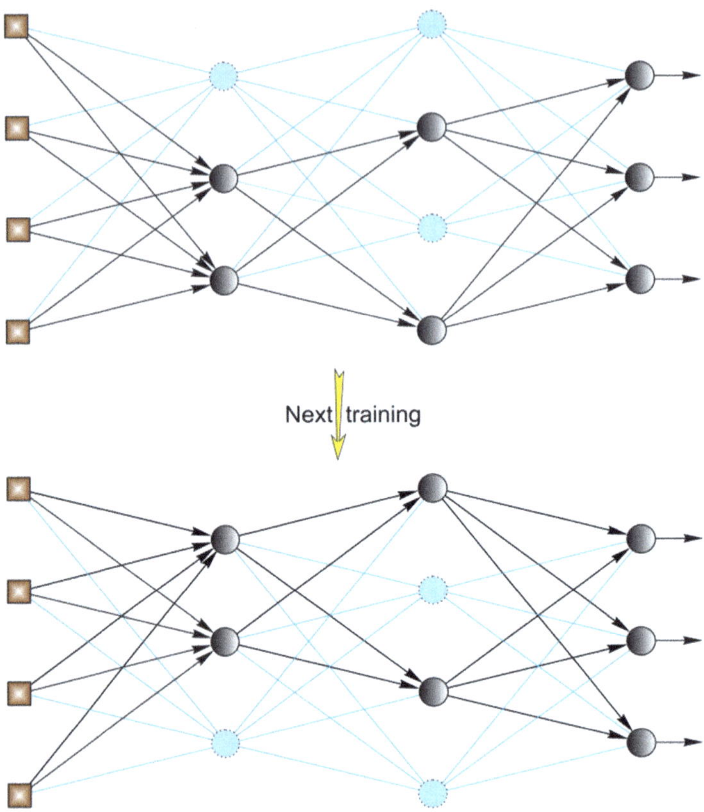

Fig. 5.68 Dropout is where some nodes are randomly selected and their outputs are set to zero to deactivate the nodes.

The technical realization in *Mathematica*,

```
⇒ drop=DropoutLayer[0.65,"Input" → NetEncoder["Image"],
   "Output" → NetDecoder["Image"]]
```

```
⇐ DropoutLayer[ Parameters
                  Dropuot probability:   0.65  ]
                  Method:                Dropout
                Ports
                  Input:                 image
                  Output:                image
```

The layer acts on the image 3-tensor by randomly and independently zeroing the individual color components of each pixel:

⇒ drop[ ,NetEvaluationMode→"Train"]

Fig. 5.69 Effect of Dropout in general.

5.6.2 ReLU

The gradient in this context can be thought to be similar in concept to the delta of the back-propagation algorithm. The vanishing gradient in the training process with the back-propagation algorithm occurs when the output error is more likely to fail to reach the further nodes. The back-propagation algorithm trains the neural network as it propagates the output error backward to the hidden layers. However, as the error hardly reaches the first hidden layer, the weight cannot be adjusted. Therefore, the hidden layers that are close to the input layer are not properly trained. There is no point of adding hidden layers if they cannot be trained, see Fig. 5.70.

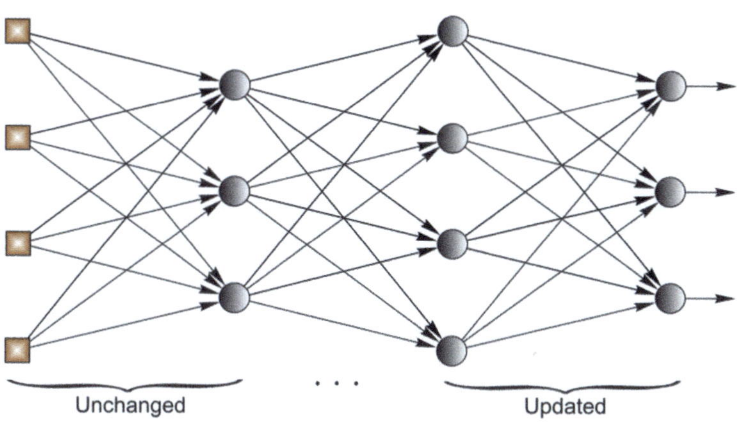

Fig. 5.70 The vanishing gradient.

The representative solution to the vanishing gradient is the use of the *Rectified Linear Unit* (ReLU) function as the activation function. It is known to better transmit the error than the sigmoid function. The ReLU function is defined as $f(x) = \max(x, 0)$. In *Mathematica* there is a built-in function, see Fig. 5.71,

```
⇒ Plot[Ramp[x], {x,-2,2}]
```

⇐ 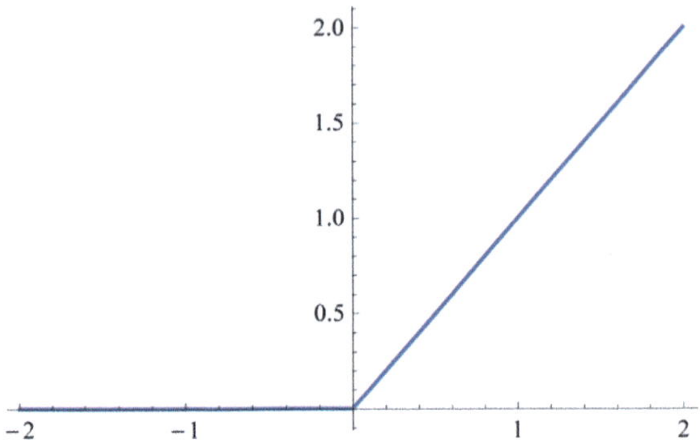

Fig. 5.71 The ReLU function Ramp(x).

ReLU(x) relates to the UnitStep(x) function, namely

```
⇒ Simplify[Ramp[x]-x UnitStep[x]]
⇐ 0
```

5.6.3 Softmax Layer

In case of classification, the probability vector produced by *Softmax Layer* of the network is converted to an actual class,

$$v = \begin{bmatrix} 2 \\ 1 \\ 0.1 \end{bmatrix} \Rightarrow \varphi(v) = \begin{bmatrix} \dfrac{e^2}{e^2 + e^1 + e^{0.1}} \\ \dfrac{e^1}{e^2 + e^1 + e^{0.1}} \\ \dfrac{e^{0.1}}{e^2 + e^1 + e^{0.1}} \end{bmatrix} = \begin{bmatrix} 0.6590 \\ 0.2424 \\ 0.0986 \end{bmatrix}$$

Mathematica

```
⇒ softmax=SoftmaxLayer["Input" → {3}]
⇐ SoftmaxLayer[ Level: -1
                Output: vector (size: 3) ]

⇒ softmax[{2,1,0.1}]
⇐ {0.659001,0.242433,0.0985659}
```

5.6.4 Cross Entropy Loss

Let us consider a new loss function *H*(*y*, *d*) called the *cross entropy function*. Technically we can compute this function in the following way,

$$H(y, d) = -d \text{ Log }(y) \; - \; (1-d) \text{ Log}(1-\text{y})$$

Fig. 5.72 shows this function, where *d* is the sample output, while *y* is network output ranging between 0 and 1. That is why this objective loss function is teaming with the Softmax layer.

```
⇒ Plot[{-Log[y],-Log[1-y]},{y,0,1},Frame → True,FrameLabel → {"y"},
    PlotLegends → {"d = 1 → -Log(y)","d = 0 → -Log(1-y)"}]
```

⇐

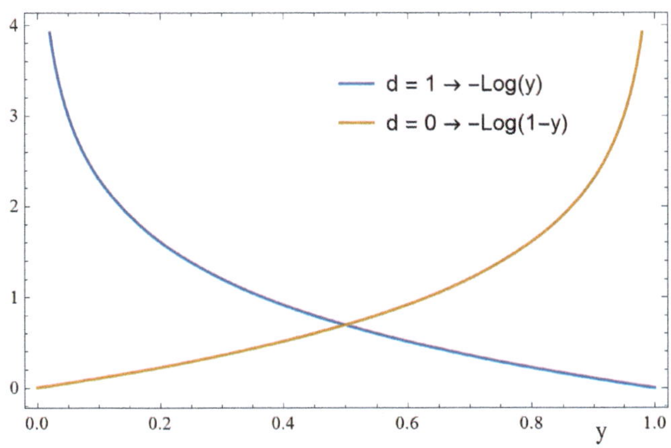

Fig. 5.72 Cross entropy function *H*(*y*, *d*) in case of *d* = 1 and *d* = 0.

If the network has *M* output nodes,

$$H = \sum_{i=1}^{M} \left(-d_i \ln\left(y_i\right) - \left(1-d_i\right) \ln\left(1-d_i\right)\right)$$

Mathematica

```
⇒ loss=CrossEntropyLossLayer["Probabilities"]
⇐ CrossEntropyLossLayer[ Target form:   Probabilities
                         Input:         tensor of rank ≥ 1
                         Output:        tensor of rank ≥ 1 ]

⇒ loss[<|"Input"  →  {0.2,0.7}, "Target" → {0.3,0.8}|>]
⇐ 0.768171

⇒ loss=MeanSquaredLossLayer["Input" → {2}]
⇐ MeanSquaredLossLayer[ Input:    vector (size: 2)
                        Target:   vector (size: 2) ]
```

```
⇒ loss[<|"Input" → {0.2,0.7}, "Target" → {0.3,0.8}|>]
⇐ 0.01
```

The cross entropy loss function is more sensitive for the error than RMSE and mainly can be employed in classification.

5.6.5 Stochastic Gradient Descent

The back propagation techniques employ gradient method to modify the weights of the network for every training sample,

$$w_{i+1} = w_i - \left(\frac{dR(w)}{dw}\right)_{w=w_i} \qquad \xi = w_i + \Delta w_i$$

where R is an error function. The algorithm sweeps through the training set, but not sequentially, randomly picking up the actual sample. We call it *stochastic learning*. Another technique is, when the algorithm sweeps through the training set, it performs the above update for each training example. However the samples can be randomly shuffled for each pass. It is called *stochastic gradient method* (SGD). In both cases at every single sample the error is computed and the weights updated immediately.

5.6.6 Batch

One sweep through the whole samples in the training set is called one epoch. In the *batch method*, each weight update is calculated for all errors of the training data, and the average of the weight updates used for adjusting the weights. This method uses all of the training data and updates *only once during one epoch.* The update of the *i*-th weight after an epoch is

$$w_{i+1} = w_i + \frac{1}{N}\sum_{j=1}^{N}(\Delta w_i)_j$$

The ADAM method (*Adaptive Moment Estimation*) is an improved version of the SGD method + *Momentum technique*. In this optimization algorithm, running averages of both the *gradients and the second moments of the gradients* are used.

5.6.7 *Mini Batch*

The mini batch method is a blend of the SGD and batch methods. It selects a part of the training dataset and uses them for training in the batch method. Therefore, it calculates the weight updates of the selected data and trains the neural network with the averaged weight update. For example, if 20 arbitrary data points are selected out of 100 training data points, the batch method is applied to the 20 data points. In this case, a total of five weight adjustments are performed to complete the training process, an epoch, for all the data points (5 = 100/20).

The mini batch method, when it selects an appropriate number of data points, obtains the benefits from both methods: speed from the SGD and stability from the batch. For this reason, it is often utilized in Deep Learning, which manipulates a significant amount of data.

5.6.8 *GPU*

1) The computationally intensive part of neural network is made up of multiple matrix multiplications, which can be evaluated in parallel,

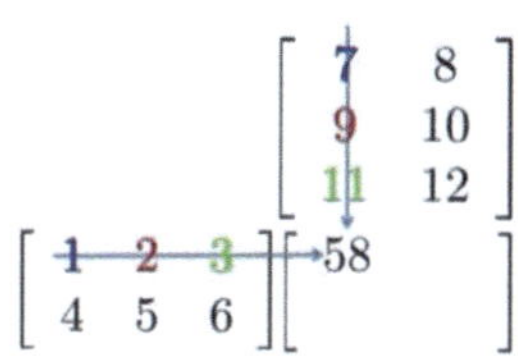

$$1 \cdot 7 + 2 \cdot 9 + 3 \cdot 11 = 58$$

A CPU core processes data in a serial one task at a time, so one is limited to simultaneous processing only up to the limit of the number of cores in general max 12 while in case of GPU there can be many thousands.

2) The CPU is designed for more general task than GPU, which has more simple architecture using basically floating-point representation for just a few types of algorithm.

5.6.9 *Classifying Double Spirals*

Let us solve the double-spiral classification problem employing deep neural network (Hu and Yuan 2016).

```
⇒ spirals=Import["M:\\spiral.dat"];
⇒ n=Length[spirals]
```

```
⇐ 194
⇒ s1={};s2={};
⇒ Do[If[spirals[[i,3]]==0,AppendTo[s1,{spirals[[i,1]],
    spirals[[i,2]]}],AppendTo[s2,{spirals[[i,1]],
    spirals[[i,2]]}]],{i,1,n}]
```

Visualization of the double spirals, see Fig.5.73

```
⇒ S1=ListPlot[s1,PlotStyle → {RGBColor[1,0,0],
    PointSize[0.02]},AspectRatio → 1];
⇒ S2=ListPlot[s2,PlotStyle → {RGBColor[0,0,1],
    PointSize[0.015]},AspectRatio → 1];
⇒ pspiral=Show[{S1,S2},PlotRange → All]
```

⇐

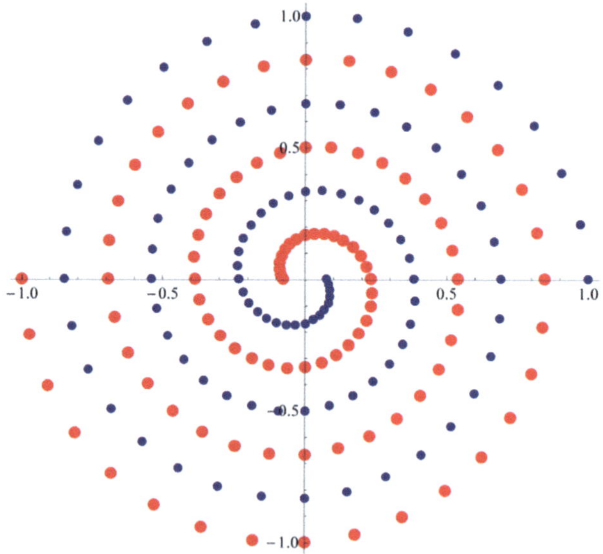

Fig. 5.73 Double spirals.

Mathematica

The elements of training set are the coordinates as input and the labels (0 or 1) are the output values,

```
⇒ S1=Map[# → {0}&,s1];S2=Map[# → {1}&,s2];
⇒ trainingdata=Join[S1,S2];
```

The network,

```
⇒ net=NetChain[{16,Ramp,16,Ramp,16,Ramp,16,
    ElementwiseLayer[1+Tanh[#]&],1},"Input" → {2},"Output" → {1}]
```

```
⇐ NetChain[
     Input        vector(size: 2)
   1 LinearLayer  vector(size: 16)
   2 Ramp         vector(size: 16)
   3 LinearLayer  vector(size: 16)
   4 Ramp         vector(size: 16)
   5 LinearLayer  vector(size: 16)
   6 Ramp         vector(size: 16)
   7 LinearLayer  vector(size: 16)
   8 1+Tanh[x]    vector(size: 16)
   9 LinearLayer  vector(size: 1)
     Output       vector(size: 1)
 ]
```

```
⇒ net=NetChain[{16,Ramp,16,Ramp,16,Ramp,16,
   ElementwiseLayer[1+Tanh[#]&],1},"Input" → {2},"Output" → {1}]
⇒ NetGraph[net]
```

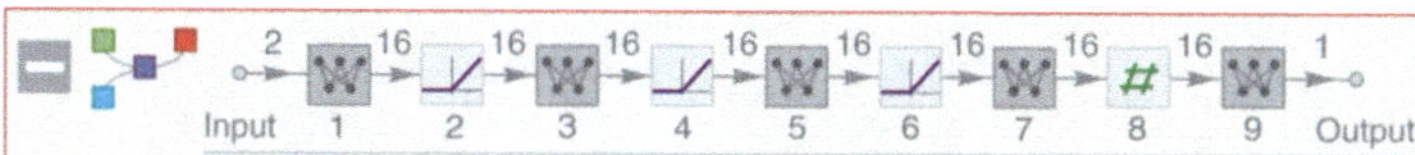


```
Inputs                    Outputs
Input: vector (size: 2)   Output: vector (size: 1)
```

```
⇒ net=NetInitialize[net];
```

We train the network with RMSProp method, which is a stochastic gradient descent using an adaptive learning rate derived from exponentially smoothed average of gradient magnitude,

```
⇒ trained=NetTrain[net,trainingdata,BatchSize → 32,
   MaxTrainingRounds → 2000,Method → "RMSProp",
   TargetDevice → "GPU"];
⇒ trained[s1]
⇐ {{0.00577259},{0.00549066},{0.00632614},{0.00683176},{0.00553894},
   {0.0056929},{0.00560635},{0.00548553},{0.0054264},{0.00556308},
   {0.00656038},{0.00603347},{0.00658998},{0.00693081},{0.00613347},
   ⋮
   {0.0036745},{0.00316484},{0.00299972},{0.00296793},{0.00299105},
   {0.0031051},{0.00339055},{0.00375449},{0.00428277},{0.0051856}}

⇒ trained[s2]
⇐ {{1.01079},{1.00865},{1.00976},{1.01042},{1.00991},{1.01131},
   {1.00825},{1.00825},{1.00989},{1.00782},{1.00986},{1.01162},
   {1.0114},{1.01157},{1.0099},{1.00786},{1.01065},{1.00889},
   ⋮
   {1.01245},{1.01273},{1.01338},{1.01434},{1.01556},{1.01702},
   {1.01867},{1.0187},{1.01828},{1.01813},{1.01826},{1.01874}}
```

The boundary points of the two classes, see Fig. 5.74.

```
⇒ Show[{pspiral,Quiet[ContourPlot[trained[{x,y}]=={0.5},
   {x,-1.01,1.01},{y,-1.01,1.01},ContourStyle → {Thin,Green}]]}]
```

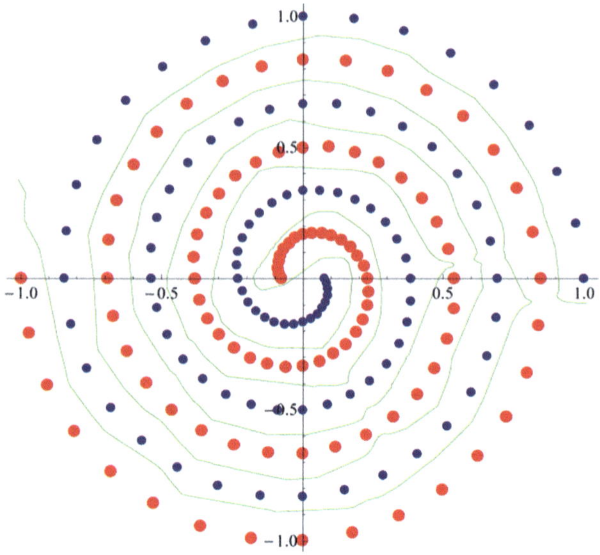

⇐

Fig. 5.74 Classified double spirals.

Python

Export data for *Python*

```
⇒ Export["M:\\spiralom.csv",spirals];
```

We employ here Keras package (Sharma 2017)

```
import numpy as np
import pandas as pd
np.random.seed(4375689)
```

```
from keras.models import Sequential
from keras.layers import Dense
from keras.optimizers import RMSprop
```

Loading the data

```
train_data = pd.read_csv('M:\\spiralom.csv').values
train_X = train_data[:,0:2]
train_y = train_data[:,2]
```

Our network structure

```
model = Sequential()
model.add(Dense(16, activation='relu', input_shape=(2,)))
model.add(Dense(16, activation='relu'))
model.add(Dense(16, activation='relu'))
model.add(Dense(16, activation='relu'))
model.add(Dense(1, activation='sigmoid'))
```

```
model.compile(loss='binary_crossentropy',optimizer=RMSprop(),
    metrics=['accuracy'])
```

Training the network

```
history = model.fit(train_X, train_y, batch_size=32,
     epochs=2000, verbose=0)
```

```
half_whascore = model.evaluate(train_X, train_y, verbose=0)t
```

```
score
```

```
⇐ {0.086479,1.}
```

```
u=model.predict(train_X)
u
```

```
⇐ {2.0923×10^-6,0.999854,0.000068496,0.993133,0.0000125849,
  0.995349,3.65539×10^-6,0.996501,3.36894×10^-6,0.999897,3.90718×10^-6,
  0.999797,0.00007199,0.99995,0.00268287,1.,0.000811408,0.999999,
   ⋮
  0.995914,3.40173×10¯11,0.998619,1.22405×10^-12,0.999704,2.72944×10^-12,
  0.999982,5.83677×10^-11,0.999997,4.11347×10^-7,0.995909,0.000571957}
```

```
⇒ z=Flatten[Round[%]]
⇒ {0,1,0,1,0,1,0,1,0,1,0,1,0,1,0,1,0,1,0,1,0,1,0,1,0,1,0,1,
  0,1,0,1,0,1,0,1,0,1,0,1,0,1,0,1,0,1,0,1,0,1,0,1,0,1,0,1,
  0,1,0,1,0,1,0,1,0,1,0,1,0,1,0,1,0,1,0,1,0,1,0,1,0,1,0,1,
  0,1,0,1,0,1,0,1,0,1,0,1,0,1,0,1,0,1,0,1,0,1,0,1,0,1,0,1,
  0,1,0,1,0,1,0,1,0,1,0,1,0,1,0,1,0,1,0,1,0,1,0,1,0,1,0,1,
  0,1,0,1,0,1,0,1,0,1,0,1,0,1,0,1,0,1,0,1,0,1,0,1,0,1,0,1,
  0,1,0,1,0,1,0,1,0,1,0,1,0,1,0,1,0,1,0,1,0,1,0,1,0}
```

```
⇒ zz=Drop[Table[spirals[[i,3]],{i,1,n}],{1,1}]
⇒ {0,1,0,1,0,1,0,1,0,1,0,1,0,1,0,1,0,1,0,1,0,1,0,1,0,1,0,1,
  0,1,0,1,0,1,0,1,0,1,0,1,0,1,0,1,0,1,0,1,0,1,0,1,0,1,0,1,
  0,1,0,1,0,1,0,1,0,1,0,1,0,1,0,1,0,1,0,1,0,1,0,1,0,1,0,1,
  0,1,0,1,0,1,0,1,0,1,0,1,0,1,0,1,0,1,0,1,0,1,0,1,0,1,0,1,
  0,1,0,1,0,1,0,1,0,1,0,1,0,1,0,1,0,1,0,1,0,1,0,1,0,1,0,1,
  0,1,0,1,0,1,0,1,0,1,0,1,0,1,0,1,0,1,0,1,0,1,0,1,0,1,0,1,
  0,1,0,1,0,1,0,1,0,1,0,1,0,1,0,1,0,1,0,1,0,1,0,1,0}
```

```
⇒ Norm[z-zz]
⇒ 0
```

Perfect!

5.7 Convolutional Neural Networks

Basic theory

The convolutional neural network (CNN) is a relative new machine learning technique integrating the traditional shallow type neural network and the computer vision mapping methods as feature extraction.

5.7.1 Problems in Computer Vision

Basic problems in computer vision: identification of images in general as classification- supervised and clustering unsupervised as well as object identification and localisation on a single image

⇒

Fig. 5.75 Classification of images. What are on the pictures?

⇒

Fig. 5.76 Clustering of images. What pictures are close to each other?

Fig. 5.77 Object localization and identification on images. What kind of objects are on the picture and where are they?

How can we solve these problems via neural network?

5.7.2 Feature Extraction via AutoEncoder

How can the input images be represented in a network? We have seen the details of the dimension reduction methods in the first chapter. As already stated, an AutoEncoder neural network is an unsupervised learning algorithm that applies backpropagation, setting the target values to be equal to the inputs, see Fig. 5.78.

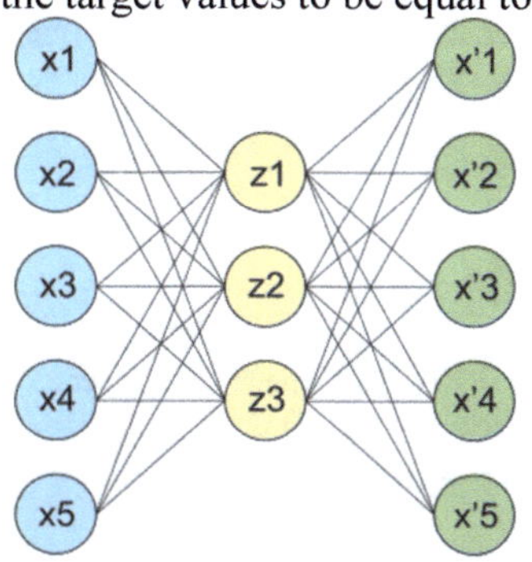

Fig. 5.78 The principle of the AutoEncoder training.

While the network represents the identity function between the input and output layers, $(x_i \to (x')_i)$ the *hidden layer* represents the reduced dimensional feature of the input, see Fig.5.79.

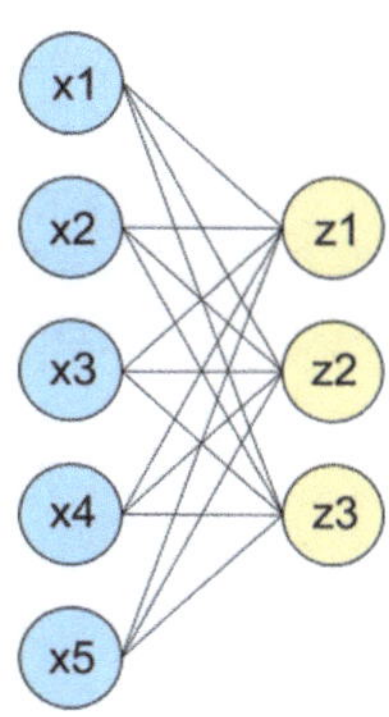

Fig. 5.79 Employing AutoEncoder.

Let us consider the following two clusters of images:

```
⇒ persons=Map[ImageResize[#,{28,28}]&,
```

{, , , , , , , ,

Fig. 5.80 Images of persons.

```
⇒ flowers=
```

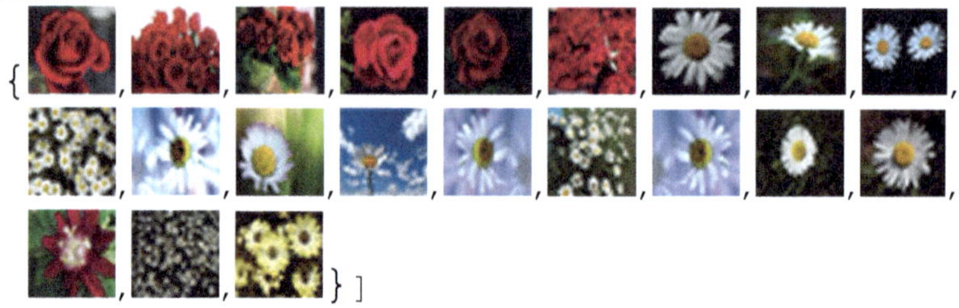

Fig. 5.81 Images of flowers.

```
⇒ mixed=Join[persons,flowers];
```

We try to characterize the features of the images with a numeric vector of size of two dimensions, which means there are two knots in the hidden layer of the AutoEncoder network. We employ built-in function

```
⇒ reduced = DimensionReduce[mixed, 2, Method → "AutoEncoder"];
```

Let us visualize the reduced dataset, see Fig.5.82.

```
⇒ ListPlot[MapThread[Labeled[#1, #2] &, {reduced, mixed}],
    PlotStyle → PointSize[0.001]]
```

⇐

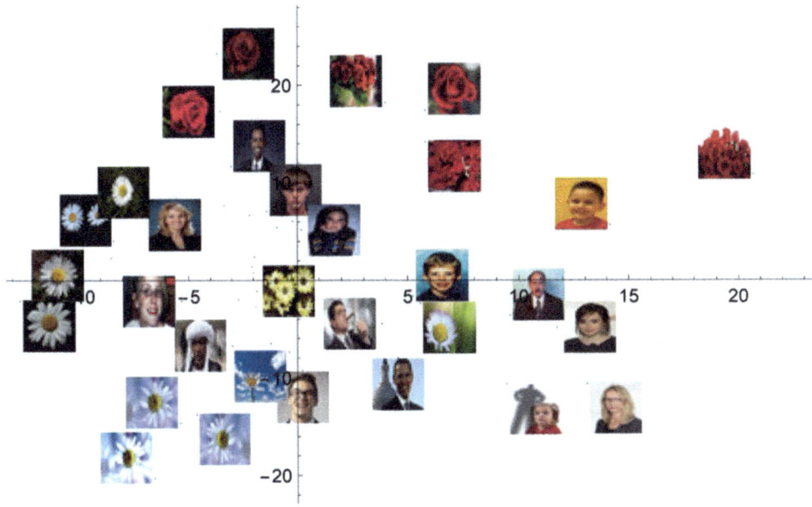

Fig. 5.82 Clusters of images represented by reduced dimension vectors.

In order to make the separation of the two clusters easier let us employ polar coordinate system,

```
⇒ reducedPolar=
    Map[ArcTan[#[[1]]/√#[[1]]²+#[[2]]²],√#[[1]]²+#[[2]]²]&,reduced];
```

Fig. 5.83 shows the transformed clusters of images

```
⇒ Show[{ListPolarPlot[MapThread[Labeled[
    #1, #2] &, {reducedPolar, mixed}],PlotStyle → PointSize[0.001],
    AspectRatio → 0.7],Plot[2.2x-20,{x,0,17},
    PlotStyle → {Red,Thick}]},PlotRange → {-17,17}]
```

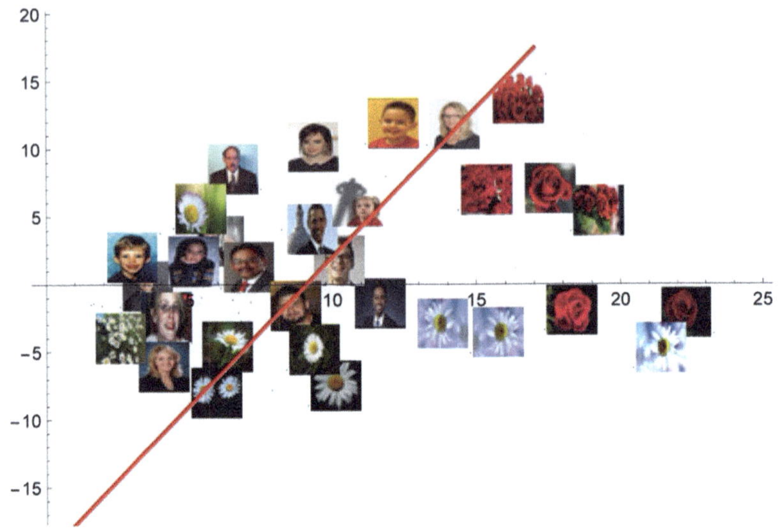

Fig. 5.83 Clusters of images transformed into polar coordinate system.

Now nearly linear separation is possible! However there are some misclustered images in both clusters.

There is another way to characterize the feature of the digital images, too.

5.7.3 Respective Fields

On the one hand, the work by *Hubel* and *Wiesel* in the 1950s and 1960s showed that cat and monkey visual cortexes contain neurons that individually respond to small regions of the visual field called receptive fields whose sizes and locations vary systematically across the cortex to form a complete map of visual space, see Fig. 5.84.

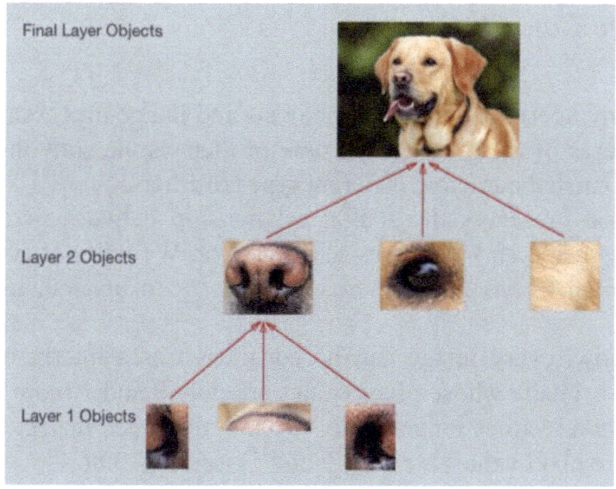

Fig. 5.84 Illustration of respective fields.

On the other hand, in computer vision, traditionally the feature extraction is based on pixel representation. It is difficult, however, for a computer to understand the meaning of raw sensory input data, such as this image represented as a collection of pixel values.

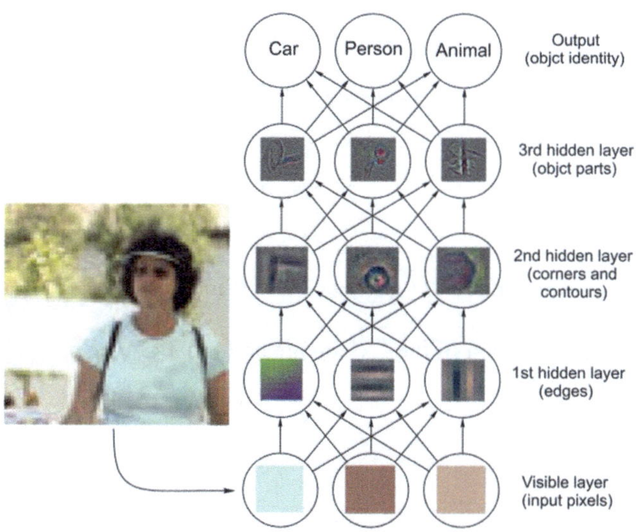

Fig. 5.85 Image feature representation at different level of the image structure (object types at different levels).

These two facts motivated the application of complex mapping to represent features of digital images in deep learning.

Convolution neural network resolves this task by breaking the desired complicated mapping into a series of nested simple mappings.

5.7.4 Image Convolution

There are many operations in image analysis and image processing for detecting different features of a digital image. One of them is the convolution operation, which can be carried out using different types of filters.

Convolution preserves the spatial relationship between pixels by learning image features using small squares of input data. We will not go into the mathematical details of convolution here, but will try to understand how it works over images.

As we know, every image can be considered as a matrix of pixel values. Consider a 5×5 image whose pixel values are only 0 and 1 (note that for a gray-scale image, pixel values range from 0 to 255, the green matrix below is a special case where pixel values are only 0 and 1), see Fig. 5.86.

1	1	1	0	0
0	1	1	1	0
0	0	1	1	1
0	0	1	1	0
0	1	1	0	0

Fig. 5.86 Binary image matrix.

Also, consider another 3×3 matrix shown below:

1	0	1
0	1	0
1	0	1

Fig. 5.87 Convolution filter (kernel matrix).

Then, the convolution of the 5×5 image and the 3×3 matrix can be computed as shown in Fig. 5.88.

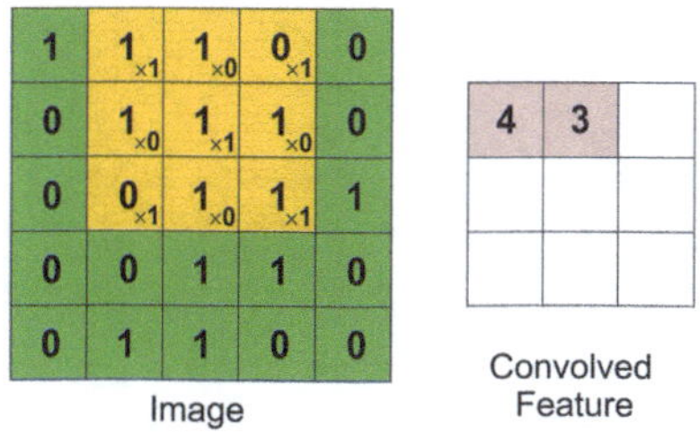

Fig. 5.88 Illustration of convolution process. Considering the multiplication and summation of the corresponding image and convolution filter elements, i.e

$4 = 1\times1 + 1\times0 + 1\times1 + 0\times0 + 1\times1 + 1\times0 + 0\times1 + 0\times0 + 1\times1$ $\quad$ $3 = 1\times1 + 1\times0 + 0\times1 + 1\times0 + 1\times1 + 1\times0 + 0\times1 + 1\times0 + 1\times1$

Let us consider an example for the convolution in case of different filters. The kernel matrices of the filters are: *edge detect*: M1 and *sharpen*

$$\Rightarrow \text{M1}=\begin{pmatrix}0 & 1 & 0\\ 1 & -4 & 1\\ 0 & 1 & 0\end{pmatrix};\text{M2}=\begin{pmatrix}0 & -1 & 0\\ -1 & 5 & -1\\ 0 & -1 & 0\end{pmatrix};$$

```
⇒ imB=Import["L:\\BudapestCNN.jpg"]
```

⇐

Fig. 5.89 Original image.

```
⇒ ImageData[imB]//Dimensions
⇐ {267,474,3}
```

Let us employ different kernels.

Edge detection,

```
⇒ im1=ImageConvolve[imB,M1]
```

⇐ 

Fig. 5.90 Detected edges.

Sharpening,

```
⇒ im2=ImageConvolve[imB,M2]
```

⇐ 

Fig. 5.91 Sharpened image.

Blurring,

```
⇒ im3=Blur[imB,5]
```

⇐ 

Fig. 5.92 Blurred image.

```
⇒ Map[(ImageData[#]//Dimensions)&,{im1,im2,im3}]
⇐ {{267,474,3},{267,474,3},{267,474,3}}
```

The result of convolution operation, the output matrix is called *Convolved Feature* or *Feature Map*.

Remark

In case of training CNN, the elements of the kernel matrix are also parameters, and should be computed like the weights of the network.

5.7.5 Spatial Pooling

Another mapping operation is the *Spatial Pooling* (also called *sub-sampling* or *downsampling*). This operation reduces the dimensionality of each feature map but retains the most important information. Spatial Pooling can be of different types: Max, Average, Sum etc.

In case of *Max Pooling* (most frequent), we define a spatial neighborhood (for example, a 2×2 window) and take the largest element from the rectified feature map within that window.

Instead of taking the largest element we could also take the average (Average Pooling) or sum of all elements in that window. In practice, Max Pooling has been shown to work better.

Figure 5.93 below shows an example of Max Pooling operation on a *Rectified Feature* map (obtained after convolution + ReLU operation) by using a 2×2 window.

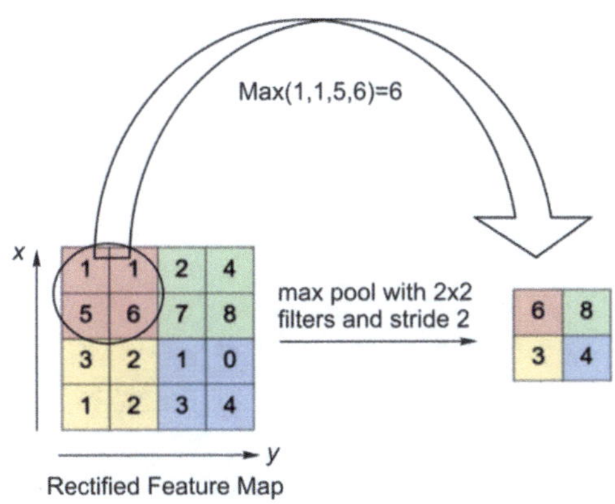

Fig. 5.93 Binary image matrix.

Let us consider the following image matrix:

$$\Rightarrow \mathrm{M} = \left\{\begin{pmatrix} 1 & 1 & 2 & 4 \\ 5 & 6 & 7 & 8 \\ 3 & 2 & 1 & 0 \\ 1 & 2 & 3 & 4 \end{pmatrix}\right\};$$

Application pooling via *PoolingLayer* in *Mathematica*,

```
⇒ pool=PoolingLayer[{2,2},2,"Input" → {1,4,4}]
⇐ PoolingLayer[ Parameters
                KernelSize:       {2,2}
                Stride:           {2,2}
                PaddingSize:      {0,0}
                Function:         Max
                Dimensionality:   2          ]
```

Resulting image matrix after pooling

```
⇒ Flatten[pool[M],1]//MatrixForm
```

$$\Leftarrow \begin{pmatrix} 6. & 8. \\ 3. & 4. \end{pmatrix}$$

Let us apply this pooling operation to the original Budapest image, Fig. 5.89. Its original size is

```
⇒ ImageData[imB]//Dimensions
⇐ {267,474,3}
⇒ pool=PoolingLayer[2,2,"Input" → NetEncoder["Image"],
   "Output" → NetDecoder["Image"],"Function" → Max]
⇐ PoolingLayer[ Parameters
                KernelSize:       {2,2}
                Stride:           {2,2}
                PaddingSize:      {0,0}
                Function:         Max
                Dimensionality:   2          ]
```

The result can be seen in Fig. 5.94.

```
⇒ imRP=pool[imB]
```

⇐ 

Fig. 5.94 The result of max pooling.

The size of the reduced image is

```
⇒ ImageData[imRP]//Dimensions
⇐ {64,64,3}
```

5.7.6 *Feature extraction via CNN*

These concepts introduced above are integrated into the convolutional neural networks, while traditional networks basically carry out classification separately from the feature extraction which is carried out for example via pixel based, Fourier as well as wavelet transformation, see Fig. 5.95.

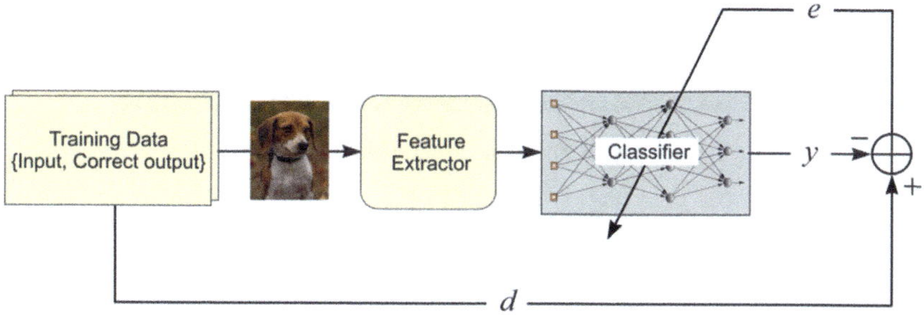

Fig. 5.95 Traditional solution of image classification via neural network. Feature extraction and classification are separated.

The convolution layer generates new images called feature maps. The feature map accentuates the unique features of the original image, see Fig. 5.96.

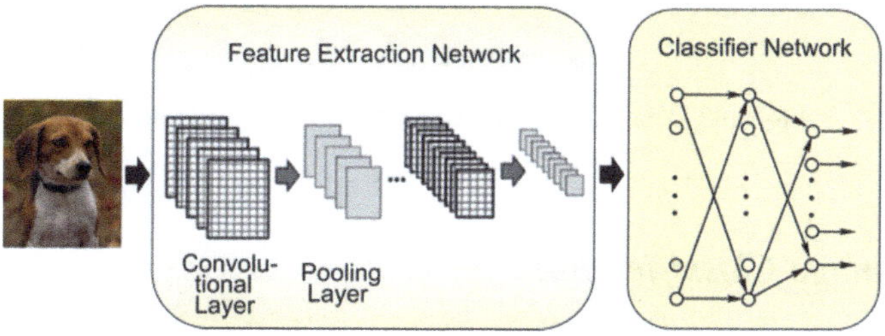

Fig. 5.96 Image classification via CNN.

Fig. 5.97 below illustrates the series of feature mappings as well as feature vectors of the fully connected classification layers (FC).

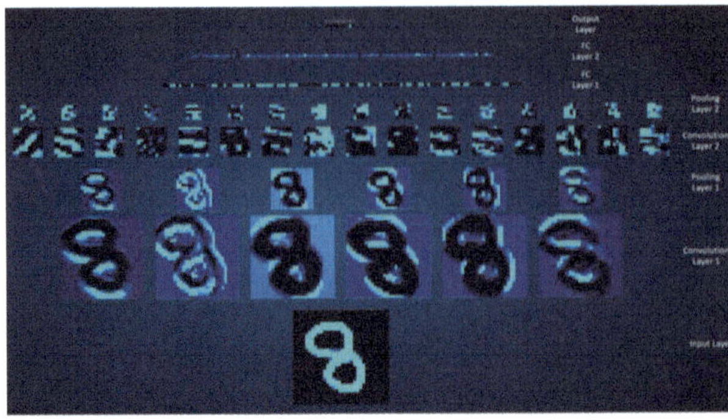

Fig. 5.97 Image go-through the CNN.

Fig. 5.98 illustrates the detailed structure of a CNN,

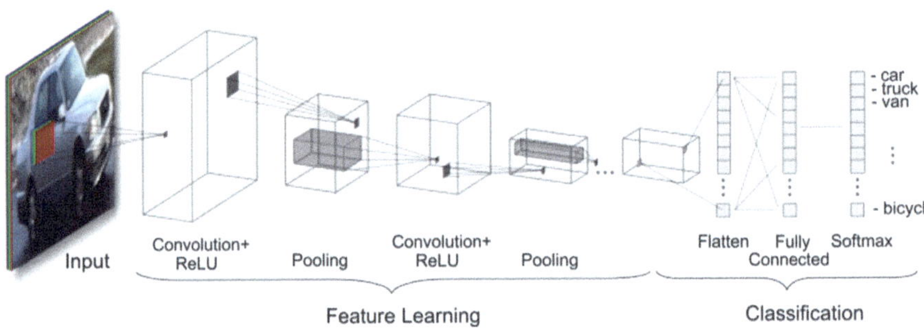

Fig. 5.98 Detailed structure of a CNN for image classification.

5.7.7 *Image Classification*

Let us see an example for the CNN classification. We assume there are three clusters: balls, persons and flowers, see Fig. 5.99 - 5.101.

```
⇒ Clear[persons,flowers,balls];
```

Let us label the elements of these clusters,

```
⇒ balls=Map[ImageResize[#,{32,32}] → ball &,
```

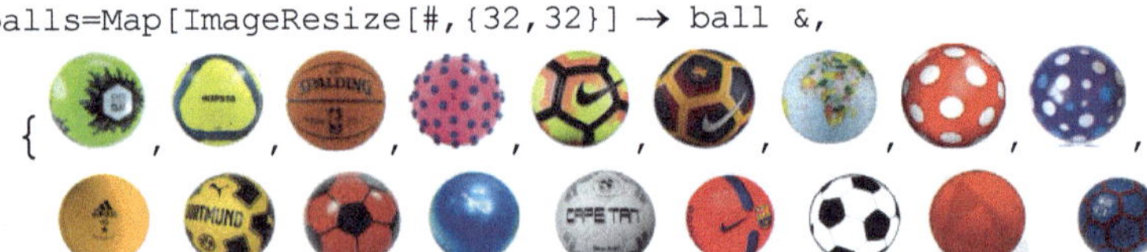

```
}]
```

```
⇐ {→ball, →ball, →ball, →ball, →ball,
```

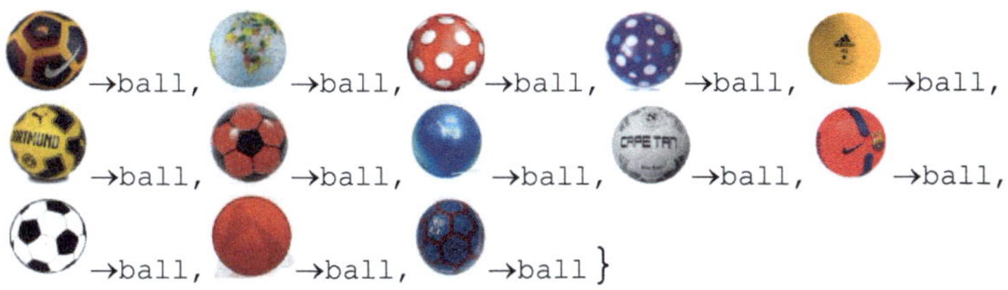

Fig. 5.99 Cluster of balls.

⇒ persons=Map[ImageResize[#,{32,32}] → person &,

Fig. 5.100 Cluster of persons.

⇒ flowers=Map[ImageResize[#,{32,32}] → flower &,

Fig. 5.101 Cluster of flowers.

The training set is

```
⇒ trainingData=Join[flowers,persons,balls];
⇒ classes = Union@Values[trainingData]
⇐ {ball,flower,person}
```

We employ a simple CNN

```
⇒
lenet = NetChain[
{
ConvolutionLayer[20,3],  (*first convolution ⇒ 20 feature images*)
ElementwiseLayer[Ramp],  (*activation function (ReLU) ⇒ non-linearity, sparsity*)
PoolingLayer[2,2],       (*max pooling ⇒ downsampling*)
ConvolutionLayer[30,3],  (*second convolution ⇒ 50 feature images*)
ElementwiseLayer[Ramp],  (*activation function (ReLU) ⇒ non-linearity, sparsity*)
PoolingLayer[2,2],       (*max pooling ⇒ downsampling*)

FlattenLayer[],          (*flattening ⇒ images to vector*)
DotPlusLayer[400],       (*first fully connected layer ⇒
                               feature vector from image features*)
ElementwiseLayer[Ramp],  (*activation function (ReLU) ⇒ non - linearity, sparsity*)
DotPlusLayer[3],         (*second fully connected layer ⇒ class prediction*)
SoftmaxLayer[]           (*normalization*)
},
"Input"->NetEncoder[{"Image",{32,32}}],      (*encoder ⇒ image to tensor*)
"Output"->NetDecoder[{"Class",classes}]      (*decoder ⇒ tensor to class*)
]
```

```
⇐ Netchain[
                                          image
      Input                               3-tensor(size: 1×15×63)
   1  ConvolutionLayer                    3-tensor(size: 20×13×61)
   2  BachNormalizationLayer              3-tensor(size: 20×13×61)
   3  Ramp                                3-tensor(size: 20×13×61)
   4  PoolingLayer                        3-tensor(size: 20×12×60)
   5  ConvolutionLayer                    3-tensor(size: 15×10×58)
   6  BachNormalizationLayer              3-tensor(size: 15×10×58)
   7  Ramp                                3-tensor(size: 15×10×58)
   8  PoolingLayer                        3-tensor(size: 15×9×57)
   9  FlattenLayer                        matrix(size: 135×57)
   10 TransposeLayer                      matrix(size: 57×135)
   11 GatedRecurrentLayer                 matrix(size: 57×19)
   12 GatedRecurrentLayer                 matrix(size: 57×19)
   13 NetMapOperator[LinearLayer]         matrix(size: 57×19)
   14 SoftmaxLayer                        matrix(size: 57×27)
      Output                              ctcbeam search
]
```

Let us train the network,

```
⇒ lenet=NetInitialize[lenet];
⇒ classifier = NetTrain[lenet, trainingData,MaxTrainingRounds → 200]
```

```
⇐ Netchain[ Input port:          image
            Output port:         class
            Number of layers:    11   ]
```

The test sets are,

```
⇒ testDataBalls = Map[ImageResize[#, {32, 32}] &,
```

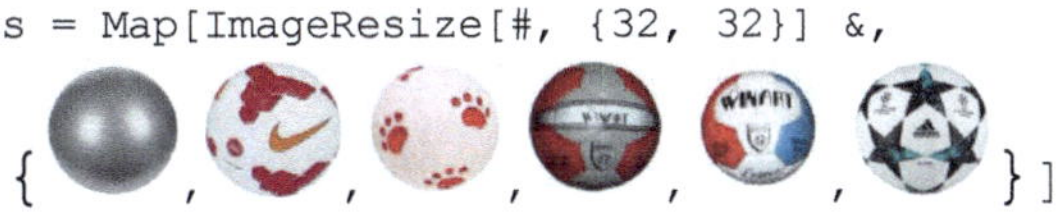

```
⇒ testDataPersons = Map[ImageResize[#, {32, 32}] &,
```

```
⇒ testDataFlowers = Map[ImageResize[#, {32, 32}] &,
```

```
⇒ testData=Join[testDataPersons,testDataFlowers,testDataBalls]
⇐
```

Fig. 5.102 Test set containing objects from all of the three different classes.

Let us employ the CNN classifier,

```
⇒ classifier[testData]
```

```
⇐ {person,person,person,person,person,person,person,flower,flower,
  flower,flower,flower,ball,person,person,ball,ball,ball}
```

We can see the detailed output of the *SoftMax* layer, namely

```
⇒ probabilities[i_Image] := Dataset[MapAt[Style[#, Bold, Red] &,
   #, Position[#, Max[#]]] &[classifier[i, "Probabilities"]]]
```

For example

⇒ `probabilities[`  `]`

⇐

ball	0.000014616
flower	0.0000032324
person	**0.999982**

Let us create a *classifier function* for the test cluster,

⇒ `TestDataL ={`

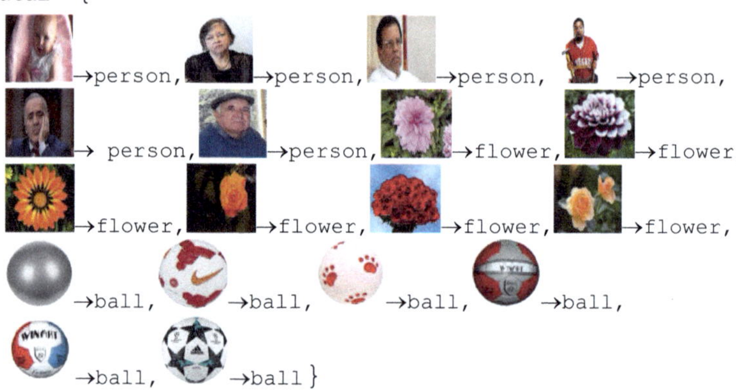

Fig. 5.103 Test set of the three sets.

```
⇒ cm = ClassifierMeasurements[classifier, testDataL]
⇐ ClassifierMeasurementsObject[ Classifier: Net
                                 Number of test examples: 18 ]
```

The overall accuracy of the CNN on the test cluster is,

```
⇒ cm["Accuracy"]
⇐ 0.777778
```

Special statistics also can be provided,

```
⇒ Dataset[
   Transpose@AssociationThread[{"Precision","Recall","FScore"},
    cm[{"Precision","Recall","FScore"}]/.
     {x_/;x>.75: → Style[x,Darker@Green],
      x_/;x<.5: → Style[x,Darker@Red]}]]
```

⇐

	Precision	Recall	FScore
ball	1.	0.666667	0.8
flower	1.	0.666667	0.8
person	0.6	1.	0.75

The confusion matrix explores the classification results graphically.

```
⇒ cm["ConfusionMatrixPlot"]
```

⇐

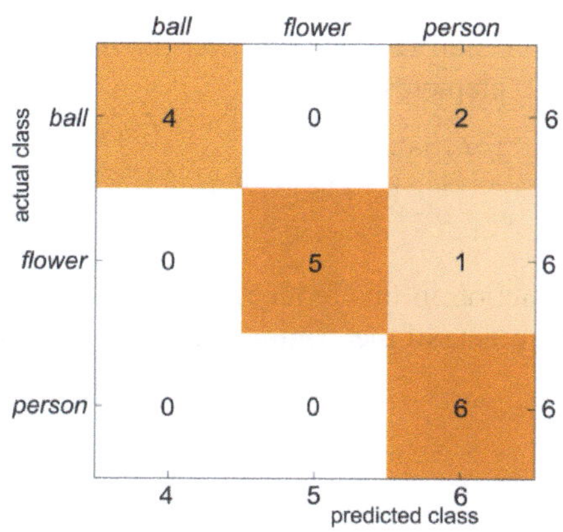

Fig. 5.104 Confusion matrix of the results of the test set classification.

5.7.8 *Image Clustering*

Clustering is an unsupervised technique as we have learned. Now we consider a special case of it. We have a cluster and an element which was not included in it. Let us find the element of the cluster which is "closest" to this excluded image.

Let us consider the data set of dog images,

```
⇒ dataset=
```

Fig. 5.105 Images of dogs for clustering.

Let us train a dimension reducer function for this dataset. Now we employ numerical vector of 10 elements,

```
⇒ dr=DimensionReduction[dataset,10]
⇐ DimensionReducerFunction[ Input type: Image
                            Output dimension: 10 ]
```

We generate a function in this reduced space which can find the closest element to one of the elements of the original data set,

```
⇒ nf=Nearest[dr[dataset] → Automatic]
⇐ NearestFunction[ Data points: 21
                   Input dimension: 10 ]
```

Using this, the function that displays the nearest image of the dataset:

```
⇒ nearestdog=dataset[[First@nf[dr[#]]]]&;
```

Use this function on images that are not in the dataset:

⇒ nearestdog[]

⇐ 

⇒ nearestdog[]

⇐ 

⇒ `nearestdog[`  `]`

⇐ 

Fig. 5.106 Images of dogs and the nearest ones provided by the *Nearest* function.

5.7.9 Object Localization and Identification

An object detection problem can be approached as either a classification problem or a regression problem. As a classification problem, the image is divided into small patches, each of which will be run through a classifier to determine whether there are objects in the patch. Then the bounding boxes are assigned to locate around patches that are classified with a high probability of presence of an object.

In the regression approach, the whole image will be run through a convolutional neural network to directly generate one or more bounding boxes for objects in the images.

Here, we introduce an object detector using the tiny version of the *You Only Look Once* (YOLO) approach (Redmon 2017).

```
⇒ Hyperlink["https://arxiv.org/pdf/1506.02640.pdf"]
⇐ https://arxiv.org/pdf/1506.02640.pdf
```

The tiny YOLO v1 consists of 9 convolution layers and 3 full connected layers. Each convolution layer consists of convolution, *leaky ReLU* and max pooling operations. The first 9 convolution layers can be understood as the feature extractor, whereas the last three full connected layers can be considered as the "regression head" that predicts the bounding boxes.

There is no native leaky ReLU layer in *Mathematica*, but it can be constructed easily using an `ElementwiseLayer`.

```
⇒ leayReLU[alpha_]:=ElementwiseLayer[Ramp[#]-alpha*Ramp[-#]&]
```

with this, the YOLO network can be constructed as

```
⇒ YOLO=NetInitialize@NetChain[{ElementwiseLayer[2.*#-1.&],
    ConvolutionLayer[16,3,"PaddingSize" → 1],leayReLU[0.1],
    PoolingLayer[2,"Stride" → 2],ConvolutionLayer[32,3,
    "PaddingSize" → 1],leayReLU[0.1],PoolingLayer[2,"Stride" → 2],
    ConvolutionLayer[64,3,"PaddingSize" → 1],leayReLU[0.1],
```

```
PoolingLayer[2,"Stride" → 2],ConvolutionLayer[128,3,
"PaddingSize" → 1],leayReLU[0.1],PoolingLayer[2,"Stride" → 2],
ConvolutionLayer[256,3,"PaddingSize" → 1],leayReLU[0.1],
PoolingLayer[2,"Stride" → 2],ConvolutionLayer[512,3,
"PaddingSize" → 1],leayReLU[0.1],PoolingLayer[2,"Stride" → 2],
ConvolutionLayer[1024,3,"PaddingSize" → 1],leayReLU[0.1],
ConvolutionLayer[1024,3,"PaddingSize" → 1],leayReLU[0.1],
ConvolutionLayer[1024,3,"PaddingSize" → 1],leayReLU[0.1],
FlattenLayer[],LinearLayer[256],LinearLayer[4096],leayReLU[0.1],
LinearLayer[1470]},"Input" → NetEncoder[{"Image",{448,448}}]]
```

```
⇐ NetChain[                                image
              Input                        3-tensor (size: 3×448×448)   ]
          1   -1+2x                        3-tensor (size: 3×448×448)
          2   ConvolutionLayer             3-tensor (size: 16×448×448)
          3   ElementwiseLayer             3-tensor (size: 16×448×448)
          4   PoolingLayer                 3-tensor (size: 16×224×224)
          5   ConvolutionLayer             3-tensor (size: 32×224×224)
          6   ElementwiseLayer             3-tensor (size: 32×224×224)
          7   PoolingLayer                 3-tensor (size: 32×112×112)
          8   ConvolutionLayer             3-tensor (size: 64×112×112)
          9   ElementwiseLayer             3-tensor (size: 64×112×112)
          10  PoolingLayer                 3-tensor (size: 64×56×56)
          11  ConvolutionLayer             3-tensor (size: 128×56×56)
          12  ElementwiseLayer             3-tensor (size: 128×56×56)
          13  PoolingLayer                 3-tensor (size: 128×28×28)
          14  ConvolutionLayer             3-tensor (size: 256×28×28)
          15  ElementwiseLayer             3-tensor (size: 256×28×28)
          16  PoolingLayer                 3-tensor (size: 256×14×14)
          17  ConvolutionLayer             3-tensor (size: 512×14×14)
          18  ElementwiseLayer             3-tensor (size: 512×14×14)
          19  PoolingLayer                 3-tensor (size: 512×7×7)
          20  ConvolutionLayer             3-tensor (size: 1024×7×7)
          21  ElementwiseLayer             3-tensor (size: 1024×7×7)
          22  ConvolutionLayer             3-tensor (size: 1024×7×7)
          23  ElementwiseLayer             3-tensor (size: 1024×7×7)
          24  ConvolutionLayer             3-tensor (size: 1024×7×7)
          25  ElementwiseLayer             3-tensor (size: 1024×7×7)
          26  FlattenLayer                 vector (size: 50176)
          27  LinearLayer                  vector (size: 256)
          28  LinearLayer                  vector (size: 4096)
          29  ElementwiseLayer             vector (size: 4096)
          30  LinearLayer                  vector (size: 1470)
              Output                       vector (size: 1470)
```

Training the YOLO network is time-consuming. We will use the pre-trained weights instead. The pre-trained weights can be downloaded as a binary file of size 172 Mb.

Using `NetExtract` and `NetReplacePart` we can load the pre-trained weights into our model,

```
⇒ modelWeights[net_,data_]:=Module[{newnet,as,weightPos,rule,
  layerIndex,linearIndex},layerIndex=
  Flatten[Position[NetExtract[
  net,All],_ConvolutionLayer|_LinearLayer]];
  linearIndex=Flatten[Position[NetExtract[net,All],_LinearLayer]];
```

```
as=Flatten[Table[{{n,"Biases"} → Dimensions@NetExtract[net,
{n,"Biases"}],{n,"Weights"}->Dimensions@NetExtract[net,
{n,"Weights"}]},{n,layerIndex}],1];
weightPos=#+{1,0}&/@Partition[Prepend[Accumulate[
Times@@@as[[All,2]]],0],2,1];
rule=Table[as[[n,1]] → ArrayReshape[Take[data,weightPos[[n]]],
as[[n,2]]],{n,1,Length@as}];
newnet=NetReplacePart[net,rule];
newnet=NetReplacePart[newnet,Table[
{n,"Weights"} → Transpose@ArrayReshape[NetExtract[newnet,
{n,"Weights"}],Reverse@Dimensions[NetExtract[newnet,
{n,"Weights"}]]],{n,linearIndex}]];
newnet]
```

```
⇒ data=
BinaryReadList["G:\\Pink\\yolo-tiny.weights","Real32"][[5;;-1]];
YOLO=modelWeights[YOLO,data];
```

The output of this network is a vector of 1470 elements, which contains the coordinates and confidence of the predicted bounding boxes for different classes. The tiny YOLO v1 is trained on the PASCAL VOC dataset which has 20 classes:

```
⇒ labels={"aeroplane","bicycle","bird","boat","bottle","bus","car",
"cat","chair","cow","diningtable","dog","horse","motorbike",
"person","pottedplant","sheep","sofa","train","tvmonitor"};
```

And the information for the output vector from the network is organized in the following way, see Fig. 5.107.

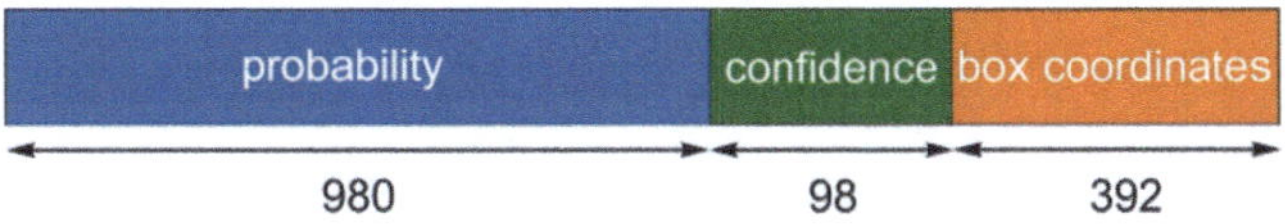

Fig. 5.107 The structure of the output vector of YOLO network.

The output vector is divided into three parts, giving the probability, confidence and box coordinates. Each of these three parts is also further divided into 49 small regions, corresponding to the predictions at each cell. Each of the 49 cells will have two box predictions. In post processing steps, we take this 1470 vector output from the network to generate the boxes with a probability higher than a certain threshold. The overlapping boxes will be resolved using the *non-max suppression method.*

```
⇒ coordToBox[center_,boxCord_,scaling_: 1]:=Module[{bx,by,w,h},
(*conver from {centerx,centery,width,height}to Rectangle object*)
bx=(center[[1]]+boxCord[[1]])/7.;
by=(center[[2]]+boxCord[[2]])/7.;
w=boxCord[[3]]*scaling;
h=boxCord[[4]]*scaling;
Rectangle[{bx-w/2,by-h/2},{bx+w/2,by+h/2}]]
```

```
nonMaxSuppression[boxes_,overlapThreshold_,confidThreshold_]:=
Module[{lth=Length@boxes,boxesSorted,boxi,boxj},
(*non-max suppresion to eliminate overlapping boxes*)
boxesSorted=GroupBy[boxes,#class&][All,
SortBy[#prob&]/*Reverse];
Do[Do[boxi=boxesSorted[[c,n]];
If[boxi["prob"]!=0,Do[boxj=boxesSorted[[c,m]];
(*if two boxes overlap largely,kill the box with low confidence*)
If[RegionMeasure[RegionIntersection[boxi["coord"],
boxj["coord"]]]/RegionMeasure[
RegionUnion[boxi["coord"],boxj["coord"]]]>=
overlapThreshold,boxesSorted=
ReplacePart[boxesSorted,{c,m,"prob"} → 0]];,
{m,n+1,Length[boxesSorted[[c]]]}]],
{n,1,Length[boxesSorted[[c]]]}],{c,1,Length@boxesSorted}];
boxesSorted[All,Select[#prob>0&]]]
labelBox[class_ → box_]:=Module[{coord,textCoord},
(*convert class\[Rule]boxes to labeled boxes*)coord=List@@box;
textCoord={(coord[[1,1]]+coord[[2,1]])/2.,coord[[1,2]]-0.04};
{{GeometricTransformation[Text[Style[labels[[class]],25,Yellow],
textCoord],ReflectionTransform[{0,1},textCoord]]},
EdgeForm[Directive[Red,Thick]],Transparent,box}]
drawBoxes[img_,boxes_]:=Module[{labeledBoxes},
(*draw boxes with labels*)labeledBoxes=
labelBox/@Flatten[Thread/@Normal@Normal@boxes[All,All,"coord"]];
Graphics[GeometricTransformation[{Raster[ImageData[img],{{0,0},
{1,1}}],labeledBoxes},ReflectionTransform[{0,1},{0,1/2}]]]]
postProcess[img_,vec_,boxScaling_: 0.7,confidentThreshold_:0.15,
overlapThreshold_: 0.4]:=Module[{grid,prob,confid,boxCoord,
boxes,boxNonMax},grid=Flatten[Table[{i,j},{j,0,6},{i,0,6}],1];
prob=Partition[vec[[1;;980]],20];
confid=Partition[vec[[980+1;;980+98]],2];
boxCoord=ArrayReshape[vec[[980+98+1;;-1]],{49,2,4}];
boxes=Dataset@Select[
Flatten@Table[<|"coord" → coordToBox[grid[[i]],boxCoord[[i,b]],
boxScaling],"class" → c,"prob" → If[#<=confidentThreshold,0,#]&@
(prob[[i,c]]*confid[[i,b]])|>,{c,1,20},{b,1,2},{i,1,49}],
#prob>=confidentThreshold&];
boxNonMax=nonMaxSuppression[boxes,overlapThreshold,
confidentThreshold];
drawBoxes[Image[img],boxNonMax]]
```

Let us test our CNN, see Fig. 5.108.

```
⇒ urls={"http://i.imgur.com/n2u0N3K.jpg",
  "http://i.imgur.com/Bpb60U1.jpg","http://i.imgur.com/CMZ6Qer.jpg"
  "http://i.imgur.com/1nEE8C7.jpg"};
  mgs=Import/@urls
```

⇐

Fig. 5.108 Test images for YOLO network.

The results can be seen on Fig. 5.109.

⇐

Fig. 5.109 The results of the YOLO network. Since the face of the white horse is not clear, the network failed to correctly recognize this object.

Further test images and their results can be seen in the following figures

k= {  }

Fig. **5.110** Test image of horses.

```
⇒ With[{i=ImageResize[#,{448,448}]},postProcess[i,YOLO[i]]]&/@k
```

{  }

Fig. 5.111 Identification and localization of horse objects.

and more test images,

⇐ {  }

Fig. 5.112 Future test images for location and identification.

```
⇒ With[{i = ImageResize[#, {448, 448}]}, postProcess[i,
   YOLO[i]]] & /@ imgs //ImageCollage
```

Fig. 5.113 The results of the YOLO network.

5.7.10 *Cat or Dog?*

To demonstrate the *Python* code for creating and employing CNN, we introduce another example, a simple binary classification problem, namely classification of cat and dog images, see Fig. 5.114.

⇐

Fig. 5.114 Simple binary classification problem using labeled photos of cats and dogs.

The detailed explanation of the Python code can be found in https://github.com/venkateshtata/cnn_medium./blob/master/cnn.py.

We have 47-47 images in the training set, and 15 -15 images in the test set.

First let us import all the required Keras packages, which we are going to use to build our CNN. Make sure that every package is installed: *Keras* using *Tensorflow* backend and using *Theano* backend.

```
# Importing the Keras libraries and packages
from keras.models import Sequential
from keras.layers import Conv2D
from keras.layers import MaxPooling2D
from keras.layers import Flatten
from keras.layers import Dense
```

Let us create the structure of our CNN, via defining the layers, activation functions etc.

```
# Initialising the CNN
classifier = Sequential()
# Step 1 - Convolution
classifier.add(Conv2D(32, (3, 3), input_shape = (64, 64, 3),
    activation = 'relu'))
# Step 2 - Pooling
classifier.add(MaxPooling2D(pool_size = (2, 2)))
# Adding a second convolutional layer
classifier.add(Conv2D(32, (3, 3), activation = 'relu'))
classifier.add(MaxPooling2D(pool_size = (2, 2)))
# Step 3 - Flattening
classifier.add(Flatten())
# Step 4 - Full connection
classifier.add(Dense(units = 128, activation = 'relu'))
classifier.add(Dense(units = 1, activation = 'sigmoid'))
```

We shall determine the type of the training algorithm as well as the loss function.

```
# Compiling the CNN
classifier.compile(optimizer = 'adam', loss =
    'binary_crossentropy', metrics = ['accuracy'])
```

Now images of the training and test sets will be loaded,

```
# Part 2 - Fitting the CNN to the images
from keras.preprocessing.image import ImageDataGenerator
train_datagen = ImageDataGenerator(rescale = 1./255,
shear_range = 0.2,
zoom_range = 0.2,
horizontal_flip = True)
test_datagen = ImageDataGenerator(rescale = 1./255)
training_set = train_datagen.flow_from_directory(
    'M:/training_setA/training_set',
target_size = (64, 64),
batch_size = 32,
class_mode = 'binary')
test_set =
    test_datagen.flow_from_directory('M:/test_setA/test_set',
target_size = (64, 64),
batch_size = 32,
class_mode = 'binary')
```

```
Found 94 images belonging to 2 classes.
Found 30 images belonging to 2 classes.
```

Then let us train the network. The test set will be employed as validation set,

```
classifier.fit_generator(training_set,
steps_per_epoch = 94,
epochs = 5,
verbose=0,
validation_data = test_set,
validation_steps = 30)
```

Let us check the result of the classification on the training set,

```
score = classifier.predict_generator(training_set)
score
```

⇐ {0.999554, 0.998548, 0.999348, 0.000058611, 8.1634×10^{-8}, 0.98943, 0.998358, 6.60737×10^{-7}, 0.000125519, 0.999842, 0.999999, 0.999987, 1., 0.99537, 0.991236, 0.999861, 0.0000295078, 0.000749577,
⋮
0.00086288, 1.17531×10^{-6}, 0.997258, 0.999844, 0.767423, 1., 0.0000296309, 0.000494728, 2.25211×10^{-6}, 0.000480905, 0.999989}

```
⇒ ntest=Flatten[%];
⇒ ntest//Length
⇐ 94
⇒ cat={};dog={};
⇒ Do[If[ntest[[i]]>=0.5,AppendTo[dog,ntest[[i]]],
    AppendTo[cat,ntest[[i]]]],{i,1,94}];
⇒ dog//Length
⇐ 47
⇒ cat//Length
⇐ 47
```

It means that the classification on the training set is perfect, 100 %. Now let us check the test (validation) set.

```
score = classifier.predict_generator(test_set)
score
```

```
⇐ {{0.999985},{1.},{1.},{0.648371},{0.999989},{0.999998},
  {0.999977},{0.999992},{0.999999},{0.0274209},{0.00695746},
  {0.000578081},{0.00398403},{0.0000143432},{0.0062359},
  {0.303308},{0.995697},{0.340169},{0.6462},{0.547773},
  {0.356563},{0.989602},{0.105532},{0.999998},{0.957598},
  {0.952294},{0.663439},{0.99315},{0.0863328},{0.418782}}
⇒ ntest=Flatten[%];
⇒ ntest//Length
⇐ 30
⇒ cat={};dog={};
⇒ Do[If[ntest[[i]]>=0.5,AppendTo[dog,ntest[[i]]],
   AppendTo[cat,ntest[[i]]]],{i,1,30}];
⇒ dog//Length
⇐ 18
⇒ cat//Length
⇐ 12
```

Which means that 6 cat images are classified as dog image. So the score of the classification is

```
⇒ (1.-6/30) 100
⇐ 80
```

namely 80 % on the test (validation) set.

Since the test set was employed as validation set, let test our CNN on further test samples:

⇒ `cat1=``, cat2=``, cat3=,``;`

⇒ `dog1=``, dog2=``, dog=,``;`

Fig. 5.115 Further images which were not involved in the training process.

Employing for image cat3,

```
# Part 3 - Making new predictions
import numpy as np
from keras.preprocessing import image
test_image = image.load_img('M:/cat_3.jpg',
    target_size = (64, 64))
test_image = image.img_to_array(test_image)
test_image = np.expand_dims(test_image, axis = 0)
result = classifier.predict(test_image)
result
```

⇐ {{0.}}

Employing for image dog1,

```
# Part 3 - Making new predictions
import numpy as np
from keras.preprocessing import image
test_image = image.load_img('M:/dog_1.jpg', target_size = (64, 64))
test_image = image.img_to_array(test_image)
test_image = np.expand_dims(test_image, axis = 0)
result = classifier.predict(test_image)
result
```

⇐ {{1.}}

5.8 Comparison of Neural Networks

Here we shall compare the performances of Hopfield Network (HN) and the Convolutional Neural Network (CNN) employing them for character recognition.

Loading Data.

Obtaining the MNIST data sets from Wolfram depository,

⇒ `testData = ResourceData["MNIST", "TestData"];`

We shall employ only a restricted size of this, namely selecting 300 digits randomly

⇒ `data=RandomSample[testData,300];`

and selecting the digits of zeros,

⇒ `zero=Select[data,#[[2]]==0&]`

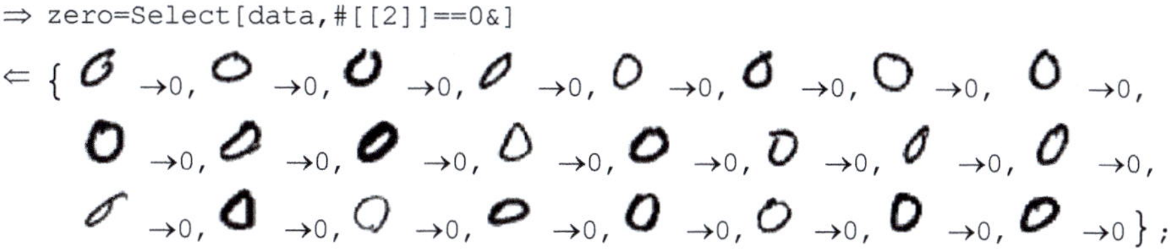

as well as the digits of ones,

⇒ `one=Select[data,#[[2]]==1&]`

⇐ 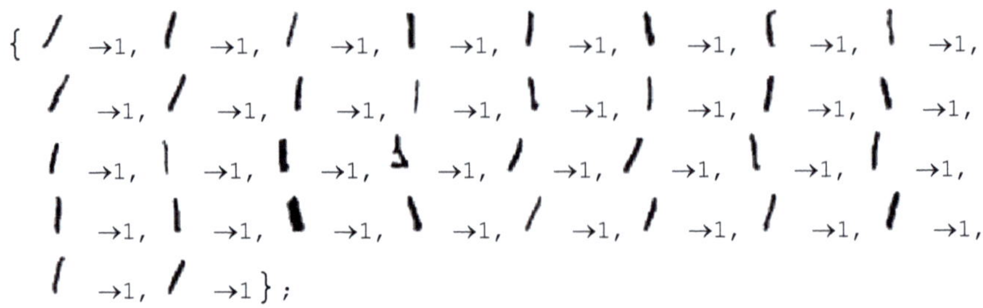

Let us take the digits separately

```
⇒ z=Map[#[[1]]&,zero];
⇒ o=Map[#[[1]]&,one];
```

Data Preparation for HN

For HN data we employ data reduction. Each digit will be represented by a vector of two dimensions,

```
⇒ dr=DimensionReduction[Join[z,o],2]
⇐ DimensionReducerFunction[ Input type: Image
                            Output dimension: 2 ]
```

To get the coordinates of the image of the digits we apply the reduction function,

```
⇒ zz=dr[z];
⇒ oo=dr[o];
```

Let us display the corresponding points

```
⇒ ListPlot[{zz,oo},Frame → True]
```

⇐ 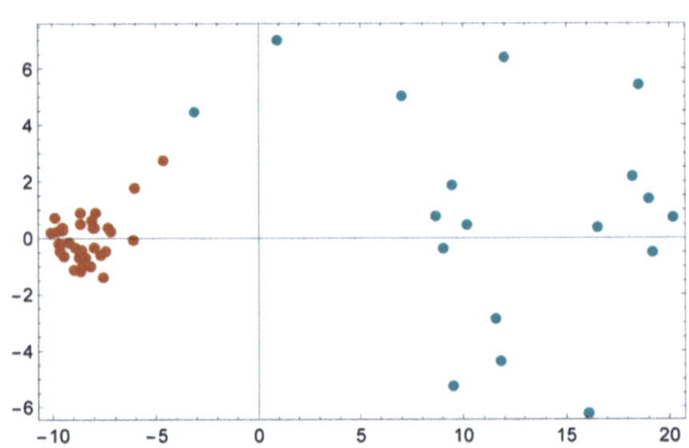

Fig. 5.116 The points of digits after dimension reduction.

Saving data for HN

```
⇒ Export["M:\\dataZero.dat",zz];
```

```
⇒ Export["M:\\dataOne.dat",oo];
```

Hopfield Solution

The next part should be evaluated in *Mathematica* version 10.3

```
⇒ ZZ=Import["M:\\dataZero.dat"];
⇒ OO=Import["M:\\dataOne.dat"];
```

We shall use 16 digits for training and 8 for testing.

```
⇒ zz=Take[ZZ,{1,16}];
⇒ oo=Take[OO,{1,16}];
```

Let us compute the average of the training samples as equilibrium points of the Hopfield network,

```
⇒ Zeq=Mean[zz]
⇐ {11.7345,0.060569}
```

```
⇒ Qeq=Mean[oo]
⇐ {-8.47825,-0.279838}
```

The training elements and the two equilibrium points

```
⇒ Show[{ListPlot[{zz,oo},Frame → True],ListPlot[{{Zeq},
    {Qeq}},PlotStyle → {Black,Black},PlotMarkers → {"✪","✪"}]}]
```

⇐ 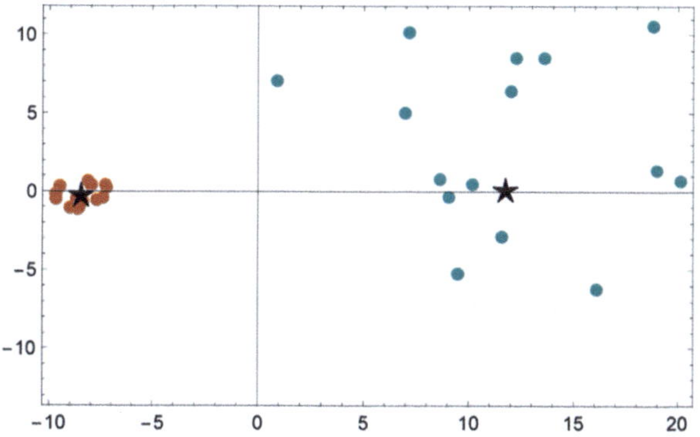

Fig. 5.117 Training elements after dimension reduction with the equilibrium points (stars).

Loading the Neural Network package,

```
<<NeuralNetworks`
```

We shall employ continuous Hopfield network with Saturated Linear activation function, which is an approximation of the tangent hyperbolic function

```
⇒ Plot[{Tanh[x],SaturatedLinear[x]},{x,-2,2}]
```

⇐

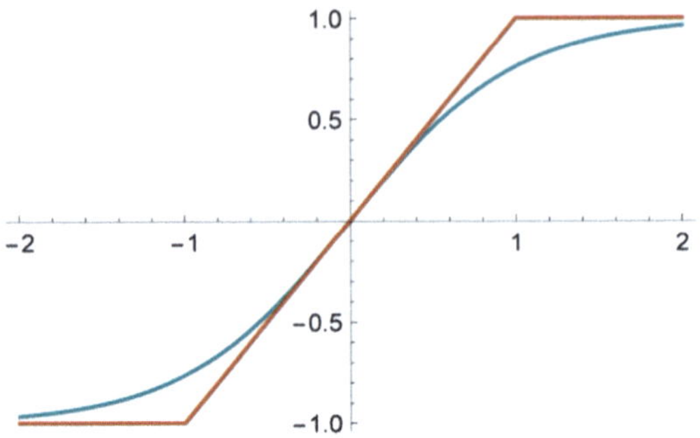

Fig. 5.118 Saturated Linear activation function (brown) compared with Tanh activation function (blue).

It means that the output of the network will be close to the values {1, 1} and {−1,−1}. Let us train the network

```
⇒ hop=HopfieldFit[{Zeq,Qeq},
    NetType → Continuous,Neuron → SaturatedLinear]
⇐ Hopfield[W,{NetType → Continuous,WorkingPrecision → 4,
    CreationDate → {2018,12,2,19,18,27.6219413},Dt → 0.00474777,
    Neuron → SaturatedLinear}]
```

Employing it for an elements

```
⇒ hop[zz[[11]]]
⇐ {{1.02087,1.0112}}
```

and

```
⇒ hop[oo[[11]]]
⇐ {{-1.02096,-0.996912}}
```

Let us visualize the trajectories of one of the two input elements. The starting point,

```
⇒ zz[[11]]
⇐ {18.7954,10.5539}
⇒ Show[NetPlot[hop,{zz[[11]]}],ImageSize → 300]
```

⇐ 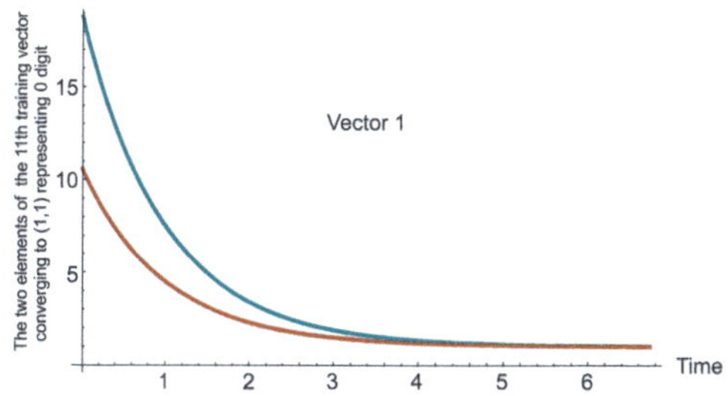

Fig. 5.119 Trajectory of 11th element of the training set of zeros.

This figure shows how the trajectories are converging to the equilibrium point (1,1) which represents the zero digit.

It is also interesting to see how the energy values of training elements are converging to the two equilibrium points (−1, −1), (1,1),

```
⇒ NetPlot[hop,Join[zz,oo],DataFormat → Surface,
   BoxRatios → {1, 0.7, 0.4},PlotRange → Automatic]
```

⇐ 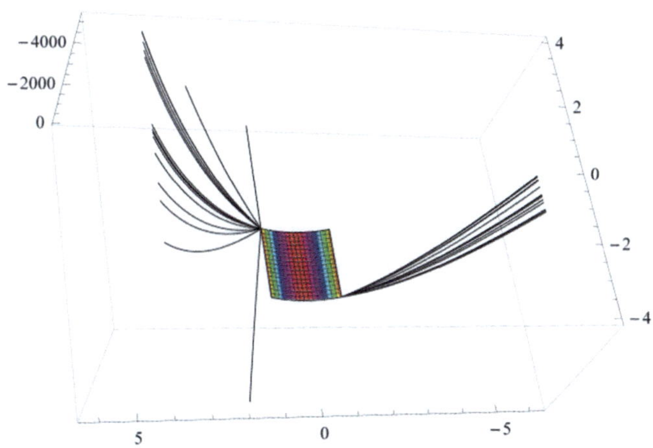

Fig. 5.120 Trajectories of the training set converging to the two equilibrium points.

How can we test digits?

Let us consider the test elements,

```
⇒ zztest=Take[ZZ,{17,24}];
⇒ ootest=Take[OO,{17,24}];
⇒ Map[hop[#]&,Join[zztest,ootest]]
```

```
⇐ {{{-1.00816,-0.979095}},{{1.02095,1.00143}},{{1.01302,0.979066}},
   {{1.02089,1.00209}},{{1.0209,0.998232}},{{1.02089,0.989609}},
   {{1.02094,0.999123}},{{1.02093,1.00526}},{{-1.0209,-0.997295}},
   {{-1.02091,-1.00132}},{{-1.02096,-0.994779}},
   {{-1.02026,-0.979135}},{{-1.02091,-0.996598}},
   {{-1.0209,-0.994331}},{{-1.0209,-0.999966}},
   {{-1.02096,-0.997844}}}
```

The first element in the zeros test set is misinterpreted, not surprisingly, see the figure below

```
⇒ zztest[[1]]
⇐ {-3.12604,4.44976}
⇒ ListPlot[{zztest,ootest,{Zeq},{Qeq}},Frame → True]
⇐
```

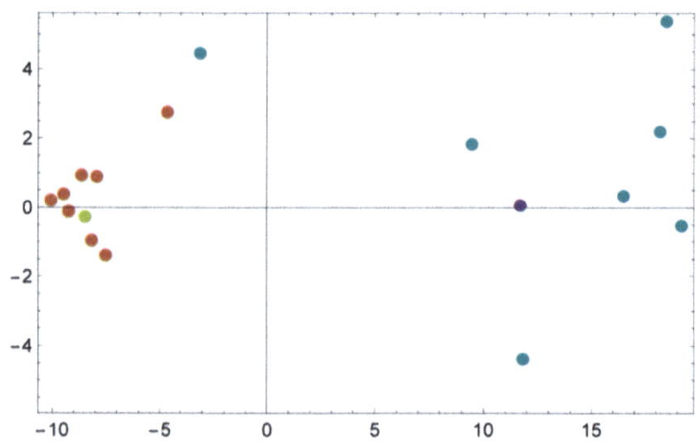

Fig. 5.121 Test set.

A zero element in the left upper quadrat (blue)

```
⇒ zztest[[1]]
⇐ {-3.12604,4.44976}
```

is too close to the equilibrium point representing one, therefore it is "misclassified"

```
⇒ hop[zztest[[1]]]
⇐ {{-1.00816,-0.979095}}
```

Convolution Network Solution

The training set

```
⇒ zero= {
```

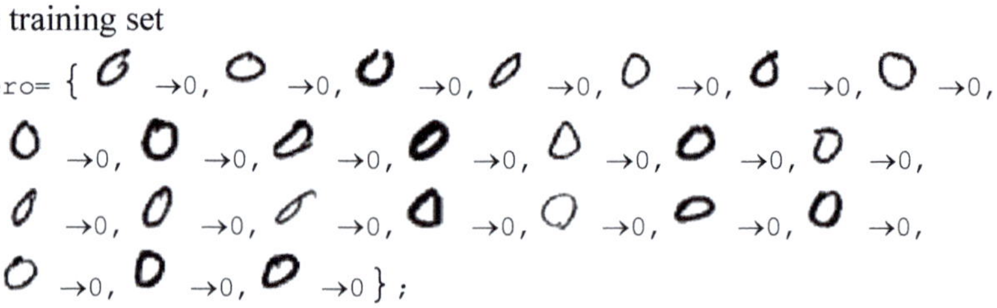

⇒ one=
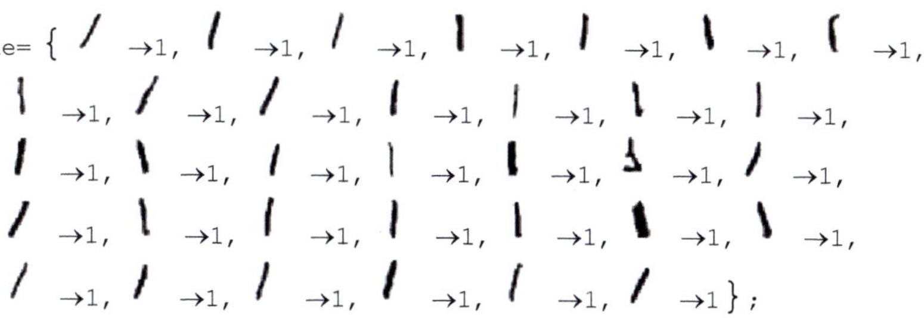

We employ the *LeNet* model,

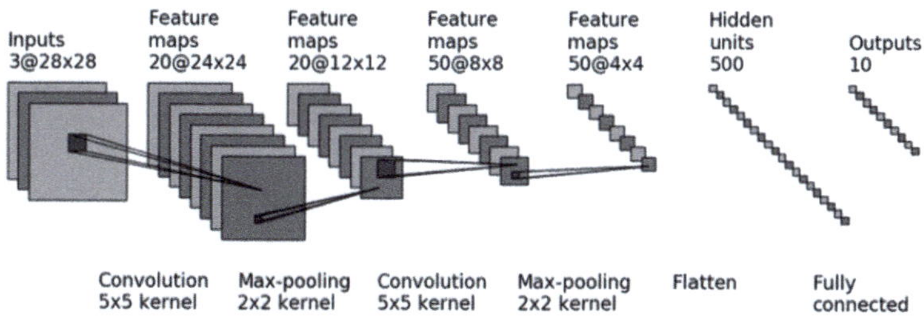

Fig. 5.122 General structure of a *LeNet* CNN mode.

```
⇒ net=NetModel["LeNet"];
```

The structure of our net,

```
⇒ net
⇐ NetChain[
                                        image
      Input                             3-tensor (size: 1×28×28)
   1  ConvolutionLayer                  3-tensor (size: 20×24×24)
   2  Ramp                              3-tensor (size: 20×24×24)
   3  PoolingLayer                      3-tensor (size: 20×12×12)
   4  ConvolutionLayer                  3-tensor (size: 50×8×8)
   5  Ramp                              3-tensor (size: 50×8×8)
   6  PoolingLayer                      3-tensor (size: 50×4×4)
   7  FlattenLayer                      vector (size: 800)
   8  LinearLayer                       vector (size: 500)
   9  Ramp                              vector (size: 500)
  10  LinearLayer                       tensor
  11  SoftmaxLayer                      tensor
      Output                            tensor
]
```

In graphics form, in two parts,

```
⇒ NetGraph[NetTake[net,5]]
```

⇐ NetGraph[
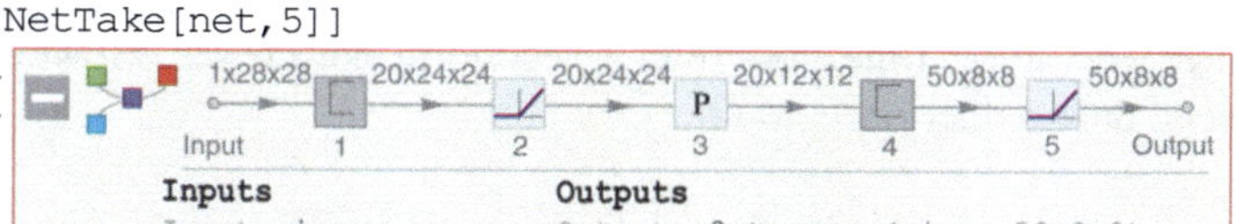

]

```
⇒ NetGraph[NetTake[net,{6,11}]]
```

⇐ NetGraph[

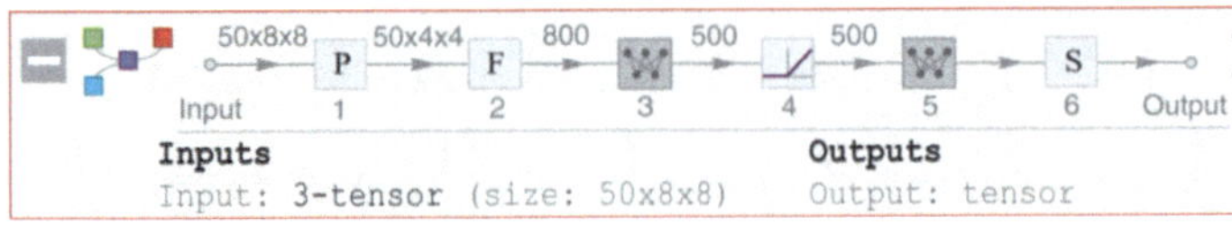

]

```
⇒ trainingset=Join[Map[Take[#,{1,16}]&,{zero,one}]]//Flatten
```

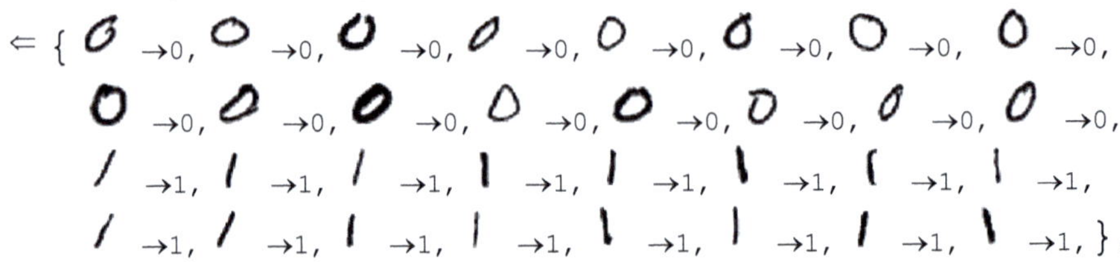

Training the network.

The parameters of the training,

```
⇒ result=NetTrain[net, trainingset,All, MaxTrainingRounds → 20]
⇐ NetTrainResultsObject
```

Total training time:	0.73 s
Total rounds:	20
Total batches:	20
Batch size:	32
Method	ADAM
Final round loss:	0.00865
Final Round error:	0%

Loss evolution plot:

Error evolution plot:

The trained network,

```
⇒ AbsoluteTiming[trained=
    NetTrain[net, trainingset, MaxTrainingRounds → 20];]
⇐ {0.867204,Null}
```

```
⇒ testset=
    Map[#[[1]]&,Flatten[Join[Map[Take[#,{17,24}]&,{zero,one}]]]]
```

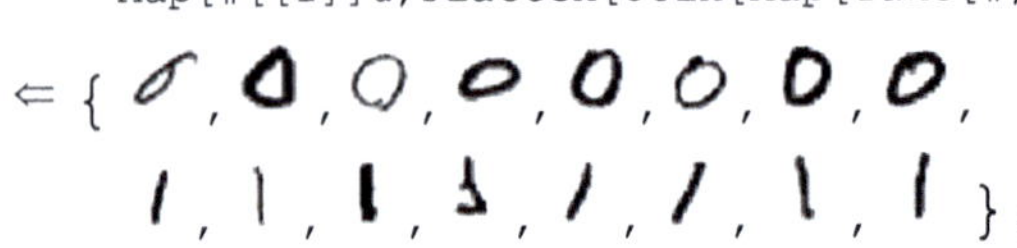

```
⇒ trained[testset]
⇐ {1,0,0,0,0,0,0,0,1,1,1,1,1,1,1,1}
```

The first element is misinterpreted here, too.

Summing up, the CNN and HN provide the same result, and the training requires very short time, less than 1 second in both cases. However CNN does not need data reduction.

References

Banerjee T (2018): Deep Learning and Computer Vision: Converting Models for the Wolfram Neural Net Repository, Wolfram Blog, https://blog.wolfram.com/2018/12/06/deep-learning-and-computer-vision-converting-models-for-the-wolfram-neural-net-repository/

Brownlee J. (2016): Time Series Prediction with LSTM Recurrent Neural Networks in Python with Keras, https://machinelearningmastery.com/time-series-prediction-lstm-recurrent-neural-networks-python-keras/

Chollet F (2017): Keras: The Python Deep Learning library, Keras Documentation, https://keras.io/

Chollet F (2018): Deep Learning with Python, Manning Publications Co., Shelter Island, NY

Deshpande M (2017): Perceptrons: The First Neural Networks, https://pythonmachinelearning.pro/perceptrons-the-first-neural-networks/

Freeman J A (1994): Simulating Neural Networks with Mathematica, Addison-Wesley, Massachusetts

Haykin S. (2009): Neural Networks and Learning Machines,Third Edition,Prentice Hall,New York, London,Sydney,

Hu X, Yuan Y (2016): Deep-Learning-Based Classification for DTM Extraction from ALS Point Cloud, Remote Sens. 2016, 8, p.730

Melnikov O (2017): Trouble fitting simple data with MLPRegressor, stackoverflow https://stackoverflow.com/questions/41069905/trouble-fitting-simple-data-with-mlpregressor

Raschka S (2018): Neural Network - Multilayer Perceptron, http://rasbt.github.io/mlxtend/user_guide/classifier/MultiLayerPerceptron/

Redmon J (2017) Mathematica implementation of YOLO, a computer vision object detection mode Object detection and localization using neuralnetwork, Mathematica Stack Exchange Network https://mathematica.stackexchange.com/questions/141598/object-detection-and-localization-using-neural-network

Rohrer B (2017): Recurrent Neural Networks (RNN) and Long Short-Term Memory (LSTM), https://www.youtube.com/watch?v=WCUNPb-5EYI

Sharma A (2017): Convolutional Neural Networks in Python with Keras,DataCamp https://www.google.hu/search?q=convolution+network+in+Python&oq=convolution+network+in+Python&aqs=chrome..69i57j0l3.14671j0j7&sourceid=chrome&ie=UTF-8

Shevchuk Y (2015): Discrete Hopfield Network - NeuPy http://neupy.com/2015/09/20/discrete_hopfield_network.html

Shevchuk Y (2017): Self-Organizing Map and Applications, NeuPy, Neural Networks in Python, http://neupy.com/2017/12/09/sofm_applications.html

Srinivasan J, Han YK and Ong SH (1993): Image reconstruction by a Hopfield neural network, Image and Vision Computing 11(5), pp. 279-282 https://www.sciencedirect.com/science/article/pii/0262885693900052

Sullivan J (2017): Neural Network from Scratch: Perceptron Linear Classifier, https://jtsulliv.github.io/perceptron/

Usama M et al (2017): Unsupervised Machine Learning for Networking:Techniques, Applications and Research Challenges. https://arxiv.org/pdf/1709.06599.pdf

Saputro DRS and Widyaningsih P (2017): Limited Memory Broyden-Fletcher-Goldfarb-Shanno (L-BFGS) Method for The Parameter Estimation on Geographically Weighted Ordinal Logistic Regression Model (GWOLR), AIP Conference Proceedings 1868, 040009 (2017); https://doi.org/10.1063/1.4995124 Published Online: 04 August 2017 https://aip.scitation.org/doi/pdf/10.1063/1.4995124?class=pdf

GPSR Compliance

The European Union's (EU) General Product Safety Regulation (GPSR) is a set of rules that requires consumer products to be safe and our obligations to ensure this.

If you have any concerns about our products, you can contact us on ProductSafety@springernature.com

In case Publisher is established outside the EU, the EU authorized representative is:

Springer Nature Customer Service Center GmbH
Europaplatz 3
69115 Heidelberg, Germany

Batch number: 09661795

Printed by Printforce, the Netherlands